BUSINESS/SCIENCE/TECHNOLOGY DIVISION
CHICAGO PUBLIC LIBRARY
400 SOUTH STATE STREET
CHICAGO, ILLINOIS 60605

D1570259

*Pierre Simon Laplace*
*1749–1827*

# Pierre Simon Laplace
# 1749–1827

*A Determined Scientist*

Roger Hahn

HARVARD UNIVERSITY PRESS
Cambridge, Massachusetts
London, England
2005

English edition copyright © 2005 by the President and Fellows of Harvard College
All rights reserved
Printed in the United States of America

Originally published as *Le système du monde: Pierre Simon Laplace—un itinéraire dans la science*
© Editions Gallimard, Paris, 2004

*Library of Congress Cataloging-in-Publication Data*

Hahn, Roger, 1932–
    [Système du monde. English]
    Pierre Simon Laplace, 1749–1827 : a determined scientist / Roger Hahn.
      p. cm.
    Includes bibliographical references and index.
    ISBN 0-674-01892-3
    1. Laplace, Pierre Simon, marquis de, 1749–1827.
2. Scientists—France—Biography.   3. Physicists—France—Biography.
4. Mathematicians—France—Biography.   5. Celestial mechanics.
I. Title.

Q143.L36H35313 2005
509.2—dc22
[B]      2005046358

*To my wife, Ellen,
with tender affection*

# *Acknowledgments*

This book fills a significant void in the biographical literature of Pierre Simon Laplace. The few other worthy treatments, by Henri Andoyer in 1922 and by the group headed by Charles Gillispie in 1978, focused principally on his scientific work, without exploring the scientist's other roles in society, his private thoughts on politics and religion, or his family life. With the help of his unpublished correspondence and manuscripts heretofore ignored, I present this eminent figure in his full historical setting. By knowing the issues and circumstances that set him along the paths he chose in his youth, we can begin to grasp how he could remain steadfast in the midst of a turbulent age.

To thank everyone individually for the moral and practical support I have received over the many years this project has taken to materialize would require several chapters. Mentors, colleagues, and students have been remarkably patient, all the while assisting me immeasurably. My intellectual debts are best measured by looking at the endnotes, but I have been more immediately sustained in the effort of writing by the encouragement of two exemplary historians of science, the late Alexandre Koyré and René Taton, who began this long trek with me a half-century ago; and by my wife, Ellen, who has endured my affair with Laplace in a ménage à trois for almost as long. More recently, I have been helped by Ran Halévi, and profited from two anonymous reviewers and a meticulous copyeditor, Julie Carlson.

# Contents

|   |   |   |
|---|---|---|
|    | Introduction | 1 |
| 1  | Norman Beginnings | 3 |
| 2  | From Scholasticism to Higher Mathematics | 16 |
| 3  | The Gifted Mathematician | 32 |
| 4  | Setting Fundamental Principles: The Philosophical Program | 48 |
| 5  | Finding the Stability of the Solar System: The Celestial Program | 64 |
| 6  | Exploring the Physical World: The Terrestrial Program | 81 |
| 7  | Revolutionary Tumult | 98 |
| 8  | The Politics of Science | 120 |
| 9  | Celestial Mechanics | 140 |
| 10 | Probability and Determinism | 168 |
| 11 | The Waning Years | 190 |
|    | Conclusion | 205 |

| | |
|---|---|
| Abbreviations | 211 |
| Draft of a Letter to Laplace by Jean-Etienne Guettard | 213 |
| Four Nonscientific Manuscripts by Laplace | 215 |
| Illness and Last Moments of M. de Laplace | 233 |
| Notes | 235 |
| Index | 303 |

*Pierre Simon Laplace*
*1749–1827*

# Introduction

For the past few decades, creative research in the history of science has focused on the great sea changes: the emergence of Greek natural philosophy, the scientific revolution of the seventeenth century, and the critical transformations in physics and biology experienced early in the twentieth century. Periods between these upheavals have generally been pictured as either anticipations or consequences of these momentous revolutions. The tacit assumption has been that the pace of scientific discoveries has mirrored the rapid change and relative stability discerned in cycles of political history.

Pierre Simon Laplace lived during a somewhat calm period of scientific history. As the arch-codifier of the astronomical sciences, he was known to his contemporaries as the Newton of France. Owing to his exemplary skills, generations of scientists were inculcated in the art of manipulating the calculus and probability theory to analyze the physical world and offered the promise that these tools would assist in further understanding it. Laplace's accomplishments confirmed the reign of classical physics prior to the advent of electromagnetism. His views on the origin of the solar system, the material theory of heat, and the determinism of the universe, while recognized as innovative, seemed to fit naturally with the classical doctrines. As a result, he was pictured as a scientific sage valued more for his magisterial syntheses than for his pioneering achievements. When his work was appraised, it was generally taken as the symbol for a glorious but tranquil era in science.

The course of Laplace's life stands in sharp contrast to this conventional picture of stability. For such a prominent figure to have flourished unscathed by the French Revolution is a feat that in itself demands inspection. The facile quips about his political fickleness do not suffice to account for his uninterrupted ascent to high offices in the French nation. In fact, the events and opportunities created by the political upheavals of the 1790s induced him to become a key science policy maker for France and the patron of a new generation. His writings also changed, from the complex scientific papers of a masterful mathematician and skillful codifier to treatises that popularized the abstruse knowledge he had helped to fashion. On a more personal level, Laplace also came to terms with the sinuous intellectual itinerary that had led him from the priesthood to the adoption of a radical view of Christianity, a view that ultimately gave way to a more temperate stance in his old age.

Through all these vicissitudes, however, Laplace remained faithful to the scientific program he had originally chartered. In the midst of change, Laplace remained steadfast in his philosophical convictions, articulating them with ever-clearer declarations that became an inspiration to the small cadre of young scientists he promoted. From the perspective of his life's experience, he can hardly be portrayed as the sedate and fundamentally conservative savant one might expect from someone who had lived between major scientific revolutions. In fact, Laplace cut a remarkably striking figure of intellectual perseverance and personal resilience in the midst of major turmoil. He clearly deserves to be understood on his own terms.

The initial publication of this work appeared in French in April 2004 as *Le système du monde: Pierre Simon Laplace, un itinéraire dans la science* (Paris: Gallimard, 2004). This English edition is a slightly amended version, with minor changes in several chapters and the addition of three appendices addressing his religious views. For the most part, translations of quoted passages are my own.

# 1

# Norman Beginnings

In the details of the growing-up years of Pierre Simon Laplace are important clues to the man—scientist, husband, father, and friend—he would become. The consistent, deliberate scholar we know through Laplace's writings, a man dedicated to articulating the laws of nature through mathematics regardless of political fashion, was shaped by both dramatic and ordinary events when he was a young child. The loss of his mother at a young age, his unrewarding relationship with his father and siblings, his education, and his decisive turn away from a career in the Church all help to explain not only his approach to science, but also his worldview.

No one present at Pierre Simon Laplace's birth would have expected him to become one of the world's greatest scientists. He was the first in his family and locality to rise to such eminence. Born at mid-century at Beaumont-en-Auge into a family of gentrified cultivators who had for generations lived off the fertile farmlands of lower Normandy, he seemed destined for no more than an ordinary life on the land. The countryside—known as the pays d'Auge—where he and his forebears grew up stretched between the rivers Dives and Touques inland to the villages of Livarot and Camembert. It was a small and tidy world, filled with lush pastures, gently rolling hills, and plentiful fruit trees. This bucolic landscape not only was the seat of his ancestors, but also remained until recently the preferred home of his distinguished heir, the marquis de Colbert, and the peaceful site where Laplace was buried.[1] Still, neither his hereditary background nor the physical and cultural surroundings suffice to explain his remarkable ascent. In his youth more subtle forces and opportunities were at work.

## Beaumont-en-Auge and Laplace's Ancestors

Some earlier writers have offered romanticized legends of Laplace as a young boy struggling to extricate himself from poverty and ignorance. But although he eventually surpassed all his forebears in wealth, position, title, and renown, it was not by pulling himself up by his bootstraps from lowly origins. His lineage can be traced in the early seventeenth century to a bourgeois of the hamlet of Bourgeauville, little more than a league west of Pierre Simon's birthplace.[2] Olivier de Laplace and his son François—direct ancestors of the fourth and third generations—were both referred to as *Maître,* a title of respect usually reserved for notables in the community. Olivier married into the Follebarbe family, which generations later produced Beaumont's notary public and mayor (in whose care our scientist would come to entrust his family affairs). The Laplaces included some of the region's most honorable citizens: another Olivier de Laplace, a royal surgeon in Bourgeauville; his daughter Marie Anne de Laplace, who married into the Le Carpentier family (whose elder sons traditionally secured royal titles); another Marie de Laplace, great-aunt of Pierre Simon, who married Doctor Robert Carrey of Lisieux in 1743;[3] and her brother Thomas François de Laplace, also a surgeon. Pierre Simon's mother, born Marie Anne Sochon, had roots in the nearby towns of Vauville and Tourgéville, two leagues to the north, and she and her ancestors moved in similar circles of small landowners, doctors, lawyers, and minor bureaucrats. After a successful career as a munitions supplier in Rouen and Lorient before the Revolution, Pierre Simon's first cousin Louis François Cordier, for example, became censor of the Bank of France.[4]

Laplace's family situation fits well with what we know about rural life in Normandy in his day from reconstructions that social historians have pieced together from local records. Around Beaumont, a village of about five hundred residents, life centered on the fertile land and its yield.[5] The Auge countryside was known then, as it is today, for its orchards, green grazing land, heady cider and powerful calvados, beef carted to Rouen and Paris, and savory cheeses. Only two leagues to the east, on the road from Caen to Rouen, was Pont-l'Evêque, the local administrative capital with a population of some 1,500, animated by the bustle of civil servants and lawyers. Six leagues to the south stood Lisieux, four times as populous and the seat of the bishopric encompassing Beaumont and its old priory. Communication between these towns was frequent, and families traveled without difficulty in the immediate area. Yet although local inhabitants mingled easily, they rarely migrated out of the region. Of marriages in nearby Annebault and Bourgeauville, 90 percent were of couples who both came from the same local area; over 75 percent of the women did not look beyond a two-league radius to find a spouse.[6]

While there were still pockets of poverty in rural Normandy and serious grain shortages in times of poor harvest, during the last half of the eighteenth century there were fewer periods of famine than earlier and no wars.[7] In the modest circles to which Laplace's ancestors belonged, life was becoming less harsh and the future more promising. Before the Revolution, the bourgeois of the pays d'Auge ate better and their children were less likely to succumb to early death.[8] Nonetheless, the population failed to grow much, possibly because women were marrying later (and because couples exercised restraint). In a detailed study of eighteen nearby parishes, Nels Mogensen found that women from landowning agricultural classes, on average, married at the late age of 25.6 and bore only 4.7 children.[9]

When he married Marie Anne Sochon at Tourgéville's parish church on 6 July 1744, Pierre de Laplace, father of our scientist, was twenty-six and one of six children. Three weeks before the wedding, the extended family had gathered in the nearby manor of Glatigny to witness the all-important signing of the marriage contract.[10] It was a grand affair witnessed by Pierre de Laplace's widowed mother, Marie Piel, and her two sons-in-law, Robert Carrey and Jacques Mabon.[11] In attendance with the future bride were her mother, Marie Le Chevalier (the only witness unable to sign her name), her mother's second husband, Olivier Le Coq, her mother's brother-in-law, Jean Jacques Sochon, and a son of her mother by her first marriage, Antoine Sochon. We find among the other signatories several Bretocqs; Marie Madeleine Isabel de Lamotte, owner of Beaumont's inn, the Belle Croix; and Marie de Lanney and Pierre Halley, who a few years hence would become Pierre Simon's godparents. The alliance of the Laplace and Sochon clans, concluded in an elegant setting belonging to local nobility, must have been the talk of the town.[12]

One of the branches of the Laplace family had already moved away from Bourgeauville. In the marriage contract, Pierre de Laplace gives his occupation as merchant of the neighboring town of Beaumont, which had been entirely rebuilt following a fire in 1689. By Christmas of the year of his wedding, he had taken out a six-year lease from Miss Isabel de Lamotte to run her tavern and inn, the Belle Croix, at the main crossroads in town.[13] It is likely that the newlyweds moved in and continued to occupy the premises for much of their married lives. Pierre de Laplace thrived in his new role as innkeeper, for within a few years he was wealthy enough to start buying nearby property, which he rented to farmers for income.[14] By 1754, he was paying taxes on a declared annual income of 750 livres derived from his property. By 1775, he had become the fifth-largest tax contributor of Beaumont, with an income that year of 1,100 livres.[15]

## First Steps

Among Pierre de Laplace's properties was a farmhouse in the township of Beaumont. Off the road from Caen to Pont-l'Evêque, near la Haie Tondue, the house was known as Le Mérisier, which has on occasion been designated as Pierre Simon's birthplace. Biographical accounts published in the 1810s, while he was still able to deny the story, specify that he was born only "near Beaumont," giving some credence to this account.[16] While there is evidence that his father occupied the farmhouse after he was widowed, one would need proof before agreeing it was the site of the famous scientist's birth. Although Pierre Simon spent time there as a child and grew attached to the property in his old age, it seems clear that he was instead born in the center of town, in a room facing rue Massue, at the Sign of the Belle Croix.[17]

Pierre Simon was the fourth of five children. An older sister, Marie Anne, was born in mid-June 1745. Twins Jacques Pierre and Julie Marguerite were baptized on 12 January 1748, but died three days later.[18] A year after Pierre Simon came Olivier, on 1 May 1750. We generally know little about the siblings and their children, none of whom figure in the scientist's later life. Marie Anne married a barely literate merchant from Périers named Henry Martinne in 1769, and they had a daughter who lived in Beaumont in the 1830s as Madame Louize.[19] Marie Anne had a second marriage to a Clarbec thatcher named Marin Simon Guernier in 1788, shortly after her father's death and the settlement of the family estate with her brother Pierre Simon.[20] Thereafter there is no indication of any special bond between Pierre Simon and his sister or her children. The last significant contact between them seems to have been in July 1788, shortly after their father passed away, when she signed away her inheritance rights to Pierre Simon in exchange for an annuity.[21] As for Olivier, who was still alive when his sister first married in 1769, he disappears completely from local records and is presumed to have died without issue.

Indeed, of the immediate family, only the father appears to have had contact with Pierre Simon past his teenage years. Uncle Louis de Laplace, deacon at Beaumont and later chaplain of nearby Criqueville, died in 1759 when Pierre Simon was a mere ten years old.[22] His own mother, who often served as godmother for local children, last shows up in baptismal records in 1757, and is listed as deceased at her daughter's first wedding in 1769.[23] We do not know when or where she died, but it seems as though she left Pierre Simon's world, creating an emotional void early in his life.[24] The absence of references to his youth in his writings or personal correspondence, and his subsequent avoidance of his native land until an advanced age, suggest that Pierre Simon wished to distance himself from those difficult critical years of development.

At the start of his days, however, Pierre Simon was well protected by immediate family and local friends. Born at home on 23 March 1749, he was strong enough to be taken to the local parish church two days later, where he was baptized by vicar Leperchey.[25] For his mother, who had lost twins a year earlier, there must have been considerable anxiety, followed by satisfaction, as the youngster flourished; for the father, pride to have a male heir on whom to fix his hopes. The choice of godparents for an eldest son was no casual matter, because they were expected to assume responsibility for him in the event he was orphaned. Pierre Simon's parents selected two local notables who shared their values and sense of well-being. Marie Magdeleine de Lanney, the godmother, came from a reliable local family that included a military officer, a lawyer, and a royal official.[26] Cousin Pierre Halley, the godfather, himself also a royal bureaucrat, was serving at the time as a minor official at one of the courts of Pont-l'Evêque.[27] The parents were ready, if fate demanded it, to place their son in the hands of local notables who shared their values and their sense of well-being. This shows the family considered itself to belong to a fortunate and worthy set of local Norman gentry. It was quite beyond anyone's dreams that this infant would someday be a marquis, peer of the realm, and that their commonplace family name would be recognized throughout the world.

The Laplaces' situation reflected the stability and moderation of their class. While not forced by economic conditions to stay at home, agricultural landowners and minor bureaucrats seldom migrated from the local countryside. The family was comfortable without being wealthy, with its members enjoying local popularity in part due to tight family connections, but none showed signs of wanting to move up the class ladder. They neither had nor sought connections with the nobility or the ruling elites in Rouen or Paris. Like most in their milieu, the men were usually literate, and many of the women could read or write, but without necessarily having had elaborate schooling.[28] None before Pierre Simon wrote anything that survived in print. One easily imagines a contented family circle, pleased with its lot in Beaumont. The scientist's father, Pierre de Laplace, is the only one who showed any particular ambition.

Perhaps it was his family connections with minor officials in Pont-l'Evêque that stimulated the elder Laplace's interest in improving his situation. The town was teeming with functionaries representing the various civil jurisdictions who seemed ready to seize opportunities for bettering their lot.[29] In addition to his connection to the court clerk Pierre Halley, Pierre de Laplace was related through his wife to the Bretocqs of Pont-l'Evêque, who had important land holdings in Beaumont.[30] More likely, it was Laplace's popularity as the successful manager of the inn that spurred him on. In provincial towns, the tavern was the focus of male society, and its owner almost automatically became the best-known individual in town. Whatever the reason, Pierre de Laplace was elected mayor of Beaumont-en-Auge shortly after his eldest son was born, a position that he held for more than thirty years.[31]

As the innkeeper, Pierre de Laplace was in an ideal position to perform his major task as mayor, to establish property rolls, and to collect taxes—all jobs that led him to encounter others from outside the immediate area, from towns such as Lisieux and Rouen.[32] Lisieux was the religious center to which his brother, the abbé Louis, owed his allegiance and where favors were dispensed by then-bishop Jacques Marie Caritat de Condorcet, whose family Pierre Simon later had to reckon with.[33] For civil matters, Beaumont was part of the *élection* of Pont-l'Evêque, itself in the *généralité* of Rouen, the capital of Normandy to which most of the property taxes flowed. But Beaumont was also an administrative center for the Vicomté d'Auge, an area that had been ceded to the Orléans family in the late seventeenth century.[34] The duke of Orléans had jurisdiction over various sales taxes, particularly on liquor and notarial transactions, and it is likely the mayor's tasks included collecting and sending such tax money to Paris. Unfortunately we do not have specific information about Pierre de Laplace's contacts, but surely they were numerous. Money and information flowed through his hands, and undoubtedly he put them to good use for his personal advancement and that of his children.

Mayor Laplace had ambitions for his eldest male heir. Since the family was blessed with neither great wealth nor influence in ruling circles, the most sensible path to success for Pierre Simon lay in education. Uncle Louis, a man of considerable erudition, took on the task of introducing the boy to the rudiments of learning. Louis had time on his hands. Though named deacon and ordained, there is no record of his performing any ecclesiastical duties in town.[35] His appointment as chaplain of Criqueville in 1752 had been a sinecure, entailing neither the care of a congregation nor residence. It is likely that this leisured abbé lived under the same roof as the youngster he started on the road to success. Uncle Louis no doubt assumed his nephew would follow in his footsteps.

## Collège de Beaumont

A more structured opportunity for education was nearby. A short distance from the center of town stood the renowned priory of Beaumont, which since the eleventh century had been entrusted to the care of the Benedictine order.[36] For many years, the local monks had been giving religious instruction to the boys of the town, always in the hope of attracting the intelligent ones into an ecclesiastical career. Following the rebuilding of the priory subsequent to a devastating fire in 1689, the duke of Orléans, who ran the priory through his appointments, institutionalized the practice by creating a school, confirmed by letters patent in 1731 and 1741.[37] The twelve monks who ran this new enterprise quickly established a solid reputation for the education of sons of Norman families, both noble and bourgeois, and even won special notice for their patient training of the deaf and dumb. Often understaffed, the *collège* called on the abbé Louis de Laplace for assistance, even though he was not himself a member of the Benedictine order. We do not know whether Pierre Simon learned to read, write, and count from his uncle at home or in the nearby school buildings, but before long, he was entirely entrusted to the care of the local Benedictine monks, under whose influence he remained throughout his youth.

Boarding students could begin study at age five, provided they would not "dishonor their family by unacceptable behavior," but generally were sent there after they turned seven. They would proceed through all the grades from *cinquième* through *rhétorique* as their individual progress warranted. Advertisements for the school display a strict set of rules for boarders, including required clothing; fixed, short vacations; and detailed regulations for when children were to go to class, study, and play. Parents paid three hundred livres annually for full room and board. In this as in other respects, the Collège de Beaumont differed little from other schools run by the clergy throughout the land. Only the special protection of the duke of Orléans singled it out, for he provided that full scholarships be offered to six needy noblemen and specified that sons of local inhabitants who lived at home could attend without cost. In recognition of his benevolence, students were required to pray for the duke daily, and "complimented" him with a formal address on every public occasion. In prospectuses, the school was known as the "collège de S.A.S. Monseigneur le duc d'Orléans." Thus from the very beginning Pierre Simon had a free and substantial education, which he owed to a distant protector in Paris. The same Orléans family was to figure often in his life.[38]

We would dearly like to know when Pierre Simon's talents were first discovered and at what age he became seriously engrossed in his schoolwork. Not only did this signal the first major turn of his life, but it occurred just as his affective life suffered its most serious setback—the death of his mother. Unfortunately, contemporary records and eulogies are of no assistance; instead we must resort to a plausible reconstruction.

Sometime after turning eight years old, Pierre Simon lost his mother. He never wrote about this loss, nor made any reference to his older sister, who might easily have been a mother-substitute. Without the presence of women in his life, he learned to rely on the company and assistance of men very early on. This pattern persisted for the first half of his life. Only at age thirty-nine, by which time he was secure in his career and respected by his male peers, did Pierre Simon marry. He took a wife twenty years his junior, treating her at first more like a child-bearing apprentice than a partner. She bore him two children in rapid succession, a boy and a girl. Sophie, born during the French Revolution, was clearly his favorite, the one on whom he lavished his affection. Indeed she seems to have been the only female to whom he was ever able to openly express his love. Her tragic death after the delivery of her first-born in 1813 was so moving an event that Laplace took seriously ill and shortly thereafter made out a will. The trauma he experienced at age sixty-four served no doubt as a painful reminder of the loss he had sustained as a child, when his mother was taken from him, and led him back to Normandy, in which he took a renewed interest at an advanced age. One senses behind this relationship with his homeland and with the opposite sex a set of circumstances that drove Laplace into an austere posture, devoid of warmth and sociability. Yet glimpses of his private life after midlife also suggest he was capable of deep emotional reactions that he had managed for the most part to keep under control.

It was in the same early and impressionable period of his life that Pierre Simon also lost his younger brother Olivier and his favorite uncle, Louis. Only his grieving father, busy with his affairs, remained. Inevitably the lad was drawn closer to his school and its activities. The regimen of the *collège* could easily fill his days, with classes and study halls beginning at eight o'clock in the morning and lasting for eight hours daily throughout the week. Not a moment was squandered, not even at mealtime, when silence was prescribed for the students while a monk would read out loud. Education was characterized by a grueling quantity of copying and endless memorization, and success required a capacity to regurgitate the teachers' lessons upon command. To receive praise, the boys had to conform to Benedictine ways and to learn without question. So concerned were the teachers about constant application that they even recommended to parents that the six-week summer vacation be skipped every other year. Today we still use the phrase "work like a Benedictine" to conjure an image of perseverance and tenacity. A contemporary of Laplace later recalled that he displayed many of these characteristics. He had a "precocious intelligence, uncommon inclinations at an age when rural children are barely embarking on the first elements of reading, and above all a prodigious memory."[39] To this must be added the habit of an unbending will to reach the appointed goal, no matter how demanding. Very early in his life, then, Pierre Simon's working habits and approach to learning were fixed by exposure to Benedictine ways. His feelings of separation from human affection in the family only drove him with more determination into the kind of behavior that would gain him attention and appreciation from his superiors, even if not their warm embrace.

One must not assume that the content of learning was as traditional as the mode of instruction. For financial reasons, the Collège de Beaumont, like other schools run by the Congregation of Saint-Maur, had to serve a variety of students, not merely those destined for holy orders.[40] Traditionally, day students like Laplace were more likely to embark on a secular career in the professions, in civil service, or in business. Sons of lesser nobility on their way to the officers' corps attended schools like Beaumont because the offerings were well suited to that calling. It is revealing that a decade after Laplace left Beaumont, when the *collège* was turned into a military school run by the same congregation, few curricular changes were required.

Latin and the classics remained the staple diet of students, particularly in the lower grades, but as the students progressed they were introduced to more modern subjects. The vernacular was used to teach geography, history, and elementary mathematics. In some Benedictine schools, the older boys were able to enroll in supplementary classes devoted to dance, drawing, heraldry, mathematics, or music to fill out their knowledge. Customarily they would display their learning before proud parents in public exercises held at the end of the school year. It is only from the programs announcing these recitations that we can gain an indirect idea of the quite respectable level of learning reached by the better students like Laplace.

Beaumont was not markedly different from other provincial schools run by the rival orders of the Jesuits or the Oratory.[41] The 1773 public exercises at Beaumont included translations of the Scriptures and various texts of Cicero, Virgil, and Erasmus; explanations of portions of the New Testament; and answers on the plot of Virgil's *Aeneid*.[42] Older boys were expected to treat the lives of holy figures and classical authors, while one class fixed its attention on Catiline, basing its knowledge on translated passages from Sallust and Cicero. Mathematical prowess was also on display; the school deemed the subject significant, particularly for those preparing for a military career, and it was a way of determining the students' progress. "The first elements of mathematics are simple and filled with certitude. The torch of truth presides over all the progressive knowledge one learns through it, with each happy discovery engendering another still more interesting."[43] Beginners answered problems in arithmetic, plane geometry, and trigonometry taken from abbé Jean Baptiste de La Chapelle's *Institutions de géometrie*.[44] More advanced students tackled algebra, providing solutions to equations of the first and second degree, and answering questions about the binomial theorem and series. They were even expected to have reached an understanding of analytic geometry, drawing on it to explicate the properties of conic sections. All these topics were treated intelligibly in La Chapelle's books, which were comparable to those of the more famous Charles Etienne Louis Camus, used at another Benedictine *collège* at Sorèze.[45] Talented boys like Pierre Simon thus had a chance to master the elements of mathematics at an early age, and to develop an interest in the subject. There were also occasional opportunities to apply book learning to outdoor exercises in surveying.

In the 1760s the school had only about sixty students, suggesting that there was a great deal of interaction between pupils and teachers. The only instructor of note with whom Pierre Simon probably came into contact during his last year at Beaumont was Dom Charles Antoine Blanchard, who was then twenty-seven years old.[46] Though born in the Ardennes, Dom Blanchard had completed his studies in the humanities at the nearby University of Caen, where he also studied philosophy and theology before being ordained. It was quite natural for promising local students to be directed to the same town to complete their training. It was equally natural that Laplace should consider going to Caen to pursue a religious career. Given that his entire education until age sixteen had been set in a religious context, beginning with his uncle Louis and ending with Dom Blanchard, these aspirations should come as no surprise; instead Laplace's eventual decision to depart from this course is the more startling outcome.

One might wonder if Laplace's mathematical aptitude, which may well have been manifest at an early age, engendered some conflict with his path to the clergy. It is not likely to have troubled him at first. In ancien régime France, the roads to all careers requiring book learning—theology, law, the military, and science—were largely indistinguishable. Given the easy mixture of mathematics with classics and Holy Scriptures in the curriculum at Beaumont, the habit of including technical subjects in religious schools, and Uncle Louis's talent with numbers, science and religion would hardly have seemed in conflict.[47] Mathematics was cultivated as much as a tool to train the mind as it was for its use in military or technical occupations. At this early stage of the young man's development, whatever talents he had displayed, they were most likely considered an asset for his entry into the clergy. He was therefore sent to the University of Caen, probably in 1765.[48]

## On to the University of Caen

While the choice of Caen is not surprising, it was not the only university available, and certainly not the most sensible had his father already recognized Pierre Simon's talents as a mathematician. Among those near contemporaries who would become colleagues in the Academy of Sciences, most moved closer to Paris after completing their elementary education.[49] Thus, the future geometers Marie Jean Antoine Caritat de Condorcet and Lazare Nicolas Marguerite Carnot, born respectively in 1743 and 1753, first attended local Jesuit and Oratorian schools, much as Laplace had attended a Benedictine one. But Condorcet completed his studies at the University of Paris, matriculating at the Collège des Quatre-Nations, an excellent school eminently suited to the fatherless nephew of the noble Bishop of Lisieux. Carnot, son of a successful bourgeois lawyer, had already opted for a military career and was sent to a preparatory school in Paris, from which he passed the competitive examinations to enter the Royal Engineers. In the capital, both thrived under the tutelage of excellent teachers. The astronomer Jean-Baptiste Joseph Delambre, born as was Laplace in 1749, studied first with Jesuits in his home town of Amiens and was sent to the Collège Duplessis of the University of Paris. Unlike Condorcet and Carnot, Delambre excelled in letters, took steps to enter the priesthood, and eked out a living as a tutor. It was only many years later, when he took advantage of public lectures in astronomy given at the Collège Royal in Paris, that he began to discover his true vocation. Yet another future colleague who studied at Caen at the same time as young Laplace was Félix Vicq d'Azyr. He too was originally destined for the clergy, but his father sent him to Paris where he began his brilliant ascent into the world of anatomy, physiology, and medicine.[50] Like three other colleagues, Antoine-Laurent

Lavoisier, René Just Haüy, and Adrien Marie Legendre, these future academicians were led into science by exposure to the most advanced education being offered in France, and their interest was reinforced by experiencing Parisians' excitement about the latest scientific discoveries. Lavoisier and Haüy, born like Condorcet in 1743, also attended the Collège des Quatre-Nations, but embraced science as a career only after trying other fields. By contrast Legendre, the mathematician three years Laplace's junior, found his vocation immediately at the same school, with the aid of a teacher who from the start recognized his outstanding gifts.

If Laplace's father had intended him to develop into a scientist, a Paris education would have made the most sense. More likely, neither father nor son recognized or acknowledged his scientific potential when he was to leave Beaumont-en-Auge. Pierre Simon's teachers must have seen in him a potential seminarian who might well follow in the footsteps of his late Uncle Louis. If this reckoning is correct, the University of Caen with its rich offerings in religious education was the most sensible option. Laplace was sent there in the expectation that he would become a priest. He stayed for three or four years that proved pivotal for his career.

# 2

# From Scholasticism to Higher Mathematics

We know very little about the young Pierre Simon's university days at Caen. In the biographical literature, there are suggestions that he went there as an auspicious student sponsored by the duke of Orléans, who paid his fees, and that he turned his back on the priesthood as soon as mathematical literature fell into his hands. But without supporting evidence, one must be skeptical of such stories. There is no solid foundation to believe either that his intellectual potential had made itself manifest to a protector before he left his hometown *collège*, or that mathematical inspiration came to him when the forbidden fruit finally appeared. My reconstruction of his crucial years at Caen, buttressed by disconnected anecdotes and incomplete university records, does not indulge in such overblown claims, although it acknowledges their symbolic meaning.

It was Pierre Simon's show of intelligence and his general success as a responsive student at Beaumont that had steered him to Caen in the first place. When Pierre Simon arrived there, he was likely to have been recommended by Dom Blanchard who, only a decade earlier, had spent seven years at the Collège du Bois, one of the three schools constituting the Faculté des Arts. Moreover, mathematics was not initially unveiled to him at Caen. Laplace had already absorbed the conventional but solid grounding in algebra and geometry provided by his Benedictine teachers in the La Chapelle text. If he was to experience something of a revelation, it was to be about the world of higher mathematics and its applications, which were quite beyond his Benedictine teachers' ken. This decisive exposure and its meaning probably grew on him during his last years at Caen, without immediately provoking a renunciation of the initial goal to enter the clergy. While his infatuation with calculus grew with time, and was authentic, Laplace was not likely to have experienced a mystical conversion. There was never anything sudden or impulsive about his actions. It seems more consistent with his later behavior to assume a gradual recognition of the attractiveness of the sciences.

At first the young teenager, confronted with a metropolis of thirty thousand, must have been at once frightened and exhilarated.[1] Even though Caen was only a day away from home by coach, the city was a long way from the tranquility and security he had experienced at the Belle Croix inn and with his monkish teachers. Students arriving in Caen often lodged at one of the many boarding houses of Caen's "Latin Quarter," in the northwestern part of town, where they pursued their studies in relative freedom from parental guidance.[2] Under a system known as *caméries,* older students at the student pensions would guide novices in their studies and introduce them to their new world. There was much gossip to be learned about teachers, assistants, and beadles; tricks to be learned for surviving in the new environment; and initiation into the usual dissipations enjoyed by university students everywhere. Most of the known complaints about them from townspeople centered on unpaid debts and excessive drinking and gambling, though the University of Caen was not especially notorious for having rowdy students.

We have no information about where Pierre Simon lived or his behavior. Given his enormous intellectual progress in a few years, it is safe to assume he was studying intensely. He was never one to indulge excessively in anything but work. Even in his later writings on probability theory, where one might expect to find games of chance as examples, he rarely featured them or railed against gambling. The only indirect indication that we have about his life was participation in Church services as an acolyte near the site of the university. Pierre Aimé Lair, a Caen agronomist who accompanied Laplace on a visit to the city in 1813, related a touching story about his Church participation. In front of the old Eglise Notre Dame de Froide-Rue, situated a few blocks from the Collège du Bois, Laplace spied a young boy doing penance for a minor infraction, and a surge of nostalgia overtook him. Addressing the priest in charge, he asked forgiveness for the lad in the name of his own service to the parish forty years earlier. "I wore a surplice in your church," he explained, "and remember it fondly."[3] No doubt his involvement in parish activities at Caen had once provided that reassuring link with the life he had left behind at Beaumont. By trading the plain black dress worn at the Benedictine establishment for a white surplice, he was marking his slow ascent up the religious ladder that would eventually lead to a degree in the Faculty of Arts, and thence perhaps to theological studies and ordination.

## Learning the Standard Curriculum

University students prepared for their examination for a master's degree by enrolling in a required two-year course in philosophy under the stewardship of a titled professor at one of the three *collèges* of the Faculty of Arts. This was normally followed by an additional year for review, during which the elder matriculants often tutored younger students. There is good reason to believe that Laplace followed this standard sequence, starting his studies in philosophy in the fall of 1766, at the age of seventeen. Records show that his fee for the master's degree was settled in mid-1768 and the examination passed a year later on 1 July 1769, with an attestation that he completed the two-year course with philosophy professor Jean Adam of the Collège du Bois.[4] Thereafter, the young student worked closely with two other faculty members, Christophe Gadbled and Pierre Le Canu, who are described by the biographer privy to local oral tradition as "more than mentors, [indeed] friends."[5] He is then reported to have tutored a young noble in the prominent d'Héricy family.

A less contemporary source, however, reports that Laplace had arrived in Caen a year earlier, which would place him in a rhetoric class prior to his course with Adam.[6] In 1765, he would have studied under François Moysant at the Collège du Mont, Jean Bouisset at the Collège du Bois, or whoever held the position at the Collège des Arts. Moysant was a man of considerable experience, having taught letters at Lisieux before obtaining a medical education in Paris.[7] He was the co-author of several compilations of surgical and historical information, and would later edit an excellent anthology of vernacular literature destined as a textbook for schoolboys. Moysant had become dean of the faculty by the time Laplace took his master's examinations in 1769. Bouisset also had connections in Paris, where he befriended Helvétius and d'Holbach, and later used his charm and conversational wit to enter the good graces of the Comte de Provence.[8] He gained a reputation as the most successful professor the Collège du Bois ever had, and was hired by the intendant Fontette as a tutor for his children. Either of these teachers would have opened up a larger world to Laplace than even the best of his Beaumont mentors could muster.

If Laplace indeed spent his sixteenth year at Caen, I suspect it was under the rhetoric teacher at the Collège des Arts, whose name is not known. Puiseux recorded the hearsay that Laplace matriculated there when he left Beaumont.[9] In addition, there is the not altogether clear notation on the payment rolls of the Faculty of Arts that Jacques Louvel, at the time professor of *seconde* at the Collège des Arts, paid Laplace's fees for the master's degree.[10] What muddies this interpretation is a further notation on the rolls showing it was paid in the name of the Collège du Mont, with which Louvel had no connection. Louvel was to be named rhetoric professor at the Collège des Arts in 1767, and later as university chair of Greek and rector of his *collège*.[11] He was also in attendance at Laplace's examination in 1769.

Whatever the case, neither Moysant, Bouisset, nor Louvel figure significantly in Laplace's career. If they were his teachers, they have now blended into the same obscurity shared by Dom Blanchard at Beaumont. Not so with Adam, Gadbled, and Le Canu, who each of whom made his mark on the education of the future scientist.

Adam and Gadbled, with their strong personalities, undoubtedly helped turn Laplace away from a religious vocation and toward a scientific one. Jean Adam represented all that was conventional and outmoded. Born in Normandy at Pierrefitte (Orne) in 1726, he earned his master's of arts degree at Caen in 1745 and continued his studies there until ordination.[12] The doctorate in theology entitled him to a teaching post at the university, and eventually to a lucrative post as a canon of Caen's leading religious institution unattached to a regular order, the Church of Saint Sépulcre. Canon Adam's appointment included lodgings close by to encourage his participation in the active life of the Saint Sépulcre assembly as one of its ten ruling officials. He often served as its spokesman on business affairs, occasionally representing the congregation before the Parlement in Rouen and royal councils in Paris, where he also held the title of "officier de la chapelle du Comte d'Artois."[13] In all these transactions, he proved himself a loyal and steadfast agent. By supporting the corporate interests of his church with detailed briefs and great fervor, he earned local notoriety and considerable contacts in high places. All this public attention was to serve him poorly when the Revolution came. He joined his colleagues in refusing to submit to a secular oath, sought refuge with his niece in the village of Verson in 1792, and was discovered there by a patrol of the National Guard in pursuit of counter-revolutionaries. He was arrested, interrogated, imprisoned, and publicly humiliated before he could slip away to the Isle of Jersey.[14] Adam died in 1795 and was interred with other Norman émigrés at London's Saint Pancras. To the end he remained a traditionalist—caught in the revolutionary storm, but unbending in his conservative convictions.

As a university professor, Adam had a similar change in fortune. Beginning as an orthodox supporter of Jesuit traditions in theology and metaphysics, he ended as the laughing stock of young students who caricatured him pitilessly as a pedant.[15] Adam has the dubious distinction of being the only Caen professor to have inspired a one-act printed farce in verse that identifies him openly as a charlatan. In *Nostradamus, ou le physicien plaideur* (1779; Nostradamus, or The Imploring Physicist), Adam is portrayed as a vain and litigious fool who proclaims himself "the best physicist of the land," all the while filling local sheets with his inane observations and flimsy theories, claiming cures by electroshock therapy, and parading himself as the wronged party in ludicrous legal battles often lost.[16] The most damning account of this would-be physicist was a remark committed to print and attributable to Laplace that he taught "lessons his pupils would be more than happy to put out of mind."[17] This passage, underlined to indicate a direct quote, appeared in a pamphlet authored in 1779 by Laplace's other significant teacher in an attempt to destroy Adam's reputation as a worthy instructor of science.

When Laplace became Adam's student in 1766, however, the conservative professor was riding high in university circles. He had been named chair of philosophy at the Collège du Bois in 1758 by the new principal, the Irish-born Milesius MacParlan.[18] Both Adam and MacParlan befriended the Jesuits of the nearby Collège du Mont, who traditionally attracted the wealthiest students in Normandy and had a well-founded reputation for producing Caen's intellectual elite.[19] One student of the Jesuits, Pierre Varignon (who was ordained), had become a prominent mathematical physicist and academician more than a half century earlier. Moreover, Jesuits had emerged as a major force in the city's life. They were the object of both admiration and envy, particularly by new teachers like Adam, who depended on student fees to supplement their paltry salaries. When the Jesuits were disbanded and chased out of the university in 1762, Adam seized the opportunity to promote his own cause. His reputation as a partisan of traditional Jesuit philosophy immediately assured him larger classes, and the purchase of their physics demonstration apparatus—which he cleverly bought at a low price in collusion with the auctioneer—made him an even more popular lecturer.[20] He quickly became the envy of other philosophy professors, who accused him of permitting students to matriculate in his course even though they had not completed their rhetoric class, as required by university rules. To add to their annoyance, Adam would swell his ranks by giving showy demonstrations and lecturing in a popular style. He was later reported to have had as many as three hundred auditors a year.[21]

In the 1760s Adam was also busy making a name for himself in print. He began by disputing the views of the venerable poet and engineer François Richard de La Londe on the feasibility of turning the Orne into a navigable river.[22] He also wrote a pamphlet entitled *Réflexions d'un logicien à son professeur* (Comments of a Logician to His Teacher), taking to task the anonymous author of the *Mélanges de littérature, d'histoire, et de philosophie* (Miscellaneous Pieces in Literature, History, and Philosophy) for metaphysical assertions that contradicted Jesuit convictions and smacked of deism.[23] Like all informed people, Adam knew the identity of the author: the chief public spokesman of "modernists" who had made considerable inroads throughout France since the publication of the *Encyclopédie*—none other than the mathematician Jean Le Rond d'Alembert, who had campaigned against Jesuits and was properly accused of advancing the cause of irreligion.

Adam's *Réflexions* are not especially interesting, but they do situate him in a particular philosophical tradition that Laplace experienced as a young man. Young Pierre Simon was educated in the midst of hotly contested issues that pitted his mentors against each other while he was still at a formative age. While we can readily accept the worthlessness of Adam's lessons, we must treat them seriously to reconstruct the intellectual milieu encountered at the time the young man moved from away from a religious vocation to science. Though Laplace was soon to dismiss Adam's teachings, he cut his intellectual eyeteeth following the debate that flared between Adam and Gadbled.

## Gadbled versus Adam

Adam's rival, Christophe Gadbled, also hailed from Normandy, having been born a few years after Adam in 1732, at Saint Martin Le Bouillant (Manche).[24] Like Adam, he was an ordained priest, a canon at the Church of Saint Sépulcre, and on the faculty of the university. He had received instruction in theology at the University of Paris, where he served as "maître de conférences de physique" at the Séminaire Saint Louis from 1757 to 1758. While a student, Gadbled had become interested in the latest features of science through his acquaintance with both the abbé Nollet, who was enticing large audiences as a physics demonstrator at the Collège de Navarre, as well as his academician colleague and chair holder at the Collège Royal, Pierre Le Monnier. These two progressive scientists represented the experimental and mathematical sides of the new science, following a path chartered by the proponents of Newton. Gadbled also was in contact with the mathematicians Etienne Bézout and d'Alembert. Undoubtedly Gadbled's commitment to the secular sciences and his adoption of many of their unorthodox philosophical views turned him into Adam's natural enemy. Professional rivalry must also have played a role.

When Gadbled returned to Normandy, he became at once a colleague and competitor of Adam. He moved to Caen in the early 1760s, where he was elected professor of philosophy at the Collège des Arts, succeeding Thomas François Le Guay.[25] In 1762, when the Jesuits were unceremoniously removed from the university and vacated their appointments at the Collège du Mont, Gadbled was temporarily named professor of mathematics.[26] He quickly secured the nomination from both the Crown and the city to become royal chair of hydrography at Caen, agreeing to resign his philosophy post and to relinquish his voting rights at the university while continuing to teach mathematics (without a salary), in exchange for retaining his lodgings at the Collège des Arts.[27] This successful career move was helped by strong backing from Parisian scientists, who counted on Gadbled to introduce the modern mathematical sciences into the university curriculum.[28]

Soon after his move to Caen, Gadbled joined the local intellectual community by gaining admission to the progressive Caen literary academy, where he would from time to time read a paper. On 5 May 1763, for example, he gave a well-publicized address entitled "Mémoire sur la haute géométrie" (Memoir on Higher Mathematics), in which he showed how a physics problem could be solved easily with the aid of integral and differential calculus.[29] A select group of students from the university would often attend these discourses, held at seven o'clock in the evening.[30] Gadbled was acquiring a local reputation as a partisan of modernist views for the study of the sciences. Indeed, he was among the earliest teachers to offer instruction in the calculus.[31] The novelty of his approach must have stimulated the young Pierre Simon, whose exposure to mathematics through the La Chapelle text had been limited to elementary topics.

It was this predilection for mathematical theory that first brought Gadbled into conflict with local townsmen and with the jealous Adam. The conflict was complicated by various alliances opposing change. Until their dismissal in 1762, the Jesuits at Caen had been in control of a Royal Chair of Hydrodynamics, originally created in 1681 to improve local navigational skills.[32] The renowned Jesuit philosopher and teacher Father Yves André had since 1726 held this post, which was affiliated with the Jesuit-dominated Collège du Mont. In accordance with the traditional instructional methods favored by the Jesuits, André focused on theory at the expense of practical training. But the Admiralty reversed these pedagogical practices in 1762 by assigning an accomplished local mariner, René Jacques Le Gaigneur, to teach hydrography. In the preface to *Le pilote instruit* (The Educated Navigator), Le Gaigneur explains that ship captains had become disenchanted by the rigorous mathematical demonstrations of the Jesuits, which they refused to follow.[33] Le Gaigneur's more direct approach pleased local pilots, who flocked to his course to learn how to sail through dangerous waters without suffering through lectures on the mechanism of the heavens. But when the retired and aged André, still the royal professor, died in 1764, the Crown replaced him with Gadbled, totally disregarding Le Gaigneur's appointment by another arm of the government.

This new selection was ratified by the university and by the city fathers despite protests by Le Gaigneur's supporters. Each appointed teacher offered a class. Bad feelings between them simmered until a major controversy erupted in 1772, when Le Gaigneur sought unsuccessfully to block Gadbled from employing the title of royal professor. Though Le Gaigneur lost his legal challenge, he continued giving, for a fee, private courses in hydrography in Caen that attracted a larger audience than Gadbled's mathematically sophisticated lectures.[34]

Into this local dispute entered Adam, who sought to increase his fee-paying audience by introducing the elements of hydrodynamics into his physics classes, given in the second year of his natural philosophy sequence at the Collège du Bois. Since the mid-1760s, he had added several lectures to his course, sprinkling them with physical demonstrations of hydraulic principles. But in keeping with his predilections, these were devoid of serious mathematical theory, filled instead with terminology and rhetorical arguments, many of which were out-of-date with current scientific knowledge.[35] By the time Adam announced that he was preparing a new treatise, he was already being scooped nationally by the renowned lectures and the expert textbook of the Parisian academician Charles Bossut.[36] Adam's old-fashioned approach to physics was no match for Gadbled's contemporary handling of the same subjects. Exposed as he was to both teachers, and well aware of their personal disputes, Laplace could not fail to compare the two. His "conversion," if one needs to call it by that name, was neither sudden nor, in retrospect, surprising.[37] Others at Caen, like Vicq d'Azyr, recognized the shortcomings of Adam's teaching and answered the call of more advanced learning. Two other students of Gadbled show how far his inspiration could lead them. Pierre Le Canu, who studied with Gadbled from 1760 to 1762, became his teaching assistant and eventually joined the Caen faculty as a professor of philosophy and medicine; and the Chevalier Jean Jacques de Marguerie had a meteoric mathematical career as a naval officer, serving as secretary of the Académie Royale de Marine at Brest before his death in a naval battle in 1779.[38]

But the greater mathematical sophistication of one mentor over another is not a sufficient explanation for Laplace's career decision. There was, after all, no intrinsic reason why he could not become an ordained priest while pursuing his scientific interests. A majority of contemporary scientists, including Gadbled himself, were trained by ecclesiastical masters or were themselves in holy orders. Many of Laplace's teachers and later his colleagues held the title of abbé, including his Uncle Louis, the abbé de La Chapelle, Nollet, Bossut, and Delambre. There was nothing incompatible about having a religious affiliation and pursuing a scientific career. Laplace's willingness in 1768 to register for the master's of arts exam suggests he considered such a dual path seriously, and does not seem formally to have abandoned it until after he passed the examination in July 1769.

There were in fact more consequential circumstances surrounding the opposition between Adam and Gadbled. From them one can presume that Laplace's faith in traditional Christian beliefs was shaken while at Caen. Gadbled and his circle may well have taught Laplace to hold a skeptical attitude not only about physics, but about God as well. When Laplace abandoned the religious path in favor of a secular one, then, he did so also on theological grounds.

## Debates over Religion

Gadbled was not the first at the university to have been called to account for the dubious religious content of his teaching. Already in 1763, Lévêque, the new philosophy professor at the Collège du Mont, had been investigated by the university councils for allegedly dictating "moral [principles] which constitute a chef-d'oeuvre of heresy and contain a doctrine dangerous for both Religion and the State." One of the main problems was over his teachings regarding the antagonism between natural law and "true liberty." According to his lessons, there was an irreconcilable conflict between freedom of the will and physical necessity. One of the professors asked by the council to examine this charge was none other than Gadbled himself, who revealed himself quite skillful at reconciling the two by not only manipulating theological sources, but also quoting from John Locke's *Treatise on Human Understanding* and some of Bernard Fontenelle's articles written a generation earlier. Gadbled's defense of Lévêque was that the offending passages from his lectures were taken out of context and contradicted in another part of his lessons. In the report to the council, Lévêque's sin was labeled as inconsistency, not as a deliberate attack against religious principles.[39]

Less than a decade later, it was Gadbled himself who was accused by his colleagues, chief among them Adam, of sowing seeds of irreligion among his students. The exact nature of his transgression is not known, but Gadbled's public admission of allowing critical remarks against traditional belief to slip into his lectures is telling. His "apology" recorded in the minutes of the council of professors is hardly convincing:

> I declare that my beliefs have always been in conformity with Church doctrine (Catholic, Apostolic, and Roman) and that my intention has never been to teach anything contrary to it; that I always attempted to expound the most decisive arguments in its favor; that I never harbored a motive other than to establish clearly the existence of a God Creator, his sovereign providence, the essential difference between good and evil (prior to civil and political institutions), the spirituality of the soul and its immortality . . . [that given] the sizable number of superior students this year in my philosophy class . . . and having had to prepare my lessons hurriedly . . . it may well be that the stream of my ideas led me to discuss lofty *[élevées]* themes perhaps less appropriate for school boys than truly educated persons; that having used a compressed style in my propositions in order to include in them more interesting notions on logic, morals, metaphysics, as well as the elements of algebra and geometry, it may well be that I omitted a few intermediary concepts that would have revealed my true sentiments.[40]

Gadbled, it will be remembered, had studied in Paris in the 1750s and was influenced by d'Alembert, a notorious and clever critic of the Establishment, civil as well as religious. In his "Preliminary Discourse" to the *Encyclopédie* and in several articles, d'Alembert knowingly supplied arguments for the brazen thesis of the abbé Jean-Martin Prades, which had been presented and then condemned by the Paris Faculty of Theology. Prades was also commissioned to write some important articles for the *Encyclopédie*.[41] This realization outraged guardians of the faith, who, led by the Jesuits, unleashed a frontal attack on the editors of this planned multivolume work.[42] At nearly the same time, George-Louis Leclerc de Buffon's *Histoire et théorie de la terre* (History and Theory of the Earth) was also censored by the religious establishment for views inconsistent with the Scriptures.[43] A decade later, with the Jesuits now barred from France, authorities feared a relapse into apostasy. The reprinting of d'Alembert's views in a revised edition of his *Mélanges de littérature, d'histoire, et de philosophie* in 1759 was a further sign of the continuing vigor of antireligious sentiment in France. When these views were aired before unsophisticated youths at the university, they became especially threatening to orthodoxy.

We know that Jean Adam courageously challenged such views in his *Réflexions d'un logicien à son professeur* (1766), but we cannot do justice to his arguments because the pamphlet has disappeared—we can only surmise from d'Alembert's work what it was that Adam may have found so objectionable. All this would be of only minor interest, were it not that Laplace was studying with Adam and that he turned to Gadbled just as the pamphlet called attention to d'Alembert's writings. Given the existing tensions in university circles, one can presume that Gadbled took the Parisian scientist's side.

In the *Mélanges,* d'Alembert rehearsed all the editorial comments found in the *Encyclopédie,* starting with its enticing preliminary discourse and the introduction to its third volume. He also included an "Essai sur les gens de lettres" (Essay on Men of Letters), which is written both to justify their need for freedom of expression and to disparage those who censored them or confuted their arguments. Most troubling, perhaps, were the comments inserted in the eulogies of several early contributors to the *Encyclopédie,* including Edme Mallet and César Chesneau Du Marsais, both ecclesiastics. D'Alembert lauded them for knowing how to confine religion to its proper domain and for demonstrating an intellectual independence from authority.[44] All this insidious moralizing most likely irritated Adam. But his ire must have been especially inflamed by the methodological principles espoused by d'Alembert in his *Eléments de philosophie; ou, Sur les principes des connoissances humaines* (Elements of Philosophy; or, On the Principles of Human Knowledge), reprinted in the fourth volume of the *Mélanges.*

There, D'Alembert effectively undermines the grounds on which traditional teachers of philosophy courses in the university justified their curriculum. In a thinly disguised call to arms for intellectuals, he maintains that the "new method of philosophizing" adopted by the editors of the *Encyclopédie* will liberate "nations from the yoke of superstition and despotism that plunge them into darkness" and will "drive them all at once from the most profound obscurity toward true Philosophy."[45] This revolution will come about only when "reason extends its empire to all the objects of our natural knowledge," excepting, of course, revealed religion.[46] While stating that the actual objects accepted by faith cannot be questioned, d'Alembert boldly asserts that the grounds for our beliefs are subject to inspection by reason as well, since "the principles of Faith are the same as those which govern historical certitude; with the difference that in matters of religion, testimony . . . must have a degree of generality, evidence and force proportional to the importance and sublimity of its object."[47] By adopting this stance, he reversed the traditional primacy of religion over philosophy, and in the process compromised the principles adopted by Adam for his lessons in the Faculty of Arts. This point of view eventually became central to Laplace's worldview.

In his printed lectures, Adam devoted as much space to logic, metaphysics, and ethics as to physics. Each topic was covered serially in over 320 pages.[48] D'Alembert, by contrast, disposed of both logic and metaphysics in twenty-three pages and focused on mathematics and physics in eighty-two pages of text and forty pages of glosses. D'Alembert repeatedly praises the sciences as intrinsically worthy of special attention. Astronomy, he says, "is one of the sciences that does justice to the human mind." And physical astronomy is held in even higher esteem as a subject "that honors modern philosophy the most."[49] Its special virtue, in contradistinction to the vague, pedantic study of metaphysics, is to enable man to derive greater certainty about truth. For d'Alembert, the study of the sciences is clearly preferable to Adam's predilection for logic and metaphysics.[50]

But d'Alembert went beyond singing the praises of the physical sciences in his fourth volume of the *Mélanges*. He also included a number of his philosophically laden introductions to his own volumes of science. As we will see shortly, none was more significant for Laplace than the introduction to the 1758 edition of the *Traité de dynamique* (Treatise on Dynamics), in which d'Alembert engages in a thinly veiled demolition of God's omnipotence. He also deliberately turns his back on the wisdom of searching for final causes, highlighting once more how peripheral to true learning was the study of theology. Despite all his professions of adhering to a deistic position, d'Alembert was keenly skeptical about the function of God in the physical universe, and even his existence.[51]

How faithful Gadbled was to each of d'Alembert's teachings is difficult to say. The modernists were not centrally organized. But one significant issue in contemporary physical astronomy is likely to have been discussed in Gadbled's classes: the possibility of devising a mechanical explanation for the origins of the solar system. In 1749, Buffon had proposed that the fortuitous apparition of a comet passing close to the sun had set it in motion.[52] More recently, one of Gadbled's best students, the

Chevalier Jean-Jacques de Marguerie, had prepared a technical paper on this very subject, published by the Académie de Marine some years after he had left Caen. In it Marguerie attempted to develop a mathematically sound mechanism for the origins of circular and rotational motion based on the mutual attraction of three bodies in disequilibrium.[53] In 1767, the idea was initially applauded by d'Alembert as a major breakthrough in dynamics.[54] But since it fell short of its goal, it was forgotten by the scientific community.[55] Nevertheless, its fleeting success cannot have failed to have an effect on Gadbled and his students. Laplace's subsequent interest in the nebular hypothesis owes something to the wish to find a naturalistic explanation for the configuration of the solar system, one that excludes or ignores the miraculous.

Because it is impossible to document when Laplace lost his conviction about the centrality of the study of theology, we cannot know what views he held during his last year at the university, after he had completed his coursework with Adam in 1767. That he began to see the futility of the endeavor, turning his full attention instead to the sophisticated mathematical lectures of Gadbled, seems very likely. It was also in his final year at Caen that he developed a close relationship with Gadbled's principal assistant, Le Canu, who had graduated only a few years earlier. By 1769, Le Canu was teaching philosophy at the Collège du Mont. Laplace must have made dramatic progress in the study of mathematics working with him that year. We can assume that he had reached a high level of competence, since he was recommended as tutor to Jacques Armand d'Héricy, the son of a young nobleman, who gave a public demonstration of his aptitude on 19 April 1769.[56] The folio-sized poster announcing the thesis d'Héricy was to defend shows it was an exercise in both differential and integral calculus, followed by a few elementary applications to tangents and cycloids.[57] The mathematical section of the poster was preceded by a three-paragraph philosophical review of the history of the calculus, which mirrors the modernist views one could find in the writings of Fontenelle, d'Alembert, and his contemporaries. In it Descartes is given credit for raising man from the "ignorance, errors, and superstitions" of the Dark Ages; and Newton was acclaimed as the genius who "banished conjectures from physics and, guided by experience and a sublime form of geometry, endowed philosophy with its proper constitution."[58] Another thesis defense of the same year by a Le Canu student covers the principles of dynamics, leading to an exposition of the law of conservation of *vis viva* (live forces). It ends with a passage entirely congruent with d'Alembert's philosophy, that "one must thoroughly reject vague or uncertain principles," using them only "after having rigorously demonstrated them."[59] By now, Laplace was fully initiated into the "new philosophy" and had become an adept of its main tool for progress, the mathematical sciences.

Nevertheless, Laplace appeared before his examiners Adam, Le Canu, Pierre Lelièvre, Louvel, and Moysant on 1 July 1769 to earn his master's degree in the Faculty of Arts.[60]

## Leaving the Path to the Priesthood

What should be made of Laplace's willingness to submit to this traditional rite of passage? It could be an indication of his desire to satisfy the aspirations of his father and Benedictine teachers that he become a priest; it could also be a reflection of his profound spiritual indecision. More likely, it was his instinctual reaction against rash personal behavior that contemporaries would later label "opportunism." Even if one assumes that he already harbored serious doubts about orthodox Christian beliefs, choosing to pursue the steps leading to an ecclesiastic career was the safest course. It was in character with what one would expect from the son of a Norman commoner. It was also why his father had sent him to Caen in the first place.

André Siegfried captured this mentality when he described the stereotypical Norman villager as openly cautious, but inwardly stubborn.[61] A profound skeptic, unwilling to commit himself in advance to an extreme position, the peasant's favorite expression was "Peut-être bien qu'oui, peut-être bien qu'non" (Maybe yes, maybe no).[62] His typical posture was to wait until it was clear which side would prevail before taking a stand. As a young man witnessing the hopeless dissension between his mentors, Laplace outwardly chose the cautious avenue, while plotting a parallel strategy. He temporarily secured his career path in the Church without renouncing his ardor for mathematics.

Becoming a cleric in Old Regime France was an accepted way for a bright commoner with intellectual aspirations like Laplace to secure a safe and socially acceptable occupation. It had been the route pursued by his late Uncle Louis. Many others followed, faced with the same dilemma. For example, the mathematics professor at Caen's école centrale after the Revolution, François Quesnot, had been pushed into an ecclesiastical career for the sake of his family despite his lack of aptitude. It was reported that as a needy lad he had struggled against this choice: "The Church of France possessed resources that seemed within sure reach [for men like Quesnot] . . . They had become the object of profane transactions for all classes of society. Entry into the Church was often the consequence of calculated self-interest or family arrangements, in which no one cared about the pure, angel-like qualities required [to perform the mission of a priest]."[63]

Laplace could well have continued on this path of least resistance. Instead, his newly discovered passion for mathematics drove him into a daring adventure. Armed with a letter of recommendation from his teachers, he traveled to the Mecca of science, knocking at d'Alembert's door with a mathematical essay in hand.

# 3

# The Gifted Mathematician

The story of a country bumpkin arriving at d'Alembert's doorstep with a letter of recommendation was recounted by Baron Jean-Baptiste-Joseph Fourier, who delivered the official eulogy for Laplace for the Academy of Sciences.[1] As resistant as he was to solicitation, d'Alembert is said to have refused to meet with the supplicant until he supplied evidence of his mathematical ability. When Laplace returned in a few days with a brilliant paper on the general principles of mechanics, so the story goes, d'Alembert was so impressed that he immediately took the young man under his wing and launched his career. Fourier claimed that Laplace was still able to recite parts of this paper many years later.[2]

This charming vignette, rhetorically well suited for the tribute, was not totally farfetched. Laplace did come to Paris announced by a letter from Le Canu; he did diagnose a knotty scientific issue for d'Alembert; and the influential patron immediately took Laplace in as his protégé. In a series of exploits that sound more like a fairy tale than a verifiable event, Laplace won over Europe's most accomplished kingmaker on his own, and the encounter transformed his life. The moral of the story that Fourier told was that merit rather than patronage prevailed. But once merit was established, patronage regained its central role. In all walks of French life, merit was beginning to be a factor, but only if recognized and reinforced by strong supporters.[3] The role of influential patrons was particularly important in technical domains like mathematics, where expertise was the province of the few. In 1769, d'Alembert was the French magnate of math.

Laplace was only twenty years old when he arrived in Paris. The assurance sent by Le Canu that he had discovered a highly talented student in the provinces was apparently not sufficient to move the fifty-two-year-old patriarch. Instead, Laplace's four-page paper did the trick.[4]

Fourier exaggerated when he said the paper was a treatise on the foundations of mechanics. It was in fact the close examination of a specific facet of a fundamental principle of physics. In his *Opuscules mathématiques,* d'Alembert had claimed to have produced a mathematical demonstration of the principle of inertia.[5] If his demonstration stood, it would remove inertia from the list of simple axioms, situating it as a necessary consequence of more fundamental principles. Already in the *Traité de dynamique* (1743), d'Alembert had replaced Newton's three laws with his own three laws of motion that bypassed Newton's concept of force (which d'Alembert had rejected because it had to be grounded on empirical evidence).[6] Now d'Alembert wanted to establish even his own laws on more solid, rational grounds. He wrote a paper entitled "Mémoire sur les principes de la mécanique" (Memoir on the Principles of Mechanics) that appeared the same year that Laplace met d'Alembert. At this point, then, this matter was close to the senior academician's heart. Laplace's willingness to engage the leading French theorist of the era on his own terrain was bold enough to gain his attention. But it was the cogency of Laplace's analysis that made him stand out as a highly talented natural philosopher.

The paper demonstrated that Laplace had grasped d'Alembert's position sufficiently to extend one of its arguments. By examining the case of a body moving through a resisting medium, he was able to raise d'Alembert's argument, which considered only a body moving in vacuum, to a more general level. In addition, Laplace was able, by expanding a function into a series, to dispose of a contradiction that d'Alembert feared.

At this point in the paper, and in its reading, it had been a good performance. But Laplace was not content to stay within the bounds of the piece he was examining. In a move that must have flattered d'Alembert, Laplace indicated that there was no better statement of the inertial law than the one d'Alembert had himself previously given in his *Traité de dynamique.* In a few well-chosen sentences, the young man both showed that he had understood d'Alembert's meaning, and raised some probing objections to d'Alembert's new assertions. Laplace even ventured to speculate that the law of inertia might itself be only an approximation of physical reality, since motion might actually be slowing down, albeit at an imperceptible rate. He ends these daring remarks by suggesting that were this the case, it could offer an explanation for the unaccounted changes in the mean motion of planets.

To justify this ingenious proposal, Laplace assumes an epistemological stance that he was to maintain for his entire career, one that eventually put him at odds with d'Alembert's own position. While agreeing that inertial motion must be treated as a metaphysical principle, he insists that it is disclosed to man only through his senses:

> If a body moves uniformly upon leaving a hand that throws it in space, it will continue to move the same way if nothing comes in its way; but nothing determines that this initial motion must be uniform if one does not possess the notion we derive from our senses about matter that it cannot [by itself] change its speed. Otherwise I see nothing that would require it to conserve its movement. We have not the least notion about the essence of bodies, and the nature of motion is manifested only through its effects, hence experience is the only guide we have for it. I therefore consider the law of inertia to be an assumption whose consequences are in close accord with the phenomena of nature.[7]

This display of mental gymnastics about fundamental issues was sufficient to convince d'Alembert that his provincial friends had indeed introduced a new, and potentially promising, way of thinking. It was evident that this young man, trained in theological debates, could argue clearly about philosophical issues in science as well.[8] The champion of the new philosophy would find this promising young man a job, and by bringing him closer to the hub of scientific activity in Paris, draw him into his scientific circle. Though still quite young, Laplace was to join the loose-knit intellectual community of scientists surrounding d'Alembert in Paris, which included Pierre Le Monnier, Etienne Bézout, Charles Bossut, Jacques Cousin, Joseph-Jérôme Lalande, and Marie Jean Antoine Nicolas Caritat de Condorcet.

## The Ecole Militaire

Wherever the mathematical sciences were involved outside of Church institutions, d'Alembert was the key figure in Parisian circles. As the editor of the scientific part of the *Encyclopédie,* he had contacts everywhere. In particular, he had enlisted two teachers from the Ecole Royale Militaire to author articles about grammar and convinced the head of the school, Paris de Meyzieu, to write about the new institution.[9] The Ecole Militaire was an experiment in secular education for sons of impoverished nobles. Its doors opened in 1753 and three years later the school moved to a sumptuous building, then on the outskirts of Paris, that still stands as one of the masterpieces of the architect Jacques Ange Gabriel. Originally intended to serve five hundred students, it rarely admitted more than half that number. The highly disciplined military institution, designed at first to satisfy the educational aspirations of the nobility with a curriculum centered on the humanities, was transformed in 1769 into a preparatory school for military officers with a new emphasis on technical subjects. The new curriculum was based on a program of elementary mathematics, which was considered a formative discipline that would help youngsters develop "order in their ideas, and clarity and precision in their thinking."[10]

The very year Laplace was to leave Caen, the school tripled its mathematics personnel, reviewing the qualifications of a dozen new teachers in early September 1769 in order to have them start classes in October. Among the newly proposed instructors, Laplace is listed in second place alongside others who have hardly left a mark in the annals of mathematics.[11] It was most likely d'Alembert who was asked to supply names of competent instructors. We know that at the end of the year, when Claude Berthelot retired as mathematics instructor, it was d'Alembert who picked Cousin of the Collège Royal as his replacement.[12]

Le Canu was notified in advance that d'Alembert had secured Laplace a new position in a missive sent to Caen on 25 August, and was asked to transmit the terms of the appointment to his fortunate student.[13] The salary was generous for a beginner: 1,400 livres, board and room, as well as a provision for firewood. As a bachelor, Laplace was obligated to take up residence at the Ecole Militaire and take his meals in the new structure, and was asked to report for work to the director of studies, Bizot, on 20 September. The young professor, who was referred to by d'Alembert as the "abbé" Laplace, was told that he should appear in his clerical garb, wearing his ecclesiastical collar. This detail is the only indication we have in Paris of Laplace's earlier intention of becoming a priest, and it served him well in this instance. Since professors were to be held up as models for some 250 youngsters, no one could be hired for the position if he were not himself of high moral character.[14]

The regime at the Ecole Militaire was very similar to that which Laplace had experienced as a student at Beaumont. There were boys as young as ten years old, of every ability. Though Latin was less in evidence, students were supervised closely and their daily schedules fully prescribed. Now, however, Laplace was a teacher, and because he was asked to hold classes only in the morning, he had a lot of free time—and some freedom. Wearing the required attire, he could leave the school during the day. But the rules forbade teachers from staying in town overnight without the express permission of the governing council, and it demanded that they follow students' progress outside the classroom and provide the director with periodic grades.[15] Instructors were admonished to be gentle with students of limited aptitude. Students who later wrote about their stay at the Ecole Militaire were impressed (and some overwhelmed) by the mathematical offerings.[16] Laplace, however, was never publicly singled out for excessive praise or specific criticism by any of his students, either at this stage of his career or later.

The school was endowed with a rich library. An inventory prepared in 1776 shows it held over four thousand volumes, and included all the current mathematics and physics textbooks in French and Latin, and some in English and German. Available as well were the transactions of all the major learned societies (of Paris, London, Berlin, St. Petersburg, Turin, and Göttingen, for example) and the Leipzig journal *Acta Eruditorum* (Philosophical Transactions), along with a good number of modern philosophical texts. Theological writings were also in evidence, but in smaller proportion than Laplace would have encountered in libraries in Normandy. Some of these works had also been available in Caen, but Laplace was not then as focused on mathematics as he became in Paris.[17]

Given his studious nature and the ambitions he harbored, it is likely that Laplace spent a great deal of time in the library or with the books in his lodgings under the eaves.[18] He took his meals at a table with the fourteen other instructors in mathematics, with whom he was to coordinate the mathematics lessons. These lessons were taken from a single elementary text for all the students. Until 1772, students studied the *Cours de mathématique* (Mathematical Lessons) of the academician Charles Etienne Louis Camus.[19] Afterward, a new text by Bossut was prescribed, though another one by Bézout was allowed, depending on which branch of the military students expected to enter. Bossut was in charge of officers expecting to become royal engineers, whereas Bézout carried out examinations for the navy and the artillery. All three of these textbook authors, befriended by d'Alembert, were members of the august Académie des Sciences, and their writings were given preferential treatment.[20]

It was clear to Laplace from the outset that academicians composed the highest possible class of scientists in Paris. They constituted the officially recognized experts to whom everyone turned for approval. Biweekly meetings of the Académie were where new ideas were tested.[21] In order to stand out among the many other teachers of mathematics, Laplace realized he would have to do more than win d'Alembert's support: he would have to have his work accredited by the Paris Académie to prove himself.

The young mathematician shared the aspirations of many who migrated to the capital. In the words of a contemporary: "Self-interest and ambition which govern all in this world; the few resources offered by the provinces to industry and talents; the attraction of multiple pleasures concentrated in large towns; etc.: that is what attracts the multitude of expatriates who bring with them their industry, their talents, their machinations, their indolence, their languages, their virtues and vices, their passions, and their fortune or indigence."[22] Laplace took an immense risk by abandoning a safe vocation in the clergy in Normandy for the promise of a career in the capital. By choosing to leave for Paris, the young man was undoubtedly renouncing any support he might once have expected from his family or his ecclesiastical mentors. Now he needed to demonstrate to himself, his father, and his teachers that this bold move would pay off. The challenge was no small task for a young man with no resources beyond his own intellectual abilities. By moving to Paris, the young mathematician effectively cut himself off from his family as well as his Norman origins. Beyond the effects of the new occupational choice, the move marked a psychological divorce from his youth.

Laplace's primary pursuit became to make a mark in the mathematical sciences. The company he had kept at the Ecole Militaire had showed him how far up the ladder he had to go to reach the top. Among his students were boys of no particular aptitude whose future as officers was assured by their noble status in society. It was also true that some could rise to the most desired military corps by virtue of their mathematical prowess. It was Laplace's task as a teacher to set the best minds on the road to success by preparing them for the competitive examinations. But

Laplace could not envisage a military career for himself because of his commoner origin. He was there to serve others, and through the institution, the Crown.

Laplace could also measure himself against his colleagues, most of whom were content with a limited vocation as teachers of mathematics. Few of them stood out. When Laplace joined the faculty, the most accomplished was the observational astronomer Edme Sébastien Jeaurat, who had recently been elected to the Académie with the backing of Lalande.[23] His forte was computation, and he retired from teaching elementary mathematics at the Ecole Militaire just as Laplace joined the staff. In 1760, Jeaurat had established a modest observatory at the school, which continued to function for many years.[24] Another older teacher of mathematics at the school, Pierre d'Antelmy, worked with Jeaurat to send some astronomical observations to the Académie in 1763. D'Antelmy also translated several Italian mathematical handbooks by Maria Agnesi and Antonio Cagnoli. The other noteworthy colleague who retired shortly after Laplace's arrival was Claude François Berthelot, at heart an instrument builder and mechanical inventor. He had attempted to write a new, simplified textbook for the school, entitled *Cours de mathématiques à l'usage de MM. les élèves de l'Ecole Royale Militaire* (1762; Mathematical Lessons for Students of the Ecole Royale Militaire), which failed to win the assent of the governing council.[25] These colleagues were not whom Laplace chose to follow.

Berthelot was replaced by the only colleague worth emulating. Jacques Cousin had already gained a reputation in Paris as the chair of mathematical physics at the Collège Royal.[26] Ten years older than Laplace, he was the only serious research-oriented scientist among his colleagues. Moreover, Cousin's prowess was in fields that Laplace had mastered at Caen with Gadbled and Le Canu. At the Collège Royal, Cousin had lectured on both integral calculus and celestial mechanics. He was a protégé of the academician Pierre Le Monnier and appreciated by d'Alembert. There is every reason to expect the two teachers exchanged ideas and learned to respect each other.

### Academic Ambitions

The young Laplace set about to expand his mathematical education. With the likely guidance of d'Alembert and Cousin, he spent hours poring over the available mathematical literature at the school library. Textbooks like Leonhard Euler's *Introductio in analysin infinitorum* (1748; Introduction to Infinitesimal Calculus) shored up his understanding of the calculus he was teaching students, and the proceedings of the learned societies brought him up to date on current research. Two foreigners in particular struck him with their versatility and inventiveness. Euler had recently published what Laplace later acknowledged as the most suggestive texts in calculus, the *Institutiones calculi differentialis* (1755; Foundations of Differential Calculus) and the *Institutionum calculi integralis* (1768–1770; Foundations of Integral Calculus). His name was repeatedly evoked in the transactions of the Berlin and St. Petersburg academies, and with envy by d'Alembert. Joseph-Louis Lagrange was also much admired for his elegant mathematical solutions to all sorts of mechanics problems that were of concern to the mathematical community. Laplace's respect for these two luminaries is confirmed by his correspondence and by the frequency with which he cites their work.[27]

Three other contemporary French mathematicians are likely to have captured Laplace's attention during this period. The elderly Alexis Fontaine had written outstanding papers published by the Académie des Sciences on integrating partial differential equations with several variables;[28] Etienne Bézout, a regular visitor to the Ecole Militaire to test students intending to become naval or artillery officers, was an innovative algebraist;[29] and the Marquis de Condorcet, one of d'Alembert's closest disciples and nephew of the Bishop of Lisieux, was furiously publishing his ideas on calculus in the three years that Laplace was a student at Caen.[30] Of the three, only Condorcet was later acknowledged by Laplace. Another innovator whose performance kindled his interest was the precocious student Adrien Marie Legendre, whose public success in 1770 at the Collège Mazarin brought him an approving nod from the academicians attending his thesis defense.[31] They were all, or would eventually become, members of the Académie des Sciences. Very soon after he was settled in Paris, Laplace realized he wanted his prowess acknowledged by this august body.

To obtain such recognition, a scientist would submit a paper through one of the Académie's members, and the secretary would place it on the agenda to be presented to the assembly. A reading, sometimes in abbreviated form and with the author in attendance, would be made at one or more sessions. The secretary would then appoint a committee of at least two academicians to prepare a report to the full body, suggesting what should be done with the submission. Some offerings were quickly dismissed as unworthy. Those that required further investigation were sent back to the author. Still others would be praised, but without promise of publication by the learned society. The best ones would appear in one of the Académie's journals. A special series, whose volumes were published as needed, was reserved for nonmembers, or *"savans étrangers."* Academicians' writings would appear in the annual *Mémoires* of the society.[32] Unpublished book-length manuscripts deemed worthy would appear under the printing *privilège* of the Académie with the full laudatory report by the committee printed at the end of the book. It was in this manner that the French scientific community formulated and announced its collective opinion of new work being produced.

## Election to the Académie

Whenever a vacancy occurred in the Académie, candidates for admission would submit their latest creative endeavor to impress the academicians about to vote in the coming election. In each disciplinary section of the Académie (called a "class"), there were three or four times a year when committees would work overtime to provide their colleagues opinions about the quality of the work of contenders prior to the balloting. Laplace wasted little time trying to make his mark. His first offering to the Académie is dated 28 March 1770, probably unsynchronized with any vacancy. But starting with his next paper, presented on 18 July 1770, his submissions to the learned society seem coordinated with deaths, retirements, or promotions that would ultimately lead to an opening. Eventually he was to bombard the learned body with thirteen separate papers in the three years leading up to his election. We have independent evidence that Laplace was considered an active candidate in the balloting on six separate occasions before he succeeded.[33] The order of his presentations is therefore more likely to be related to his ambition to become an academician than to the succession of his ideas.[34]

Laplace's initial presentation of March 1770 was on the manipulation of maxima and minima in a family of functions. It was undoubtedly stimulated by Laplace's readings in the Turin Academy's transactions, where Lagrange had given an analytic demonstration of one of Euler's propositions.[35] The academic committee chosen to evaluate Laplace's first paper was composed of two experts on calculus, Condorcet and Jean Charles Borda, and Borda wrote the extensive report.[36] Borda pointed out that Laplace had arrived at the same conclusions as Lagrange using a different approach, and had repeated results obtained by Borda himself years earlier. But he also praised the young man for having forged his own path, and for having developed some ideas that went beyond current research: "M. Delaplace has provided . . . a less direct and rigorous new demonstration, but one that is simpler and rather elegant . . . It seems to us that M. Delaplace's paper reveals more mathematical knowledge and more intelligence in the manipulation of the calculus than is ordinarily found at his age." The committee recommended publication, though Laplace was told to abbreviate the section that was not original, and to employ standard notation. Though not a total success, his first attempt had at least caught the attention of older and more accomplished mathematicians.

Laplace's next dozen presentations were superb, and ultimately landed him the prize he was seeking, election to the Académie. Condorcet later wrote that the Académie "had not seen anyone so young offer so many important papers on such diverse topics and of such difficulty."[37] They constitute a body of original work that falls into three interconnected categories that are more easily understood topically rather than chronologically: mathematics, celestial mechanics, and probability theory. It was precisely in these three domains that Laplace was to make his most significant contributions to the world of science in the next half century. But before looking at them in some detail, one must consider the magnitude of his career achievement, which profoundly shaped his behavior and personality.

There were precedents to election at such a young age. Laplace was elected an *adjoint* (assistant) in March 1773 when he was twenty-four. Alexis-Claude Clairaut had been welcomed into the Académie in 1731, at the age of eighteen. But the 1770s was a period of flourishing mathematical talent in France. Laplace's competitors were a formidable lot, and his truly remarkable success was due to a combination of extraordinary abilities and perseverance, as well as the backing of d'Alembert. Those who won the nominations that he coveted for three years included a much older mathematician of great talent who had been a candidate for years, Alexandre Vandermonde; Jacques Cousin, his colleague at the Ecole Militaire, who benefited from high visibility as a professor at the Collège Royal; and two experimentally oriented physicists who were well appreciated in academic and administrative circles, the abbé Alexis-Marie Rochon and Nicolas Desmarest.[38] More telling is the list of competitors he beat out: the other mathematics professor at the Collège Royal, Antoine René Mauduit; the abbé de La Chapelle from whose textbook Laplace had been initiated into mathematics at Beaumont; Mathurin Georges Girault de Kéroudou, Condorcet's talented teacher in Paris; and Laplace's superior at the Ecole Militaire, Pierre d'Antelmy, who was "inspecteur des classes de mathématiques." Several other notable mathematical scientists were also in the wings, submitting papers and books to the Académie: the Jesuit Father Ruđer Bošković; the astronomer Charles Duvaucel; the military engineer Tinseau d'Amondans; and three local mathematics teachers, the abbé Joseph François Marie, the abbé Jean Saury, and François Marie Fyot. Even more remarkable were those he bested the year he was finally chosen: Jean Jacques de Marguerie, a fellow student of Gadbled's who won second place in the election, as well as three who would eventually become Laplace's colleagues at the Académie: Gaspard Monge, Adrien Marie Legendre, and Jacques Charles.

Laplace produced hundreds of pages of original work in these three years. In the process, he completed his scientific education by mastering the current literature, most of which was available at the Ecole Militaire library. He worked with great intensity, and was initially frustrated at being rebuffed in his attempts to become an academician.[39] In 1772, his disappointment led him to explore possibilities of emigrating to Prussia and Russia, emulating the careers of Euler and Lagrange. He talked the astronomer Lalande into introducing him by writing a letter for Euler in St. Petersburg in the hopes of landing a position there.[40] D'Alembert also wrote on Laplace's behalf to Lagrange in Berlin. The letter reveals how seriously unhappy Laplace had become:

> There is a young man here named Laplace, currently professor at the Ecole Militaire where I placed him. This young man has shown much zeal for mathematics and I think enough talent to make a mark. He wants to devote his entire time to it, and since his post as mathematics professor is time-consuming, he is seeking another which would leave him entirely free. Our Académie could not satisfy him because pensions come after many years, sometimes only after twenty-five years of waiting; and in any case he is not an academician, having been unjustly preferred despite my backing and that of most of our mathematicians

by a person much his inferior who, as professor at the Collège Royal, assembled considerable support from [other] academicians ... Is there actually a position at the Berlin Academy where he could immediately expect a revenue of about three to four thousand livres?[41]

None of these stratagems worked.[42] Laplace remained in Paris at the Ecole Militaire, which awarded him two commemorative medals upon his election to the Académie and continued to employ him as an instructor.[43] When the Claude Louis Comte de St. Germain's reforms brought about a change in the school's curriculum in 1776, Laplace was relieved of his duties and given a pension of six hundred livres "to permit me to follow my work as a member of the Académie."[44] A year earlier, the Académie had assigned him an annual stipend of five hundred livres from one of its special funds to provide him with some relief.[45] He had no other resources because his father had long ago abandoned him.

Laplace, living until this time at the Ecole Militaire, led a spare existence in Paris, centering his emotional energy on his career probably to the exclusion of an extensive social life. Although he would have had access, Laplace does not seem to have participated in any of the renowned salons in Paris, nor have been involved deeply with the Enlightenment debates raging around him. One might have expected him to befriend the Holbach circle or the Freemason lodge of the Neuf Sœurs, given his theological inclinations. Nevertheless, he is not cited in contemporary correspondences nor noticed in newspapers. Foreign visitors to the cultural capital of France are likewise silent on Laplace. Outside of scientific circles, he remained largely unknown.

His scientific output while a teacher was stunning. He presented some twenty papers to the Académie, half of which appeared in print by 1777.[46] These compositions chartered the research directions that Laplace was to pursue throughout his life. Indeed, with the exception of his interest in imponderable fluids (heat, light, electricity, and magnetism), which came later, at the time of his collaboration with Lavoisier, these papers could stand as a blueprint for his entire scientific career. His willingness to pursue the same course of activity in spite of major upheavals in his life and the nation is testimony to his determined, resolute, and almost obsessive character. It is rare to find a mind so focused on his professional work for over fifty years. Laplace showed no decline in endurance or tenacity until he was well past the age of seventy-five.[47]

Even though his initial presentations to the Académie won him immediate encouragement, Laplace never reached the top tier of creative mathematicians of his day. Leaving aside temporarily the domain of probability theory, where his role became pivotal, one must in retrospect place Laplace among a cluster of gifted mathematicians who pushed the research front forward without making sweeping changes to its character. When compared to the sheer mathematical genius displayed by Lagrange, Carl Friedrich Gauss, or Legendre, Laplace remains a figure of second rank. He was fully in command of contemporary literature, willing and able to tangle with the masters in the field, to correct their errors or to improve on their methods, but he rarely led them to new ways of conceiving mathematical problems. Ultimately the motivation for writing about mathematics came from his conception of it as an instrument rather than an object of aesthetic pleasure.

Contemporaries like Condorcet and Cousin, themselves talented mathematicians who were centrally involved in the research front, noted:

> The successors of Newton enriched infinitesimal analysis with a good number of deep and ingenious methods that they applied with success to rather complicated problems, but calculus in itself was rarely prepared to lead the object of their research . . . the greatest physicists of each century, those to whom we owe the discoveries of most utility, were also profound mathematicians . . . The study of higher mathematics trains the mind for profound deliberation; it provides it with a sense of comprehensiveness and lucidity; it prepares it to follow truth through its most twisted convolutions without losing its way.[48]

In keeping with these attitudes, Laplace learned to devote his attention to mathematical problems when these gave promise of providing superior power for dealing with issues of dynamics. His first essay of 1769, written to win him d'Alembert's support, had already demonstrated this mode of operation.

The immediate problem for consumers of the calculus was to determine conditions of exact integrability for functions encountered in analyzing celestial dynamics. Laplace's March 1770 paper on maxima and minima, and the one following it in July on integration of finite differences, were stimulated by recent writings of Euler and Lagrange, but also were aligned with part of d'Alembert's attempts to resolve fine points in the "system of the world."[49] As a mathematician, Laplace always lived in the shadow of these three innovators. His choice for publication of the July 1770 paper in the proceedings of the Turin Academy indicates a desire to become part of their ongoing dialogue. The mathematical section of the first four volumes of the *Miscellanea philosophico-mathematica Taurinensis* (Philosophical and Mathematical Miscellany of the Turin Academy) was dominated almost entirely by exchanges among these three figures, one of whom, Lagrange, was a founder of the Turin Academy who lived in Savoy until 1766.

Laplace was not alone in following their lead. Condorcet, his senior by six years, had earlier made his mark as a precocious mathematician. In 1765 when he was only twenty-two, Condorcet collected a set of mathematical problems that were published as the treatise *Du calcul intégral* (On the Integral Calculus).[50] These were followed by another set entitled *Essais d'analyse* (1768; Essays on Mathematical Analysis), which eventually opened the doors of the Académie for him a few years ahead of Laplace. It would seem that the young Laplace patterned his subjects of inquiry after reading closely the works of Condorcet, whom he both admired and envied. The strategy he developed was to enlist himself in a productive enterprise that d'Alembert admired, while also showing his technical superiority over figures like Condorcet. These two contemporaries, Laplace and Condorcet, while always remaining on the surface respectful and cooperative, were in fact competitors for the attention of the Parisian mathematical world. It was fortunate for both of them that Condorcet adjusted his career in recognition of the superior mathematical facility of the younger mathematician, taking advantage of his own aptitude for philosophy. Shortly after his election to the Académie in 1769, Condorcet began to pen philosophical essays comparable to d'Alembert's contributions to the *Encyclopédie,* while continuing to work in mathematics. With d'Alembert's encouragement, he saw himself as a successor to Jean Paul Grandjean de Fouchy as "perpetual" secretary of the Académie. He prepared eulogies of the founding members of the Académie, published in 1773, and took over the duties of secretary as the assistant of the aging Fouchy. In that capacity, he was to be the interpreter of Laplace's work to the public, writing introductory sections of the Académie's *Mémoires*.[51]

The comparison between Condorcet and Laplace has often been drawn, but their personal relationship remains somewhat unclear.[52] Condorcet lacked the clarity and concision of expression in dealing with

Physionotrace of the young Laplace signed by Madame Laplace, from the author's collection

the mathematical sciences that Laplace quickly demonstrated. Laplace, for his part, was never willing to engage in the extended and elegant discussions of metaphysical and philosophical issues that became second nature for Condorcet. It was not until Condorcet's death during the Revolution that Laplace attempted more popular treatments of his own achievements. Even then, they lacked the sophistication and fervor of Condorcet's essays. In the historical sections, Laplace clearly benefited from Condorcet's example, but he waited for Condorcet's demise before publishing them. Condorcet kept his own revised treatment of calculus in manuscript, some of which has only lately been recovered for study, as if he were conscious of its incompleteness in the face of contemporary advances. Their parallel and symbiotic relationship was one of the most important catalysts for Laplace's intellectual career.

# 4

# Setting Fundamental Principles: The Philosophical Program

Despite the intellectual distance separating the University of Caen from the academic milieu in Paris, the two environments shaped Laplace's scientific career in similar ways. Neither Gadbled nor Le Canu, each well versed in the intricacies of calculus, chose to cultivate the subject for its own sake. For them both, it was a language to master in order to appreciate, and if possible to contribute, to the advancement of the physical sciences. Among their students only one besides Laplace, the hapless naval officer de Marguerie, wrote an original piece in pure mathematics.[1] Laplace himself was momentarily tempted to indulge in research about the theory of numbers, but never pursued this inclination.[2] In Paris, other skillful users of the language, like Condorcet and Cousin, preferred to wield it as a tool for advancing the physical sciences. The same attitude initially motivated Laplace's other talented mathematical colleagues Monge and Legendre. It took an unusual amount of independence for each of them to separate his mathematical inventiveness from its applications. Not until the era of the French Revolution were Monge's descriptive geometry and Legendre's theory of numbers considered on their own merits. Before the Revolution, in the 1770s, attention was clearly focused on what was then called physical astronomy, a position generally acceptable to everyone, whether in Paris or in the provinces.

D'Alembert had set the nation's scientific agenda by asserting that completing the Newtonian program was the most virtuous and urgent activity for fellow scientists to pursue. In the *Mélanges* that Adam had remonstrated against in 1766, d'Alembert reiterated a view he had expressed in an earlier scientific work: "If astronomy is one of the chief sciences that ennobles the human mind, physical astronomy is one of those that honors modern philosophy the most. The search for causes of celestial phenomena, which has made such progress in our times, is no mere sterile speculation, but [an activity] whose value is gauged by the extent of its object and by the formidable effort required to comprehend it." He went on to explain what it would take to fulfill the goals set for the study of astronomy: "One could presume to have identified the authentic causes of the motion of planets once one could derive consequences from these causes that match what observation has disclosed."[3] In France, modern scientists had made great strides in that direction since the 1740s, when Cartesian principles had been gradually abandoned. D'Alembert emphasized that progress was directly linked to the preference that Continental scholars had shown for calculus (the analysis of infinitesimals) over Newton's attachment to geometric methods (synthesis). In his judgment, the change in mathematical language was largely responsible for the advancement of physical astronomy. Nonetheless, the object of the researcher was not the language used, but what it enabled him to say about the universe.

Two kinds of issues at opposite ends of a spectrum confronted those attempting to join d'Alembert in his enterprise to complete the Newtonian program. In his magisterial way, d'Alembert had mastered them both. On the technical side, there were still major obstacles to fitting theory and data together. Finding solutions to these problems constituted the agenda for mathematical physicists—which he, Euler, and Lagrange were continually discussing in their scientific productions. At the philosophical end, there were major ontological and epistemological issues that remained unstipulated or incompletely considered. Among them was the proper way to weigh the importance of material substances to explain phenomena as opposed to mathematical formulations to describe nature.[4] Whereas there was a small army of specialists competent to deal with the technical issues, very few of his colleagues had either the opportunity or his talent to address effectively this philosophical concern. D'Alembert, both *philosophe* and *savant*, possessed the means to speak to both sides in different registers. The *Encyclopédie* and the Académie Française offered him a forum to air the philosophical aspect of the task, whereas the technical dimensions could be worked out in his scientific treatises as part of the normal activities of learned societies.

By the 1770s, since d'Alembert had a fixed view on most philosophical issues, he urged that they be tabled until more progress had been made in understanding the details of the system of the world. He often referred to "metaphysics" with contempt, decrying it for its obscurity, uncertainty, and sterility. As indicated earlier, he ridiculed the excessive emphasis that traditional teachers like Adam accorded this dimension of scholastic learning. Laplace had equally been distressed by the idle verbiage used by Adam to buttress his antiquated stands on natural philosophy. The trouble with metaphysics was that it could be criticized only by more unverifiable philosophical discourse. This was not the way to advance the Newtonian program championed by his mentor. Nevertheless d'Alembert had to justify his philosophical position by contravening older authorities on their own grounds, and as a result found himself engaged in metaphysical arguments. It could not escape his younger colleagues that, despite his disdain for barren philosophical discourse, he frequently made important and debatable pronouncements about it. One example has already been indicated with reference to his unproven law of inertia. Men like Condorcet and Laplace, trained as they had been in religious disputations, could hardly resist studying—and amplifying—these pronouncements. Although the balance of their attention was devoted to technical issues concerning the system of the world, both scientists chose to react with their own philosophical formulations, which far from constituting a diversion from their professional activities, became the underlying framework for their own scientific pursuits. It is ironic that for all their rejection of metaphysics, young scientists found it necessary to engage in fundamental philosophical discussions.

In 1768 Condorcet penned a little-noticed public letter entitled "Le Marquis de Condorcet a Mr. d'Alembert, sur le systême du monde et sur le calcul intégral" (A Letter to Mr. d'Alembert from the marquis de Condorcet on the System of the World and on Integral Calculus), in which he attempted to summarize the current situation in these domains. It was carefully scrutinized by Laplace when he arrived in Paris, because in some ways it was meant to replace d'Alembert's introduction to his *Traité de dynamique* as the baseline from which a student of nature would begin his inquiries. In the years that followed, Laplace often found the younger Condorcet to be a more congenial expositor of philosophical principles than their patron d'Alembert. Laplace, for example, was to side with Condorcet—and against d'Alembert—in his assessment of the merits of probability theory in science. There were other issues on which Laplace patterned his views after Condorcet's rhetoric. The most significant instance had to do with the status of the laws of nature.

## The Berlin Academy Prize

In 1756, the Berlin Academy offered a prize for the best answer to the question of "whether the truth of the principles of statics and mechanics is necessary or contingent."[5] D'Alembert's opinions were laid out in the preface to the second edition of the *Traité de dynamique* (1758), where, in an effort to avoid taking an overtly antireligious stance about God's power to intervene in the very natural laws God was credited with establishing, d'Alembert insisted that what we could infer from science could never settle the issue of God's existence: "It is not a matter of deciding whether the Author of nature could have provided it with laws other than the ones we observe; as soon as one admits an intelligent being capable of acting on matter, it is evident that this being can at any moment move or stop it at will, or according to uniform laws, or laws that differ at every instant and for every unit of matter."[6] Unlike Newton, who had explicitly asserted that the nature of God could be illuminated through the study of nature, d'Alembert tried to disengage the domain of science from that of faith.[7] His attempt to remove theological implications from scientific discourse was only partly successful. Both Condorcet and Laplace saw through this maneuver and were led to reopen the question.[8]

Condorcet agreed with d'Alembert that two kinds of essentially different laws existed, the comparison of which was the principal business of scientists.[9] Some were laws based on metaphysical assumptions about the nature of matter; others were laws based on human observation. Condorcet identified the former with "mechanics," portrayed as a mathematical exercise that, when played by the rules of logic, must yield necessary truths. If the premises about the nature of matter were correct, then the consequences derived from them would be equally valid. The other set of laws, those derived from human observation, were empirically generated, and imperfectly known. They constituted the "system of the world." According to Condorcet, they were "the product of the free will of an intelligent Being who has willed the world to be as it is rather than any other way."[10] Because the system of the world was contingent upon this Being, its manifestations could not be considered inevitable. Necessary laws were, in other words, the byproduct of human rationality, whereas contingent laws were within the purview of God. The convergence of these two types of laws might well give man confidence in his abilities to divine God's laws, but the system of the world would always remain the handiwork of this intelligent Being.

Though clear about the ontological situation, Condorcet was not entirely sure what epistemological meaning should be attached to this conclusion: "Perhaps these laws differ among themselves only because, according to the current relationship between things and us, we require more or less sagacity to know them."[11] It was thus with regard to man's ultimate ability to arrive at a full understanding of God's empirical laws that Condorcet seemed to deviate from his mentor. He proposed a hypothetical situation that would strike Laplace as a most fertile way of thinking about the problem:

> One could conceive [of the universe] at any instant to be the consequence of the initial arrangement of matter in a particular order and left to its own devices. In such a case, an Intelligence knowing the state of all phenomena at a given instant, the laws to which matter is subjected, and their consequences at the end of any given time would have a perfect knowledge of the "System of the World." This understanding is beyond our capabilities, but it is the goal toward which all the efforts of philosophical investigators *[géomètres philosophes]* must be directed, and which they will continually approach without ever expecting to attain.[12]

It was this distinctive language, used by Condorcet to explore his epistemic quandary, that led directly to Laplace's celebrated stand in favor of a deterministic universe.[13] His views were given a different formulation than those of his Parisian colleagues. Embedded in Laplace's formulation were a complex of theological and philosophical issues intimately tied to his practice of science. He must have struggled with these issues for months before taking a stand in a paper submitted to the Académie, at the latest on 10 March 1773. The views eventually appeared in print in 1776, unobtrusively sandwiched between two sections of an article entitled "Recherches sur l'intégration des différentielles aux différences finies et sur leur application à l'analyse des hasards" (Research on the Integration of Differentials and Finite Differences and

Their Application to the Analysis of Chance).[14] To position this now momentous philosophical creed in the middle of a mathematically technical paper was not only indicative of the subsidiary role that Laplace assigned to metaphysics; it also announced to his colleagues that he was to be considered foremost a professional mathematical physicist, not a natural philosopher—a self-image that would change dramatically after the French Revolution. The same philosophical conception was repeated with minor variations in the 1812 *Théorie analytique des probabilités* and in the 1814 *Essai philosophique sur les probabilités,* the latter of which has become the locus classicus of his important dictum about determinism.

It was not sufficient for d'Alembert and his followers to embrace the distinction between the certainty of deductive systems derived by humans and the contingency of knowledge secured empirically. Their half century of experience with the Newtonian program also raised new ontological dilemmas, of two related sorts. On the one hand, the repeated empirical verification of the principle of universal gravitation was turning it into a nearly certain principle. What then was to be its status? Was it to be considered a fundamental law of nature, or simply a unitary and uniform cause? On the other hand, the remarkable success registered by the use of the calculus to portray the mechanics of the heavens was impelling natural philosophers to adopt a mechanistic attitude in which God's role was relegated to that of the Creator, or first cause. This combination of criticism of the voluntarist position and a compensating emphasis on causality forced contemporary philosophers of all persuasions to reexamine their views. For instance, Laplace's literary colleague at the Ecole Militaire, the abbé Charles Batteux, penned a historical review of the dilemmas faced by his contemporaries the very year our mathematician moved to Paris.[15]

## Intellectual Sources

Laplace's interest in these central metaphysical questions, though directly inspired by Condorcet and d'Alembert and possibly echoed by Batteux, began much earlier in Caen, which sported a literary tradition of its own. Because Laplace rarely cites his intellectual sources, there is no direct evidence of the influence that specific authors or texts may have had on his thinking. But similarities between his concerns and those of some predecessors active in Brittany and Normandy are suggestive. It ought not to surprise us that some of these provincial authors still cherished Cartesian values. One must recall that the triumph of Newtonian physics, while soundly established for dealing with the dynamics of the heavens, remained problematic for many who were more philosophically inclined. As late as the decade when Laplace entered the Académie, the learned society had been asked to review the writings of mediocre authors like Gaspard Forbin, Joseph Bertier, and Jean Saury, each of whom clung to certain aspects of Descartes's worldview. These minor enthusiasts posed little threat to the new Newtonian consensus in the Académie. More troublesome outside the scientific circles were philosophers clinging to a Christianized version of the mechanical philosophy, which was based on Cartesian physics.

Among these philosophers were two Bretons whose writings Laplace may well have known: Charles Hercule de Keranflech and Father Yves André.[16] At the University of Caen, André had held the royal professorship of mathematics and hydrography immediately prior to Gadbled. Like de Keranflech, André embraced the ideas about God and nature held by the highly respected academician Nicolas Malebranche.[17] Even if Laplace had no direct contact with them, their reputation in Caen would have become known to him. Another writer living in the Lowlands whose work reflects similar and well-articulated preoccupations is the Huguenot David Renaud Boullier, who was both familiar with some of Leibniz's views on issues of the day and a partisan of Pascal.[18]

The significance of each figure as a mediator between these older philosophers and the young mathematician is unknown and thus will have to remain problematic; but it is evident that Laplace showed an interest in the ideas they broached. The peculiar blend of the theological, metaphysical, and scientific that they all exhibited in their writings was something Laplace was conversant with from his own student days, even if it now caused him some discomfort. Though he consciously tried to disengage one domain from the other, following d'Alembert's counsel, his temperament was moored in traditional preoccupations. In the end, Laplace consciously repressed his true disposition in order to present to the world a "modern" scientific persona that seemed to ignore metaphysical issues. By opting for this posture, he openly sided with d'Alembert, adopting what may be called a pre-positivist mentality.[19] Nevertheless, Laplace never totally gave up mulling over these older and more metaphysical issues, particularly those touching on the nature of

God. Somewhat later in his career, the task of discerning the function of God in the secular world was to preoccupy him in ways that one might not initially suspect.

At this early stage in his life, what stands out is Laplace's personal resolution of some fundamental questions. The exercise led him to articulate a new and seemingly paradoxical philosophy that combined the uncertainty of human knowledge and the determinism of nature with a remarkably confident attitude about the value of the scientific enterprise.[20] This philosophy ultimately gave meaning to his life's work and endorsed in his own eyes the decision he had made to abandon the theological path in favor of a scientific career. It also validated his entire scientific program, which was devoted equally to celestial mechanics and probability theory.

Laplace arrived at this conclusion by turning away from the ontological problem raised by the Berlin Academy prize in favor of resolving an epistemic proposition. He reasoned that the task of the *géomètre philosophe* was to worry less about the nature of laws—that is, whether such laws are necessary or contingent—than about the degree to which they were to be taken as established truths. Like d'Alembert, he understood that if there were a God, he could countermand any law of nature. But the quarrel between Gottfried Leibniz and Nicolas Malebranche, later echoed in the correspondence between Leibniz and Samuel Clarke about the kinds of laws God had legislated, was an issue on which science offered little enlightenment. Whether God was omniscient, having created perfect laws to which he was always subject, or omnipotent, and thereby able to break them at will, seemed unresolvable. Laplace could do little but take refuge in the Augustinian position, echoed by Pascal, that since man can never fully know the essential nature of matter, he can say little with any assurance about the nature of God.

What man can do with his own wits, however, is to perfect his knowledge of the laws that seem to govern the entities placed in the world. The quick march of scientific research in this direction was striking. In the previous century, Descartes's speculations had been rapidly supplanted by Newton's codification of the laws of physics and his postulation of the principle of universal gravitation. For Laplace, like for d'Alembert and his coterie, the fruitfulness of this principle was to be found in transforming it from a plausible principle into a verified set of laws of nature.[21] What was singularly impressive was its unifying and reductionist character, which made it analogous to a first cause.[22] Like the principle of least action, it seemed fundamental because, with a single antecedent cause, it explained so much.[23] But unlike this principle, which its proponent Pierre-Louis Maupertuis customarily linked to the nature of God, the law of gravitation was adopted because it was derived from phenomena that are solely within the province of human understanding. Laplace looked to Newton's law as a highly likely universal truth to which he, as a layman, could subscribe with confidence, yet without determining its relationship to God. In this way he could justify to himself the discordance he created when he chose to abandon religion for science.

Universal gravitation, in line with the older tradition of metaphysics that Laplace had absorbed and criticized in Caen, seemed initially to Laplace to be able to stand as an all-encompassing truth. Throughout his life, Laplace expected that attraction would operate not merely in the macroscopic realm of the solar system, where it had shown its potency, but in terrestrial physics as well and perhaps even as the central constituent of the laws governing chemistry, mineralogy, and the life sciences. The very phrasing of the original formulation of his deterministic credo in 1773 suggests that he meant it to apply to all beings, not merely to astronomical entities: "The present state of the system of nature is evidently a result of what it was in the preceding instant, and if we imagine an Intelligence who, for a given moment, encompasses all the relations of beings in this Universe, It will also be able to determine for any instant of the past or future their respective positions, motions, and generally their disposition *[les affections de tous ces êtres]*."[24] The use of terms such as *êtres* and *affections* is reminiscent of the writings of authors like Boullier, André, and de Keranflech, who had variously argued for and against Descartes, Malebranche, and Leibniz. This brief passage, which he later tempered, was the only overt concession that Laplace made to the intellectual environment he had encountered as a theology student in Caen.[25] The remainder of the deterministic credo spoke directly to his Parisian colleagues in the Académie, justifying his agenda as a physical scientist. Taking on a d'Alembertian voice, Laplace wrote:

> Physical astronomy, that subject of all our knowledge most worthy of the human intellect, offers an idea—albeit imperfect—of what such an Intelligence would be. The simplicity of the laws that move celestial bodies, and the relationship of their masses and their distances, permit

us to follow their motion up to a certain point with the use of calculus [*analyse*]. To determine the state of the system of these large bodies in past or future centuries, it is enough that observations provide the mathematician with their position and speed at a given moment. Man derives this capacity from the power of the [mathematical] instrument he uses and the small number of links [*rapports*] they include. But our ignorance of the various causes that produce these events, as well as their complexities taken together with imperfections in the calculus, prevent him from making assertions about most phenomena with the same assurance. For him therefore there are things that are uncertain, and some that are more or less probable.[26]

On all these points Laplace was in agreement with his close academic colleagues. What stamped this deterministic credo as innovative were the sentences that immediately followed. They explicitly linked the future progress of knowledge about nature to the application of a mathematical tool that had been principally used to solve problems of chance, a totally different activity. For Laplace, a more evolved probability theory would provide the key to the epistemological puzzle: "Given the impossibility of [total] knowledge, man has compensated by determining their different degrees of likelihood; so that we owe to the frailty of the human mind one of the most delicate and ingenious of mathematical theories, namely the science of chance or probabilities."[27]

## Probability Theory

Laplace immediately explained away the Democritean connotations of the word "chance" or *hasard*. The term was not meant to suggest that chance ruled the world, as pre-Socratic atomists and their many disciples asserted. Chance was not the absence of cause in nature, but a sign of the absence of man's knowledge of that cause. The word "chance," he said, corresponds to no reality in nature; "it is merely a term that designates our ignorance of the way the different aspects of a phenomenon are arranged among themselves and in relation to the rest of nature."[28]

The concept he expressed was not new.[29] Since the beginning of the century, probabilists like Jacob Bernoulli, Pierre Rémond de Montmort, and Abraham de Moivre had been codifying this notion and giving it operational meaning. Laplace's central contribution was to insist that the proper use of this new tool was to help the scientist out of his epistemic dilemma.[30] Though one could not infer certainty in most instances, one could evaluate the degree of certainty of a particular cause for given consequences. This would offer the scientist an objective way of preferring one causal hypothesis over another, and would give him logical grounds for selecting one of them. It was a strategy for dealing with the unavoidable uncertainties stemming from the nature of the contingent system of the world. Firmly believing that nature followed inflexible, necessary laws, Laplace recognized that it was the scientist's special challenge to discover them, or to come as close to determining them as possible. In making this point explicitly, Laplace was projecting his personal intentions onto the whole community. Like d'Alembert and Condorcet before him, he defined the role of the scientist for himself, and then proclaimed it as a universal truth.

Laplace's central assertions were not entirely rash, nor were they difficult for contemporary scientists to absorb. Many of the ideas had a generic relationship with those expressed by leading figures of his day. The epistemic issue had been part of an ongoing philosophical dialogue in which Maupertuis and Laplace's patron d'Alembert played a central role. The notion of an Intelligence with vast calculating powers was familiar to Leibniz, Maupertuis, and Condorcet. Determinism itself was far from novel, having been expressed in various ways by Leibniz and the Jesuit scientist Bošković before him.[31] It was perhaps for these reasons that Laplace did not dwell on them further in print. To his mind, he was merely reformulating accepted positions for his own use. The only part of his program requiring elaboration was the notion of probability. It was in fact the only one to which he eventually devoted an entire text, bearing the title *Essai philosophique sur les probabilités* (Philosophical Essay on Probabilities).

In retrospect, we know how difficult the enterprise was, and that Laplace did not completely resolve it.[32] One of the difficulties lay in distinguishing between the epistemic and the aleatory nature of probability. Laplace's transposition of Leibniz's concept of insufficient reason into the idea of equipossibility did not immediately help to clarify the misunderstanding. One can appreciate the historical situation by recalling that the techniques for dealing with probability were invented by mathematicians initially fascinated by games of chance. They solved problems relating to betting on coin tossing or rolling the die. In these instances, it was fitting to adopt the notion of equally likely outcomes and to use combinatorial arithmetic to devise a prognosis. The assumption was always that chance, rather than a specific cause, ruled the outcome. For this reason, the science that we now call probability was initially named the doctrine or calculus of chance. Laplace, by contrast, was concerned with determining laws in the physical world and ascertaining causes. He meant to do this by focusing on instances in nature where there was a marked and recurrent departure from randomness, singling out for analysis instances when the principle of sufficient reason was in operation. He wanted to bring the new calculus to bear on cases where reason rather than hazard ruled.

Laplace did this by codifying and adapting practices already employed by contemporary users of the doctrine of chance, notably Daniel Bernoulli. By insisting that likelihood was measured by the ratio of actual to possible outcomes, he harnessed to his own epistemic concerns techniques devised by mathematicians to analyze games of chance. In so doing, he borrowed procedures from a new and respectable mathematical tradition, placing them at the service of the physical sciences. In other words, in Laplace's hands, the doctrine of chance became a science of probability. The key discovery was divulged in his lengthy technical paper entitled "Mémoire sur la probabilité des causes par les évènemens" (Memoir on the Probability of the Causes of Events), which was most likely written in mid-1773, shortly after he was inducted into the Académie. It has recently been the subject of considerable attention, often presented under the label of Bayesian inference.[33]

Many questions still remain about how Laplace was led to this major innovation. It seems unlikely that he was at all familiar with the English mathematician Thomas Bayes's views on inference, which had been published posthumously in the 1763 issue of the *Philosophical Transactions of the Royal Society*.[34] Though the concept of inversion from effect to cause bore similarities, the mathematical notation and approach that Bayes adopted were quite distinct. A more likely scenario is that Laplace's interest in the technique for integrating finite partial differences, which caught his attention because of its relevance to celestial mechanics, led him to de Moivre's *Doctrine of Chances*. Laplace's progress with integration turned out to offer solutions to issues that de Moivre had encountered.[35] So fascinated with this seminal work was Laplace that he once contemplated translating it, even though he did not know English.[36]

Laplace encountered other situations in which a knowledge of combinatorial calculations was critical. There was the local and mathematically trivial problem of understanding odds in the lottery held at the Ecole Militaire where he taught.[37] More interesting were issues he faced in closely studying the resolution of problems in celestial mechanics. Two of these in particular made him realize that he needed to gain mastery over the aleatory science. One was in finding an appropriate, mathematically justifiable way to select one observation in a specific situation when three or more observations of the same event were available. Choosing the "best" mean value, as the problem has come to be called, was quite a challenge. Laplace admits that what egged him on was learning that Daniel Bernoulli and Lagrange had tackled it themselves.[38] Competition was clearly part of what drove him to pursue the question; Laplace would boast that he had arrived at a solution before knowing theirs. Bošković and Johann Heinrich Lambert had also tried their hand at it, but in ways that were unsatisfactory.[39] The second issue centered on knowing when and why to discount or discard observations that were discordant. As a budding physical astronomer, Laplace, like his colleagues, was continually called on to deal with such "errors."

In both of these cases, Laplace's approach was to reason deductively about what ideally should happen, and to fix attention on the differences between the theoretical and the observed values.[40] The remainder of the mathematical problem would focus on formulating this difference as a differential expression and finding the conditions for minimizing the function. To solve this problem in error theory, the least-square method was only decades away, but alas, in the future. Curve fitting was also more an art than a science, and easily led to major disputes. Laplace was shortly to tangle with Bošković over the proper means to calculate the path of comets given three close observations, a topic given currency by the former Jesuit scientist, who was in Paris presenting his views to the Académie during the very years that Laplace was knocking at its doors.[41] Though this dispute did not directly turn on probability issues, reflecting on the work of contemporaries like Bošković may have given Laplace the extra thrust he needed to push through to a solution. As an eager and alert young scientist, Laplace was all too aware of his competitors, and keen to show his mathematical superiority. D'Alembert had shown him the way, and now Daniel Bernoulli, Lagrange, Lambert, Bošković, and Condorcet spurred him on. Clearly he owed them a debt for providing the stimulus. But he could also be proud of the originality of his mathematical solutions. The accusation later made by the mathematician and biographer Augustus De Morgan that Laplace often failed to cite scrupulously his sources is appropriate, but the implication that he was in this way minimizing their contributions to swell his own is unwarranted.[42] Laplace was an original thinker, but one who did not operate in an intellectual vacuum.

## Challenges to Probability Theory

The status of probability theory as Laplace came to conceive it was also challenged by the major thinkers Buffon, d'Alembert, and Condorcet, whose visions for this technique were at variance with Laplace's. All of these formidable colleagues in one way or another concerned themselves with the distinction between pure mathematics and rational human behavior based on the use of this theory.[43]

Buffon was convinced that in addition to degrees of mathematical certainty, one had to account for "moral certitude." He expressed these views cogently in a letter following a reading of Laplace's 1774 *Mémoire,* and later set his views down in print in his *Arithmétique morale* (1777; Moral Arithmetic).[44] Laplace chose to ignore this objection, perhaps because the limits that Buffon selected for the introduction of moral dimensions in his calculations were patently arbitrary.[45] Unlike Condorcet, who in his eulogy came close to ridiculing Buffon for his mathematical ineptness, Laplace kept his assessment to himself.

The challenge from d'Alembert was much more serious and pointed. Already in the *Mélanges de littérature,* which Laplace presumably already knew in Caen, d'Alembert had reiterated his doubts about the usefulness of probability calculations to determine what rational course of action one should take with regard to inoculation.[46] He developed this skeptical stand further, probably in reaction to the growing interest in probability throughout the next decade, and undoubtedly discussed it in person with Laplace. D'Alembert openly criticized his protégé in his last pronouncement on the subject in 1780.[47]

D'Alembert's difficulty lay in his not believing that mathematically sound solutions that followed the doctrine of chance and that assumed equipossibility had any value for physical reality. In line with his dichotomous treatment of the principles of mechanics and of empirical laws, d'Alembert asserted that the two domains could not be reconciled. In the end, he capitulated before the ontological differences that separated mathematics from physics. More concerned with methods for ascertaining knowledge, Laplace transcended this viewpoint by claiming that the two could come asymptotically close to one another, as long as choices made in the handling of observational data used to deepen our knowledge of laws of nature were based on rational grounds. His great innovation was to master statistical inference and put it to work as a viable technique for working with empirical data.

Condorcet joined Laplace in countering their common patron, though with considerably more respect than he showed Buffon. Most of his rejoinders to d'Alembert were published after the master's passing, and conceded the intelligence of d'Alembert's questions, if not their persuasive force.[48] As Keith Michael Baker has shown, Condorcet benefited from Laplace's penetrating research in probability theory as he formulated his arguments, to the point that it is difficult today to separate their individual contributions.[49] It may even be the case, as Baker suggests, that Laplace's important but pithy insertions into his technical papers, discussed earlier, were part of Condorcet's campaign to satisfy d'Alembert's philosophical objections.[50] As is true for all truly symbiotic relationships, it is difficult to figure out which one of the partners should be credited for each innovation.

This close intellectual relationship with Condorcet pulled Laplace in a direction that he had not anticipated. In part through his association with Turgot, Condorcet was becoming increasingly concerned with the use of probabilistic techniques to illuminate social rather than natural problems.[51] In a letter to Turgot in September 1774, Condorcet even suggested that Laplace would be willing to devote himself fully to producing a treatise on the application of these techniques to the political economy, provided that Turgot would relieve him of his teaching chores at the Ecole Militaire.[52] By developing a new use for the old doctrine of chance, Condorcet extended its domain to include the human sciences. Just as understanding the system of the world could be inferred from data, so the argument went, knowledge about human behavior could also be gained, if probability theory were properly wielded. Condorcet took it upon himself to stress the potential value of Laplace's discoveries for his own philosophical enterprise, in the spirit of the Enlightenment. Nowhere is this clearer than when, as secretary of the Académie, Condorcet characterized Laplace's memoirs in the historical introduction of the volume where they were published. "One sees that this question encompasses all the applications of the doctrine of chance to the uses of ordinary life, and that it is the only component of this science of general utility deserving the serious attention of *philosophes*. Ordinarily this calculus is employed only to yield probabilities in games of chance and lotteries, and it [nevertheless] fails even to dissuade men from these amusements, which are harmful to their industry and mores."[53]

There is no record of how Laplace reacted to this slanted characterization of his interest in probability theory. On the one hand, Condorcet could capably explain to French society the value of the pursuit of scientific knowledge, something Laplace was reticent or unable to do. He did learn very quickly that there was merit in trumpeting the potential utility of abstruse research. Indeed Laplace cleverly chose to illustrate the application of his newly won techniques with examples drawn from life as well as from nature. (In the early memoirs, he takes up cases of gambling, lotteries, and demographic predictions.) Nonetheless, he most wanted to advance the subject so that he could use it to determine the laws of nature for physical astronomy.

Given his steadfast and calculating character, Laplace chose to entertain and even to participate in the new usages for probability theory, but without relinquishing his initial goal. He was willing to assist in studies championed by Condorcet that centered on population statistics, voting patterns, and insurance and annuity plans, each of which required sophisticated manipulations of probability theory. But Laplace did so without fully sharing in Condorcet's expectation of developing a truly sound, mathematically based social science.

During the feverishly busy years in which Laplace established himself in the Parisian world of science, he managed to carve out an innovative and coherent plan for his intellectual itinerary. The three lines of research he had been pursuing converged in a unified philosophical program. The advancement of calculus, probability theory, and celestial mechanics promised to offer the scientifically adept an organized understanding of the system of the world. Man could finally hope to apprehend fundamental causes of nature's mechanisms, coming asymptotically close to ultimate reality without having to be concerned with the nature of the Creator. Laplace's hypothetical Intelligence, who could see backward and forward in time, was the intrepid calculator whose wits were sharpened by a superior understanding of the mathematical tools of the trade. It is not hard to imagine which scientist Laplace assumed would be the worthiest candidate for this task.

# 5

# Finding the Stability of the Solar System: The Celestial Program

Laplace was not a *philosophe* in the sense that Enlightenment figures like d'Alembert and Condorcet were; nor can he be considered a philosopher in the classical tradition of Thomas Hobbes, John Locke, Gottfried Leibniz, or Etienne de Condillac. It is quite consonant with his personality that his philosophical program could be found in a mere two pages inserted in his early and voluminous technical mathematical productions.[1] As a young adult, he was reticent to engage in protracted discourse on metaphysics and never ventured into public debate about fundamental matters that, in retrospect, we deem important intellectual issues. Instead, since Laplace had earned his career from his prowess as a professional scientist, he needed to exercise his mathematical craft on a daily basis. His archives are full of sheets of calculations attesting to his feverish and continuing activity as a physical scientist. The habit of working hard and with quill in hand never left him, not even in old age.

Nevertheless, the overarching design governing Laplace's career was set by a philosophical blueprint from which he hardly deviated once it had been internalized early in his mature life. Moreover, these two levels of activity were not treated by him as distinct. The technical operations were pointless without the underlying philosophical program; and this program depended on careful examination of data that could be fitted to establish natural laws. Everything that he considered important in his professional life was channeled into supporting this goal. Though he was obviously talented in mathematics, astronomy, and physics, he did not regard himself as a specialist in a particular domain of science. His self-image was more like that of a modern natural philosopher blessed with a special mastery of the contemporary language of mathematics. It explains why he thought nothing odd about being elected to the Académie as a member of the mechanics section instead of the geometry or astronomy sections. As long as he could deploy his skills, he was at least initially gratified.

## The Young Academician

As a junior academician, Laplace was expected to sit on various commissions that examined papers submitted to the royal arbiter of scientific innovations. Many of the submissions to the Académie were below the standards he set for himself, a situation that reinforced his own self-esteem. Indeed a number of contemporaries considered him to be arrogant.[2] During his first dozen years in the learned society, he was appointed to over twenty commissions dealing with windmills, paddle wheels for water mills, diving suits, contraptions to scrape mud off of river basins and ports, pumps, fire ladders, stage coaches, and even the improvement of a keyboard instrument.[3] Most of the time his more mechanically inclined colleagues Mathurin Jacques Brisson and Jacques de Vaucanson were assigned the task of writing the reports and discussing them in academic meetings. In those instances, Laplace merely concurred with their findings, appending his signature to their report. For example, Laplace was chosen to head the committee reviewing Bonaventure Le Turc's machine for producing lace patterns and Dez's formula for estimating the liquid capacity of barrels with shapes generated by the revolution of a parabolic segment.[4] Both authors, who happened to be employed at the Ecole Militaire, obtained a favorable hearing from their faculty colleague, but only after he authored a precise and detailed account of the proposals that demonstrated the new academician was taking his job seriously. The clarity of the reports leaves little to the imagination.

Laplace was less inclined to squander time on the dozens of proposals for squaring the circle, trisecting angles, or inventing perpetual machine motions. The very first report Laplace presented to the Académie the month following his election was on one Gosset des Aunes's scheme for squaring the circle. He dismissed it with a few curt sentences.[5] Other proposals that he examined were not even reported to the Académie, presumably to save the author embarrassment.[6] In 1773, he also reported succinctly on errors committed in an elementary solution offered by Thomassin for the physics of free fall.[7] The reports he prepared were generally informative without being lavish either in praise or in length. Only when faced with a matter he deemed scientifically significant did he display the full potential of his analytic mind. One senses that he quickly grasped the difference between a routine committee assignment, in which the petitioner was likely to be declined, and one involving a serious new idea.

One of the benefits stemming from these assignments was that he became familiar with colleagues whom he otherwise knew only by sight. He served on committees with an assortment of partners, ranging from astronomers and mathematicians, whose activities he followed closely, to chemists and surgeons—of whose work he was less likely to keep abreast. Early on he expanded his set of friends, originally drawn from d'Alembert's coterie, to include prominent amateur astronomers who moved in aristocratic and conventional religious circles, like Achille Pierre Dionis du Séjour, Jean Baptiste Gaspard Bochart de Saron, Alexandre-Guy Pingré, and Jean-Sylvain Bailly. He met and impressed Lavoisier—the young partner in the tax-collecting business firm, the Ferme Générale, and a rising star in chemistry—with whose family he occasionally dined. A special affinity developed as well between the mathematician Bézout and Laplace, who shared a responsibility to ready young nobles intending to become officers, as well as providing contacts with the military establishment. For example, Laplace taught the young students at the Ecole Militaire, who were then examined in mathematics by his academic colleague Bézout. Because of a lack of available information, I cannot be more specific about how Laplace's social life unfolded. Presumably it was centered around his quarters at the Ecole Militaire and the biweekly gatherings of the Académie.

It seems likely that Laplace did not frequent the famous salons of the Enlightenment or of progressive thinkers like Holbach or Turgot. This was despite his early association with d'Alembert and Condorcet, who would have made it possible to develop a liaison with the many social reformers of the ancien régime.[8] While not openly antagonistic to them, Laplace preferred not to take advantage of those contacts, given his lack of concern with philosophical discourse. If anything, Laplace's social inclinations seem to have favored the more conventional, established social elites of Paris rather than its more radical members. One of them, Bochart de Saron, was to become a few years hence his patron. Even during the Revolution, Laplace would shy away from extremists.

But even in such refined surroundings, Laplace was not timid about expressing personal opinions. According to the Swedish astronomer Anders Johan Lexell, the Spanish ambassador once considered bodily throwing Laplace out for speaking with contempt of Spaniards as "sots et ignorants."[9] And at a dinner party held at the home of Lavoisier's relative Clément Augez de Villers, the geologist Jean-Etienne Guettard was staggered by Laplace's bold denunciation of the existence of God, a view he reiterated at the home of Dionis du Séjour.[10] If one is to take Guettard's words literally, Laplace's unambiguous expression of his atheistic belief was supported by a thoroughgoing materialism, not by any reference to the system of the world.[11] It was a deep conviction derived from his earlier repudiation of a religious life. As such, it was meant to remain a personal matter, inadvertently blurted out in private company, but not for publication. At these dinner parties, he made no attempt to link his stand on religion with any research into the heavens. Yet on scientific matters, he was never reported to have been less than judicious, if at times forceful. Ultimately, one of his rewards, granted in 1786 by the new duke of Orléans, was to be appointed a pensioned member of the duke's scientific entourage.[12] Like other men of learning who frequented the courts of Louis XVI's relatives, Laplace fit in easily in ancien régime society.

Among his peers, Laplace adopted either a diplomatic or combative posture, as circumstances dictated. Very early in his career, he was called on to adjudicate a delicate priority dispute between two of his talented older colleagues, Condorcet and Cousin, over the integration of partial differential equations.[13] The matter seems to have been settled "out of court" diplomatically. With the ex-Jesuit Bošković, however, Laplace had a major row in 1776 over the use of proper techniques to calculate orbits given three closely made observations. The quarrel reverberated in the scientific community and won him the approval of Lagrange, among others.[14] Bošković, however, was shocked by young Laplace's attack, which he regarded as unprovoked and out of keeping with academic courtesy. Laplace considered the issue to be a serious professional dispute, in which nothing less than the validity of the use of calculus against older geometrical techniques was at stake. He eventually carried the day, notwithstanding efforts by his academic colleagues to mediate the caustic disagreement.[15]

## Calculating Orbits

It was in fact the challenge of computing the nearly elliptical paths of celestial objects that made Laplace's approach to physical astronomy distinctive and led to his masterful display of the subject. While one can safely assume that d'Alembert's essays were the starting point for Laplace's general formulation of astronomical problems, the more proximate occasion for his detailed involvement came from a long text that Dionis du Séjour proposed for publication and which was assigned by the Académie for Laplace to review in 1774.[16] The *Essai sur les comètes* (Essay on Comets) subsequently appeared with a detailed and laudatory appraisal penned by its reviewer, which underscored the progress that Dionis du Séjour had made in understanding these errant bodies of the solar system.[17]

Comets were a concern for Parisians. A widely reported address by Lalande, prepared for a public meeting of the Académie in April 1773 about the perturbing effects that planets might have on paths of comets, had revived popular fears about the likelihood of a major collision of comets with the earth. In the address, which actually failed to be delivered for want of time, Lalande had attempted to alleviate superstition about what comets portend, by arguing that their insignificant mass and distance from the earth all but eliminated any threat.[18] Nonetheless, Lalande's unwillingness to rule out absolutely that such an accident could occur spread alarm among the Parisian populace.[19] Dionis du Séjour was prompted to extend the argument by making a detailed calculation of the improbability of such collisions. With Laplace as its reporter, the committee naturally underscored this innovative aspect of Dionis du Séjour's work, particularly piquing the reviewers' interest since it called on elementary notions of probability to make its case. But Laplace also seized the opportunity to focus on more fundamental astronomical issues, only a few of which Dionis du Séjour had treated.

Dionis du Séjour's treatment was ultimately fashioned for practicing astronomers, who were busy in the late eighteenth century spotting new heavenly bodies, calculating their paths, and predicting their location or reappearance. In addition, most of his colleagues in the astronomy section of the Académie were devoted to simplifying the operations of observational astronomers. This included the time-honored concern with instruments, the reduction of data to tabular form, the preparation of ephemerides, and the teaching of their use to amateurs.[20] Laplace had already encountered this operational turn of mind among Gadbled's detractors in Caen, who considered astronomical education worthy of support solely because of its use by navigators. In Paris, he met a more sophisticated set of practitioners, including his nearby colleagues Jeaurat and d'Antelmy, who had established an observatory at the Ecole Militaire. Other dedicated amateurs manned stations scattered throughout the capital and its surroundings.[21] The older academician Lalande, an early supporter of Laplace, was this group's most prolific booster, producing popular manuals, supervising the publication of the annual *Connoissance des tems* (Weather Almanac) and lecturing at the Collège Royal, where he trained scores of amateurs. All their activities were thoroughly grounded on a quantified empiricism serving the needs of the observer. Everything about astronomy, including the introduction of new mathematical techniques and the calculation of orbital elements, was eagerly appropriated for the improvement of the art of observation. It was a stand shared by most of Laplace's astronomical associates at the Académie, chief among them César François Cassini de Thury and his son Jean Dominique Cassini IV, Bochart de Saron, Bailly, Pingré, Dionis du Séjour, and Lalande, who was its relentless champion.[22]

Laplace's interest in astronomy had from the outset been markedly different. It was no accident therefore that he was never considered a serious candidate for the astronomy section in the Académie. His perspective was chiefly sustained by a desire to untangle the mechanics of the universe for the transcendent purposes noted earlier. Dionis du Séjour expressed this well in the preliminary discourse to a book appropriately assigned by the Académie for review by Laplace in 1789: "The more complicated this affair [astronomy] becomes, the more one must realize how difficult it is to deal with it in a simple fashion. But one is buoyed by the sublimity of the subject, and human intelligence, flattered by its discovery of truths seemingly beyond it, imagines it is participating in the immensity of nature." With these words, Dionis du Séjour was echoing his colleague's profoundest sentiments. For Laplace, celestial investigations were but a means to a loftier end.[23]

## The Law of Gravitation

From the very beginning of his career, Laplace had been fascinated by Newton's universal law of gravitation. He repeatedly commented on its correctness, and in his correspondence underlined with relish each confirmation of its validity. In the lengthy paper published in 1776 in which he first made his views on determinism known, Laplace states unequivocally: "There is no truth in physics more indisputable and better established by the agreement between observation and calculation than the one that 'all celestial bodies gravitate toward each other.' Newton, the author of *this most important discovery ever made in natural philosophy*, found that the observed motions of planets can be sustained only by a tendency toward the sun proportional to their masses and inversely to the square of their distance from this body."[24] At this early stage of his research, he was particularly conscious of the ways in which this fundamental law could be manipulated so that its consequences would match the observed motions. There had been unsuccessful attempts to "save the appearances" in the recent past; for instance, by assuming that the distance function bore a different exponent.[25] Leonhard Euler and Charles Bossut had just published their findings on the assumption that planets and their satellites were slowed by a small resistance inherent in the transparent "ether" that permeated the universe.[26] Discrepancies stubbornly remained.

Laplace now entertained several other possibilities. The first of these was that bodies in motion were affected differently by gravitation from those at rest; the second was that the gravitational effect operated with a time lag. Raising these issues reveals his parallel concerns, one with the metaphysical foundations of mechanics, the other with their actual consequences for the system of the world. His aim was to disclose the laws of nature that were logically plausible and acceptable, while at the same time yielding values consistent with astronomical observations. This was a goal he never tired of pursuing, from the beginning to the end of his career. With respect to the law of attraction acting on stationary and moving bodies, he questioned the Cartesian and Newtonian assumption that rest and motion were ontologically equivalent. Some years later, in intimate musings, he asserted that bodies at rest must be endowed with a "modification" when they were set in motion, either through collision or by a gravitational pull.[27] In his earliest published articles, Laplace chose to bypass such knotty metaphysical issues in favor of elucidating the mechanism by which attraction operated. Was it not the case, he asked, that if gravitation were activated by a material cause, the force of the moving body would be affected by the standard centripetal gravitational force plus the resistance to its own motion, orbital or otherwise? If that were so, he concluded that the gravitational force, in addition to depending on mass and distance, would have to include a small term similar to a secular equation, proportional to the square of the time.[28] Most likely he imagined this feature borrowing from the accepted mechanism used to explain the aberration of light. With this hypothesis, Laplace suggested that one might be able to account for the secular acceleration of the moon's longitude, a vexing problem that had not yet been resolved despite efforts by d'Alembert and Lagrange.[29] Not content with this qualitative remark, he proceeded to calculate how fast a hypothetical gravitational corpuscle would have to move in this case to account for the moon's secular acceleration. The result was an astounding 7,680,000 times the speed of light. Clearly this was a speed far surpassing man's ability to verify or falsify. Hence this intricate supposition had to be set aside.

Laplace also entertained another possibility that he was to reject for similar reasons—namely, that the transmission of the gravitational force was not instantaneous. He pointed out that "it is not likely that the attractive virtue . . . is propagated instantly from one body to another; for all that travels through a [finite] space must pass through its successive points; but given our ignorance of the nature of forces and the way they are transmitted, we must be guarded about our inferences until experience can shed some light on them. Nonetheless I would point out that even when evidence leads to a belief that the propagation is instantaneous in nature, we must not jump to that conclusion, for there is an infinitely great distance between instant propagation and one that takes an undetectable amount of time."[30]

The recurrent problem for Laplace was to find an operational test for this assertion. He never succeeded. No wonder that a decade later he would return to the issue, toying with the highly speculative system of minute, fast-moving "ultra-mundane" particles postulated by Georges Louis Le Sage to offer an explanation for the law of gravitation.[31] In the back of his mind, Laplace was always hoping to find a plausible physical mechanism for fundamental principles, but he remained extremely cautious about adopting openly any untested and untestable scheme.

We are able to peer into Laplace's mind only because he allowed himself to work through these shrewd but ultimately fruitless pathways in his publications. Motivated by an all-consuming desire to uncover nature's deepest secrets, Laplace nonetheless always controlled his urges with a strict rule to connect theories with hard evidence. In these early writings, he would share with his readers a chain of reasoning that may now appear as mere flights of fancy. Later, he would relegate such chimera to footnotes or popular treatises that might tantalize the reader or stimulate further research; but he would clearly distinguish hard-won knowledge from clever conjectures. The most famous example of this practice was to be his speculation on the origins of the solar system.

Lest one think Laplace devoted all his time to such speculative musings, one must recall that all his astronomical memoirs written before the French Revolution make plain his pains to master the standard problems for the mechanism of the solar system. Literally hundreds of pages in the Académie's *Mémoires* are filled with his attempts to grapple with the detailed mathematical consequences of Newton's gravitational law. Like other theoreticians of the era, Laplace wrote extensively on the moon's erratic motion, on the anomalies in the motion of Jupiter and Saturn and their satellites or rings, on the significance of the shape of the earth and its tides, and on comets. In the dozen years after his election to the Académie, he established himself as an equal to leading mathematical scientists like Leonhard Euler, d'Alembert, Lagrange, Daniel Bernoulli, Legendre, and Monge. The best advanced textbook of the era, Cousin's *Introduction à l'étude de l'astronomie physique* (Introduction to the Study of Physical Astronomy), published in 1787 and based on his lectures at the Collège Royal, reports on Laplace's contributions with the same respect given to older contributors. His place in the professional annals of astronomy was now well secured.

## Crediting Predecessors

It may not be worthwhile to reanalyze each of these memoirs.[32] Nonetheless several features deserve to be extracted from these exhaustive studies. The most delicate and revealing concerns his originality. Augustus De Morgan long ago complained about Laplace's annoying habit of failing to properly acknowledge his debt to other scientists.[33] More recent commentators have focused on citation analysis to temper the accusation.[34] This approach has serious pitfalls, given that the practice of meticulously referring to past literature was not standardized in his time. Laplace frequently listed names of contemporary workers whose ideas he appropriated and developed, but usually neglected to indicate exact references. He could assume that the limited number of mathematical scientists reading his complex memoirs would know which papers were germane. This practice did not always satisfy everyone. D'Alembert in particular took umbrage on several occasions and insisted that Laplace add to the printed version of papers presented orally to the Académie a section indicating the source of his inspiration. The young scientist naturally bowed to his patron's wishes, as is clear from one missive to him in 1777.[35] The addendum he offered for the printed memoir was reverential and somewhat inflated: "If one considers how difficult are the first steps in any subject, and especially in such a complex subject, [and] if one considers the immense progress accomplished in analysis since his publication, one will not be surprised that he left something for us to work on, and that, abetted by the theories we owe almost totally to him, we have made headway in the course he was the first to lay open." His letter of explanation to d'Alembert concludes with a self-effacing comment that must have gratified his mentor: "I have always cultivated mathematics by fondness rather than to establish a vain reputation about which I care little. My greatest delight comes from studying the steps taken by other inventors in order to visualize their genius working to overcome the obstacles they encountered. I put myself in their place and ask how I would have handled these same obstacles, and even though this transposition is often humbling for my self-esteem, I take pleasure in following their triumphs."[36] Laplace's submissive compliance needs no elaboration, even though one might wonder why d'Alembert craved this additional public recognition.[37] With nearer contemporaries the issue is more intricate. Two older mathematicians to whom Laplace clearly owed a debt—Condorcet and Lagrange—provide a more complex situation.

We have already noted how closely Laplace's youthful work shadowed that of Condorcet. They were natural rivals and could legitimately have been jealous of each other's reputation.[38] Fortunately for Laplace, Condorcet sought and obtained his renown in the larger Enlightenment community as a *philosophe* and public spokesman for science just as Laplace entered the scene. Laplace depended on his colleague's goodwill for the prosecution of his career and found a way to be deferent without losing his independence. It may well be that Laplace, who was the clearer and more profound mathematical technician, saved Condorcet some embarrassment. Condorcet never completed his major treatise on integral calculus, most likely because he felt outmatched by contemporary workers like Laplace and Lagrange.[39] Although we have little solid evidence about their precise personal relationship, clearly Condorcet was a major player both in Laplace's early creative life and in the progress of his career.

Laplace's relations with Lagrange are historically more important and better documented, even though they are not straightforward.[40] Laplace was Lagrange's junior by thirteen years, and by the time he came into contact with Lagrange, the Piedmontese mathematician was already well into his own career, serving as head of the mathematical section of the Berlin Academy. He had won several prize competitions and was highly esteemed by the leading scientists of his time, especially d'Alembert.[41] Initially, Lagrange noted the brash and self-serving impression that Laplace made when he inquired about obtaining a pension from the Prussian monarch, but the older scientist generously ascribed this excess to Laplace's youth.[42] After their next series of exchanges in 1775, Lagrange began to praise the young mathematician's accomplishments, treating him almost as an equal. This attitude led to the private disclosure of some notable mathematical information that spurred each one on. Each tried politely to upstage the other, such that their relationship was always tinged with both mutual admiration and jealousy. The letters they exchanged are filled with phrases that could either be taken as an expression of genuine respect, or merely conventional courtesy. Laplace wrote to Lagrange admiringly: "I received your beautiful papers and read them with the greatest pleasure . . . It would be difficult to add anything to them, except . . . Your work [on the theory of numbers] is one of the most magnificent things ever done in this branch of mathematics . . . Your excellent paper on integrals [is] a chef-d'oeuvre of analysis by virtue of the importance of the topic, by the beauty of the method, and by the elegance with which you present it."[43] But in the same letter, Laplace followed each compliment with a suggestion that someone (presumably Lagrange or Laplace himself) could improve on the brilliant work, either by generalizing the procedure or extending it to another special domain of science. Indeed their creative relationship was fed by the thought that each one could surpass the other's accomplishments. Their mathematical styles were nonetheless quite distinct. Lagrange was the "purer" mathematician, more taken with the aesthetic beauty of the discipline, whereas Laplace used mathematics to solve physical issues in celestial mechanics, often at the expense of succinctness and symmetry.[44] The two bonded even more when Lagrange moved to Paris in 1787 and through the difficult days of the Revolution, but they nonetheless remained independent thinkers. Laplace took his colleague's demise in 1813 very hard, reportedly suffering from depression.[45]

Laplace's relations with closer contemporaries were not always so harmonious. He made heavy and unabashed use of the accomplishments of two excellent French mathematicians who deservedly earned a proper reputation after Laplace had succeeded in academic circles. Monge and Legendre became his colleagues in 1780 and 1783, respectively. Perhaps because of their manifest talent, Laplace thought nothing of using their results before they were able to publish them. Any annoyance Monge may have felt was mitigated by a footnote Laplace inserted in one of his publications.[46] There nonetheless remained a professional distance between the two mathematicians throughout their careers.

Legendre, however, was bitter about the apparent usurpation of his priority, pointedly remarking in his published paper that he had presented his ideas to the Académie before Laplace had developed his.[47] He might also have indicated that Laplace earlier had been the committee chair who reviewed a number of his essays when Legendre was a candidate for admission to the Académie, and hence had ample occasion to study, appreciate, and borrow from Legendre's arsenal of innovations.[48] But in fairness, Laplace explicitly refers to Legendre when first using his formulation of elliptical integrals in the 1784 *Théorie du mouvement et de la figure elliptique des planetes* (Theory of the Motion and Elliptical Path of Planets), in the second part of the essay dealing with the shape of the earth.[49] The citation is brief, and is immediately followed by a critical comment about the limitations of Legendre's mathematical innovation. It is clear here and elsewhere that Laplace considered any new ideas expressed in the meetings of the learned society to have become public property, hence grist for his mill. It is also clear that Laplace followed others' productions very closely and used their results for his own purposes. What mattered most for him was the advancement of mathematical knowledge, not the ownership of ideas. In operating this way, Laplace was not acting differently from contemporary mathematicians like Monge, Cousin, or Charles, who were also sparing in their reference to others.

## Exploiting Others

For the progress of celestial mechanics, Laplace also learned how to exploit the services of others. To my knowledge, he never observed astronomical phenomena with his own eyes in a professional manner. He relied on the reports of authors of ephemerides, comet watchers, and workers in renowned observatories. In his published writings, there are frequent references to foreign observers like John Flamsteed, Edmund Halley, James Bradley, Nevil Maskelyne, and Johann Tobias Mayer. Locally he relied initially on Lalande, Charles Messier, and Pierre-Françoise-André Méchain, who were trustworthy reporters of observations. But increasingly, as he checked his theory against evidence, he found the need to recruit others to do his bidding. Starting around 1785, he struck an agreement with Jean Baptiste Joseph Delambre, a prized student and trusted assistant to Lalande, to verify the elements of various planets and satellites, and to reduce raw observations to test his equations.[50] At a later date Laplace enlisted the help of other competent observational astronomers, notably the Piedmontese Barnaba Oriani and the Viennese Johann Tobias Bürg. Eventually, too, he relied largely on his faithful calculator Alexis Bouvard, who stayed with him until the end.[51] From the beginning it was clear that his collaborators were never considered as equals. They were Laplace's assistants.

The case of Delambre is particularly revealing because Laplace was instrumental in shaping his career, and established a pattern for such reciprocal assistance that he was to follow later to help himself and to launch other scientists into the profession. Delambre was a great success story, eventually rising to the pivotal post of secretary of the Académie in its post-Revolutionary version at the Institut; he also wrote a still-influential history of modern astronomy. In return for Laplace's offer to sponsor his election to the Académie, Delambre helped Laplace solve the knotty Jupiter/Saturn problem.[52] Several attempts by Delambre to enter the Académie failed, but shortly before the learned society was disbanded during the French Revolution, he was finally elected.[53]

The prospect of entering the Académie was surely not the only reason that Delambre had agreed to assist Laplace with his calculations; Laplace had lured him with his ambitious overarching scientific program. While this program was clearly narrated in the language of modern analysis, Laplace used this idiom to raise the practice of astronomers to a more elevated goal, one shared by philosophical minds ever since the pre-Socratics. Delambre had been schooled in the classics and appreciated Laplace's plan to map the entire heavens in the service of natural philosophy by bolstering Newton's scheme for the system of the world. The long-range objective with which he associated himself was no less than the reconstruction of the cosmos in the language of differential calculus endorsed with impeccable evidence. It was, as Dionis du Séjour said, a subject partaking of the sublime.

Two additional features of the set of technical papers produced prior to the Revolution need special notice. The densely argued series of essays by Laplace revealed that Newton's laws, when properly manipulated, could predict the paths of all types of celestial objects, planets, their satellites, as well as comets; and that this solar system was stable. Laplace did not arrive at these conclusions easily or alone, but he provided the capstone to the progress forged in his lifetime.

A spectacular demonstration of the power of the new celestial mechanics manifested itself following the detection of a new heavenly object by the Hanoverian astronomer William Herschel.[54] His discovery, made at Bath on 13 March 1781, was at first assumed by everyone to be a comet, following a parabolic path. A host of astronomical practitioners (including Bošković, Lalande, Méchain, Oriani, Lexell, and Laplace himself) immediately busied themselves to determine its orbit, and on the basis of three initial observations came up with four possible paths. Laplace shared his frustration with this result before his academic colleagues on 13 June 1781. Further sightings of Herschel's "star" showed its eccentricity to be very slight, raising the startling possibility that it might be moving in a closed orbit. Bochart de Saron, with whom Laplace was quite friendly, determined that it was more than fourteen unit radii from the sun, lending further support to the planetary theory.[55] With the aid of new exact observations and a fourth location by Méchain, Laplace formulated a new method to calculate its elements as a planet, and triumphantly presented his results to the Académie on 22 January 1783. Shortly thereafter he was able to establish a value for the inclination of its orbital plane and the longitude of its node. He communicated his new general method to his colleague Pingré, who quickly inserted it in his *Cométographie* (which reviewed all the proposed techniques) and proclaimed Laplace's method the most satisfactory. This significant accomplishment not only confirmed Laplace's intellectual prowess, but also vindicated his earlier disparagement of Bošković's method as an inadequate procedure based on approximate geometric techniques.

### Demonstrating Stability

The demonstration of the stability of the solar system was also something of a triumph, although in retrospect we know it was short-lived because the new tables for the position of Jupiter and Saturn based on Laplace's findings did not sufficiently match the observations. In 1787 Lalande hailed the demonstration as a major step forward in planetary astronomy, and ever since historians have underscored the ingenuity with which Laplace and Lagrange "solved" the Jupiter/Saturn problem.[56] The issue arose because the perceived secular acceleration of Jupiter and secular deceleration of Saturn could apparently not be explained by Newton's laws. If that were the case, either some other law would have to be added to celestial mechanics, or the system would in the long run be unstable. For a natural philosopher who had proclaimed his faith in a deterministic universe, the situation was unacceptable.

Laplace attacked the problem several times, first in his earliest paper on the system of the world written in 1773, and then again a decade later, arriving at his brilliant solution of 1786. At first Laplace thought he had demonstrated that the secular inequalities of the mean motion of planets due to the mutual attractive force of other planets were negligible, and later that they were nil.[57] Hence he at first assumed that the Jupiter/Saturn problem might be solved by considering perturbations of the orbits due to passing comets. Given what little was known about the total history of comets in the solar system, this hypothesis gave no promise of relief. As we know, it did stimulate Laplace to investigate the calculation of orbits of comets, and to regard Dionis du Séjour's 1774 essay and Bošković's geometric method worthy of close attention. The discovery of Herschel's new planet, Uranus, also offered the possibility that this celestial body might hold the key, which as historians of astronomy know turned out instead to lead to the discovery of yet another planet, Neptune, some sixty years later. But in 1784, Laplace developed another approach suggested by Lagrange that resolved the issue to his satisfaction.

Already on 11 August 1784 Laplace indicated a concern with stability in dealing with the oscillations of a fluid body (the ocean) enveloping an oblate spheroid (the earth) when that central body was itself moving.[58] A year later, on 23 November 1785, he outlined his new strategy to solve the Jupiter/Saturn problem in a masterful paper "Mémoire sur les inégalités séculaires des planètes et des satellites" (On the Secular Inequalities of the Planets and Satellites), which was clearly inspired by two papers he had received from Lagrange, one of which had shown theoretically that "the planets . . . gradually change the form and position of their orbits, but without ever going beyond certain limits."[59] Laplace insisted that this theory needed to be applied to all planets, but especially to Jupiter and Saturn, and eventually their satellites. He offered the details of his work in two epoch-making presentations given before the Académie in May and July 1786. The gist of his argument outlined in November 1785 is that so-called secular variations of the elements of planetary orbits actually oscillate around a mean value when considered for very long time periods.[60] According to his initial calculations, the acceleration and deceleration regularly reverse themselves after 877 years.[61] In this way even the unexplained apparent anomalies assumed for shorter periods disappear in favor of a long-term equilibrium.

Laplace preferred the term "firm equilibrium" to the word "stability," which connoted a static phenomenon. The motions of planets and satellites were temporarily "perturbed" by gravitational tugs, but the entire system remained fixed under one set of laws. While others marveled at the mathematical and conceptual acrobatics he had performed in resolving the Jupiter/Saturn problem, Laplace remained most excited about its consequences for the fundamental law of gravitation. In these papers, and writings that followed on Jupiter's satellites and on the moon's motion, he returned to the theme with which he had started his astronomical career. The exultation was over the confirmation of Newton's law:

> Until now, the theory of universal gravitation could not account for these phenomena . . . I expect to show that, far from constituting an exception to the principle of gravity, they are its necessary consequence, thus offering a new confirmation of this admirable principle.
>
> The correspondence of other celestial phenomena with the theory of gravitation is so perfect and satisfying that one could not without regret see the secular equation of the moon . . . [as] the only exception to this general and simple law whose discovery, by the grandeur and variety of the objects it embraces, so honors the human mind.

> I would hope that one would be gratified by these phenomena which seemed inexplicable under the law of gravitation [being now] reassigned to this law, for which they supply a new and striking confirmation.[62]

By fitting all major celestial phenomena under a unitary principle requiring no exception or additional rule, Laplace hoped he had realized his original dream of a deterministic solar system. As if this were not enough, he raised the tone of his claim to an even higher pitch:

> So the system of the world merely oscillates around a mean state from which it deviates very little. By virtue of its constitution and the law of gravitation, it enjoys a stability that can be disturbed only by causes foreign to it. On the basis of observations from the earliest times to the present, we are certain that their effects are undetectable. This stability of the system of the world, which assures its duration, is a phenomenon most worthy of our attention in that it shows us in the heavens the same intention to maintain order in the universe that nature has so admirably followed on earth to sustain individuals and to perpetuate the species.[63]

# 6

# Exploring the Physical World: The Terrestrial Program

Any scientist of the era who was determined to reconstruct physical astronomy with the language of modern calculus would have had a full agenda. Laplace had more than proved his mettle by taming Herschel's planet, and by demonstrating the stability of the Jupiter/Saturn system. Lunar anomalies seemed well on their way to falling into place. And with the assistance of Delambre and Oriani, Laplace was closing in on the remaining puzzles of the solar system prior to writing his Promethean *Traité de mécanique céleste* (Treatise on Celestial Mechanics). Had nothing interfered, that monument should normally have appeared on the eve of the Revolution as a counterpart to the classic outstanding treatise in mechanics, the *Méchanique analitique* (1788; Analytic Mechanics), produced by his friend and new Parisian colleague Lagrange. It would then have joined the ranks of major synthetic works being produced by other Parisian academicians, among them Antoine Laurent Jussieu's massive *Genera Plantarum* (1789; A Grammar of Botany) and Lavoisier's *Traité elémentaire de chimie* (1789; Elements of Chemistry). Paris was unquestionably achieving its claim to be the preeminent site of scientific activity during the late Enlightenment.

Yet Laplace was not satisfied merely with the completion of this colossal program. He embarked on a new scientific adventure before finishing his well-chartered research agenda. Coincidentally, the French Revolution intervened, upsetting all appearance of normalcy and giving him an additional justification for not fulfilling sooner this astronomical program in celestial mechanics. As Laplace wrote in a letter sent in July 1790: "Here, minds are turning toward civic affairs, and for some time, at least, the sciences will suffer from this diversion."[1] In fact, Laplace had already deviated from his well-defined research program in astronomy. A dozen years before the Revolution, he had entered into a set of major investigations concerning the nature of the imponderable fluids of heat, light, and electricity. For him the consequences were to be far-reaching, swelling his reputation with accomplishments in terrestrial physics and chemistry, and even sparking an interest in him in animal physiology and human psychology. Today, Laplace is still remembered as a major partner in the development of heat theory, and celebrated as a valuable collaborator of Lavoisier in the fashioning of his chemical reform. His contribution to the chemistry of life, also carried out with Lavoisier, was likewise significant.[2]

Antoine Laurent Lavoisier was better known to the French public than was Laplace. He was a prominent partner of the private corporation entrusted by the government to collect taxes, a position he had attained by marrying one of his partner's young daughters, Marie Paulze. This corporation, known as the Ferme Générale, was the major source of his comfortable income, and proved to be the reason he was guillotined during the Terror along with his father-in-law. Lavoisier was clearly part of the Old Regime establishment, and despite his liberal and reformist tendencies, succumbed like others after the Jacobins took power.

A slightly older colleague of Laplace at the Académie, Lavoisier was forging ahead with his complete transformation of chemistry by challenging the standard theories of phlogiston in order to replace them with a new principle of combustion based on the function of a portion of the air we now call oxygen. During his serious collaboration with Laplace, a major stumbling block of the ongoing chemical revolution was a precise understanding of heat exchanges in chemical reactions of combustion, exchanges that occur both in calcination and respiration.

What led Laplace to interrupt his program of astronomical research to join up with Lavoisier remains a major puzzle, one that Laplace himself admitted he was at a loss to explain fully or convincingly. To Lagrange, he confessed in 1783 that he was unable "to spurn the entreaties of his colleague Lavoisier whose obliging and sagacious contributions could not be more congenial."[3] To his Genevan friend Jean André Deluc, he acknowledged not knowing "how I let M. Lavoisier lure me into working with him on heat."[4] Asserting that his true passion was for the mathematical sciences, he added somewhat disingenuously: "An irresistible penchant for mathematics and a disposition toward laziness will always prevent me from contributing to the progress of physics."[5] In the letter transmitting the now-famous "Mémoire sur la chaleur" (Memoir on Heat), Laplace conceded his dissatisfaction with the work and confessed he was tempted to "abandon this career [in physics] and confine myself to geometry alone."[6] A year earlier, he had used similar expressions and identical grounds in an attempt to extricate himself from continuing his research with Lavoisier.[7] Though it seems evident that he was angling for reassurance and compliments, these statements nonetheless imply he thought a rationale was necessary for what appeared as a detour from his previous activities.

One easy answer would be to credit the move to ambition. Around him specialists at the Académie were concerned with understanding physical phenomena that had been discovered since Newton's *Principia*: the "imponderable fluids" of light, heat, electricity, and magnetism. These phenomena were discussed in the queries to Newton's *Opticks* as open questions, ripe for a host of demonstrations and deserving of serious explanations. To contribute to this enterprise would have been a reasonable motive for Laplace to veer from his chosen trajectory. He was, after all, a member of the mechanics section of the Académie. One could even argue that he hoped to put his special mathematical talents to work in this domain, and to attach this branch of knowledge to the laws of universal gravitation. To bank solely on this explanation, however, would require us to assume that he anticipated his post-Revolutionary achievements as early as 1777. That appears unlikely. One needs to know why he turned to Lavoisier in particular and so early for this new departure.

We can begin to explain this apparent inflection in his research activities by recalling a set of circumstances in his personal life that must have contributed to it. Laplace's active partnership with Lavoisier began in 1777, peaked in 1783, and continued sporadically until the chemist's execution in 1794. But Laplace's involvement in physics and chemistry was not limited to a personal interaction with his attractive colleague. The research he carried out eventually triggered protracted association with other scientists, notably Alessandro Volta, René Just Haüy, Deluc, and Claude-Louis Berthollet, with whom he shared his thoughts long after the days of the Terror that ended Lavoisier's life. Investigations into terrestrial physics were later to constitute a major component of his scientific life, dominating as they would the weekly scientific meetings he held with Berthollet at his country home in Arcueil during the Napoleonic era. Laplace devoted much of the fifth volume of the *Mécanique*, published in 1825, to particular aspects of physics. Moreover, several of his major disciples, chief among them Jean-Baptiste Biot, François Arago, and Siméon Denis Poisson, devoted their lives to physicists' topics that Laplace deemed worthy of major attention. In the twilight of his career, he became entangled in significant disagreements with Jean-Baptiste-Joseph Fourier, Pierre Prévost, Arago, Augustin Fresnel, and other accomplished physicists. How did it all begin?

## Instruments

One can decidedly rule out the influence of Laplace's teacher Adam, whose views on physics Laplace thought it best for students to expunge from their memories.[8] Nor did his institutional association with the *mécaniciens* in the Académie immediately lead him closer to the discipline. His first significant engagement with physics actually occurred when he examined measuring instruments as a commissioner for the Académie. On those occasions, his sharp analytic mind made him an excellent critic who could on occasion offer constructive ideas to inventors. With Jean Baptiste Leroy, he reported on a new thermometer devised by a Rouen instrument-maker named Scanégatti early in 1775.[9] More consequential was the paper submitted by the Genevan physicist Deluc, which dealt with the meniscus of barometers and thermometers. Laplace was asked to examine that proposal with his colleague Bossut in the summer of 1776.[10] Since it was meant as an amendment to Deluc's thickly argued *Recherches sur les modifications de l'atmosphère* (Research on the Variations in the Atmosphere), Laplace read Deluc's whole work first, becoming engrossed by the topic. This interest led directly to a never-published paper on experiments he carried out with Lavoisier, read before the public meeting of the Académie on 9 April 1777: "Sur la nature du fluide qui reste dans le récipient de la machine pneumatique" (On the Nature of the Fluid Left in the Receptacle of the Pneumatic Instrument When a Vacuum Is Produced).[11] In this public account he compared French procedures for using the barometer to the English procedures to determine degrees of vacuum. Later that year he was appointed to report on Louis François Dellebarre's microscopes, which had purportedly been derived from Euler's theoretical work on optical aberration.[12] In 1779 Laplace wrote a report on Nicolas Fortin's pneumatic machine pumps for the review committee that included Lavoisier.[13] The next year, Laplace was sitting on a committee with Lavoisier to review Jean Honoré Robert Paul de Lamanon's plans for correcting the effects of heat on the dilation of mercury columns in barometers.[14] That same year he was considered expert enough to report on a senior colleague's physics dictionary written in the tradition of René-Antoine Ferchault de Réaumur's pioneering efforts a half-century earlier.[15]

It is likely it was this set of activities that led Laplace to Lavoisier in the first place. Laplace is recorded to have been in the delegation from the Académie sent to witness some of Lavoisier's experiments on several gases on 26 February 1777.[16] The activity was part of Laplace's routine academic obligations, and his "expertise" as a commissioner on instruments must have been uppermost in Académie members' minds when he was appointed to authenticate these critical laboratory analyses. As Henry Guerlac pointed out, Lavoisier had earlier required assistance to compare thermometers so he could determine how cold the freeze of the 1776 winter had really been.[17] While Lavoisier had sought an answer from Brisson, who owned an original Réaumur instrument on which a marker for the low readings of the 1709 winter was still visible, he also turned to other academicians conversant with difficulties stemming from using a comparable apparatus. It seems as if Lavoisier's initial professional contacts with Laplace all revolve around measuring temperature and pressure when substances miraculously change from a liquid to a gas, an issue critical for Lavoisier's ongoing program to elucidate various transformations accompanying chemical reactions. Exploring such topics was a key issue that had to be investigated before a revolution in our understanding of chemistry could occur.[18]

## New Lodgings

Two other circumstances brought Laplace and Lavoisier together. Laplace had to abandon his rooms at the Ecole Militaire when its curriculum was reformed and he was pensioned off in 1776.[19] He took up lodgings nearer the center of Paris, first residing at rue des Noyers (not far from St. Séverin), and a few years later moving close to the Palais du Luxembourg, on rue de Condé.[20] Though liberated from the obligations of teaching cadets and on the edge of Paris, he had to mind his finances carefully. Until he was named examiner for the artillery corps in 1783,

Laplace by necessity lived on a modest scale, with a meager revenue of less than two thousand livres per annum.[21] There is no evidence that he received any assistance from his family. On the contrary, Laplace learned early in 1775 that his father had been charged with a misdemeanor in Dieppe that ultimately required him to pay a sizable settlement.

Laplace's limited means stood in stark contrast to Lavoisier's substantial personal wealth. Everyone—especially Laplace—was aware of Lavoisier's good fortune to have joined and married into a profitable enterprise as a full partner of the tax collecting corporation, the Ferme Générale.[22] Lavoisier could, and did, devote part of his assets to ordering expensive instruments. Indeed, the comment Laplace makes in a letter to Lagrange about Lavoisier's wealth might refer only to his colleague's habit of commissioning apparatus that would insure accurate results.[23] Surely this modern equipment helped Laplace with his invention of the calorimeter, which garnered him celebrity and more than warranted the collaboration with Lavoisier.[24] While Lavoisier gained a partner endowed with exceptional gifts of analysis and freed from any particular chemical or physical dogmas, Laplace was extended the opportunity of working in a laboratory with a brilliant collaborator who spared no costs.

Yet there was more behind the association of Laplace and Lavoisier than mutual professional advantage. Letters they exchanged suggest that a contractual arrangement was struck between the scientists, in which some of the terms of alliance were understood rather than spelled out. Though never explicit, there is the suggestion that Laplace would gain in other ways from the partnership. For example, now that Laplace lived closer to Lavoisier's home on the rue Neuve des Bons Enfants, to the east of the Palais Royal, he knew he would often be included in Lavoisier's social circle.[25] A more tantalizing part of the bargain was a sizable loan Lavoisier made to the elder Pierre de Laplace, presumably to pay off his fine. The loan, which involved Lavoisier professionally as well as personally, settled an incident that, if publicized widely, could have redounded badly to Laplace's reputation.

## The Cider Affair

Laplace's father, Pierre, had for years exploited his Normandy land holdings by producing cider and eau-de-vie from his crop of apples and pears. He often sold his wares through his cousin Jean Mabon, who distributed kegs throughout the region. Because the cider remained in barrels for long periods, and in order to keep the liquid clear, it was common in the trade to add preservatives and precipitants to the juice, and Mabon was known to follow such practices. The wrong dosage often resulted in colics for consumers, and on occasion could prove fatal. It was such an incident in a wine shop in Dieppe, in which several barefoot Carmelite Sisters were stricken in 1775, that led to a charge of negligence against Mabon.[26] He was fined, and in turn sued his cousin for selling him adulterated cider. A bitter squabble ensued before local magistrates, in which family members accused each other of unethical behavior.[27] To settle matters out of court, Pierre de Laplace and Mabon agreed to pay damages of three thousand livres to the congregation. It was a galling episode for Laplace's elderly father, who was just withdrawing from the business and who could ill afford this expense.

The problem of adulterated cider was of immediate concern to civic authorities in Rouen, where Pierre de Laplace was known as mayor of Beaumont. Several new edicts protecting the community from such perils had been enacted by the Parlement of Normandy shortly after the Dieppe affair was aired in local courts.[28] Consequently, officials were authorized to requisition samples of suspected cider to have it analyzed by local pharmacists. These pharmacists, in turn, appealed to the local Rouen Academy for expert help. Eventually the newly created Parisian Société Royale de Médecine, of which Lavoisier was a founding member, became involved with technical disputes over adulterated cider, and turned to its own analysis of the litharge, ceruse, and minium often added to the liquids.[29] Lavoisier was clearly abreast of the Dieppe fiasco, all the more so since the organizer of the Société, Vicq d'Azyr, was a Norman colleague with whom Laplace had studied in Caen. The upshot of this convoluted affair was that Pierre de Laplace trekked to Paris and obtained a substantial loan from Lavoisier to bail him out of his financial difficulties, thereby saving him from further notoriety.[30] Since the affair stretched from January 1775 to the end of 1777, it is not clear if the loan was part of the original bargain Laplace had made to render services to Lavoisier, or if it is an example of the chemist's generosity and friendship toward Laplace. To further cloud matters, there is a clear indication that Lavoisier used his influence to promote the career of Olivier de Laplace, the brother of Pierre Simon.[31] However it unfolded, the cider episode linked Laplace ever more closely to Lavoisier. Laplace's often-expressed desire to disengage from the collaboration was thus offset by the personal bond and financial obligations he had incurred. The two scientists' ties were reinforced a decade later when Laplace invested some of his capital in a public corporation that Lavoisier supervised for the government.[32]

## Initial Collaborative Efforts

Regardless of the motives for embarking on the collaboration, it quickly came to thrive on far more essential scientific grounds. Laplace's concern with precision derived from his experience with celestial physics. He had learned through practice that analysis and probability considerations could be applied usefully only if there were quantifiable and reliable data at hand.[33] Instruments employed for this purpose needed to have sound designs and proper calibration if they were to be harnessed to search for laws. Data would have to be matched to detailed theories that could be expressed mathematically to test their reciprocal merit. He knew from listening to paper submissions to the Académie that the study of terrestrial phenomena was seriously lagging behind work in celestial physics. Many amateurs were still practicing what Condorcet would condescendingly term "physicaille," in contradistinction to the higher practice of proper natural philosophy.[34] It was Laplace's eagerness to join in a research program compatible with the more advanced mathematical sciences that made him so valuable to Lavoisier, and that drew him into full partnership with this physical chemist.

In their daily collaboration, Laplace and Lavoisier treated each other as equals. The surviving notes they exchanged show that Laplace was an active participant who often initiated measures and even dictated procedures to his older and more experienced colleague.[35] Laplace was not always present at experiments, many of which were carried out at the Arsenal laboratory early in the morning; but he was intimately involved with both their planning and with the arithmetic calculations they generated. Most significantly, he focused on how the apparatus could yield the most precise and reliable information on which to base conclusions. A telling example was his design of an experiment on the expansion of solids to be repeated under similar changes of temperature, first increasing and then decreasing the heat for equal time intervals. In this way

Laplace hoped that errors stemming from the experimental procedure would be neutralized.[36] It was this dimension of the activity that most captured his fancy, and for which he repeatedly expressed excitement in his correspondence.

When he and Lavoisier originally entered into an agreement to investigate physical phenomena associated with chemical reactions, Laplace indicated that these investigations included experiments on heat, the expansion of solids, and electricity.[37] It was their original conjecture that the production or release of these imponderable fluids significantly accompanied changes of state. By making careful measurements of the variables, it was hoped that the mechanism of the phenomena could be unlocked. Special prominence was expected from the two aspects of physics currently being explored, namely the operations of static electricity and the behavior of gases. Laplace's interests also intermittently led him to consider the role of light and magnetism, including even animal magnetism. Clarifying the domain of imponderables was a mystery deserving attention. For Laplace the only direct way to do so was by devising or refining techniques of measurement.

Laplace's impression of current chemical practices was not fundamentally different from Condorcet's view of the displays of the physical sciences currently in vogue in Paris. Popular demonstrations of new phenomena would often entertain and mystify the public rather than enlighten the scientific community. Laplace was reminded of his experience as a student of Adam, who loved to demonstrate physical phenomena without providing a reasonable understanding of their nature. Such demonstrations smacked more of magic shows than of the honest attempts of a serious investigator. To understand chemistry reasonably well, Laplace lamented that "it would be necessary for me to read through all the writings that have appeared in large numbers, especially of late; and you know that they are not often written with suitable concision, so that often a few truths are drowned in a sea of tomes."[38] What Laplace proposed was to bypass this web of unreliable traditions in order to deal directly with verifiable phenomena. He brought to his study a subtle understanding of technical difficulties in instrumentation learned from his exposure to astronomy and on academic committees. Weighing all substances including gases, which Lavoisier frequently did, was only an instance of the broader goal of calibrating all significant terrestrial phenomena. In this way the verbiage and conjectures he found so objectionable could be reduced to verified data presented in condensed mathematical format, and lead, it could be hoped, to sound theories.

After their initial collaborative effort in 1777, several years passed before Laplace and Lavoisier resumed their association in 1781. The first new experiments they published were unspectacular.[39] Laplace and Lavoisier reported on measurements of static electricity associated with the changes of state. It was the presence in Paris of the Italian professor of natural philosophy from Pavia, Volta, who had come in December 1781 to publicize and test his sensitive electroscope, that sparked their renewed curiosity.[40] Volta was in attendance as the French scientists measured the production of negative electricity when hydrogen, carbon dioxide, and nitrous oxide were generated under strict laboratory conditions.[41] Their expectation that positive electricity would manifest itself with the reverse sublimation process turned out to be inconclusive. It was a line of work that yielded little for Lavoisier's chemical revolution, though the role of electricity was still to puzzle them when a half-dozen years later water was produced from a mixture of oxygen and hydrogen ignited with an electric spark. Laplace remained in contact with Volta through intermediaries, and after the Revolution welcomed him with all honors when Volta received a sizable prize from Napoleon for inventing the battery.[42]

The inclusion of electricity in the study of gaseous evaporation had implications far beyond standard physics and chemistry. For those concerned with atmospheric sciences, including the analysis of cloud formation, lightning discharges, and weather patterns, the invention of Volta's condenser and the Neapolitan amateur Tiberius Cavallo's silver electrometer promised to yield new clues for meteorology. None was more eager to bring the subject into his orbit than the Genevan Deluc, who came to play a significant role for Laplace as a go-between with the group of physicists concerned with linking theories to solid experimental evidence. He had already authored an important treatise on hygrometry and thermometry in 1772 that was continually in need of updating to include new discoveries.[43] Deluc eagerly seized on Volta's ideas in the expectation that they would help link the various atmospheric phenomena to his favorite cosmological views derived from his mentor in Geneva, the brilliant recluse Le Sage. It was from this quarter that Laplace felt a fresh wind of ideas worthy of attention.

## An Explanation for Gravitation

In a letter to Volta, Deluc indicated that among Parisian physicists he met in January 1783, Laplace was the one most serious about Volta's theories, which Deluc proposed to air to the public by way of a detailed article in the *Journal de Physique*.[44] It was his intention to link Volta's new theories of electricity to Le Sage's explanation for gravitation and chemical affinity. In expressing this desire, Deluc showed a penchant for a causal physics closely linked to experimental observation, of the sort that Laplace favored for physical astronomy. Though his relationship with Laplace was never visible to the public, the two men seemed to enjoy each other's intellectual company. They conducted a substantial correspondence, and when Deluc visited Paris, the men spent significant time together discussing physics. The Genevan had much to contribute because he was unusually well informed through his local home contacts and his colleagues at the Royal Society—he had been "Lecturer to the Queen of England" while living at Windsor. Though an exiled Genevan, he operated like many of his fellow scientists at the crossroads of an international network, picking out the best traits from each national group. Through Deluc, Laplace began to appreciate and take seriously the contributions of Joseph Priestley, James Watt, and Adair Crawford in England; Jan Hendrik Van Swinden and Martin Van Marum in the Lowlands; Le Sage, Horace Bénédict de Saussure, and Prévost in Geneva; and Volta and Marsilio Landriani in Italy. Unlike some of his Parisian colleagues, Laplace was less parochial in his scientific outlook, willing to borrow selectively from any quarter, though he always filtered the results through his own critical sieve.

Beginning in August 1781, when he was temporarily residing in Passy on the edge of Paris and attending meetings of the Académie, Deluc was determined to make astronomers appreciate the significance of Le Sage's system. Already earlier he had pointed to Le Sage's "sublime theory," which he lamented physicists were slow in adopting.[45] One of the components of Le Sage's theory was his conjecture that the agitation of particles in elastic fluids emanated from the same cause he postulated for gravitation. Laplace was aware that Le Sage assumed the existence of tiny particles moving through space at enormous velocities to account for universal gravitation. Indeed that notion had been rejected a decade earlier, when Laplace had considered alternatives to Newton's law, because of the difficulty that experimentalists would have in detecting their existence. But now Deluc disclosed that the hypothesis might also be useful in accounting for a variety of terrestrial phenomena such as cohesion. In Le Sage's original publication, the *Essai de chymie méchanique* (1758; Essay on Mechanistic Chemistry), he had advanced the idea that chemical as well as physical phenomena were linked to his speculative theory, paralleling the grandiose plan that Bošković ultimately set forth in his *Theoria philosophiae naturalis redacta ad unicam legem virium in natura existentium* (1763; Theory of Natural Philosophy). These notions were among the few intelligent theories suitable to explain the well-known selective affinities among chemical substances, and were discussed over the years in Parisian circles, always with a great degree of skepticism. According to Deluc, Laplace was also willing to consider such speculations, but with serious reservations. Deluc alerted Le Sage that Laplace "reiterated that he had not seen anything approaching your system; and that if one were really concerned with mechanisms in the universe, it was significant . . . [But] that he continued not to believe in it except for the data it provided for mathematical problems . . . Only if we could extract some attractive problem to solve would he be tempted to pursue it."[46] Deluc persisted. He provided so many details to various academicians about Le Sage's theory concerning elastic fluids that he resolved to compose a series of open *Lettres à M. Delaplace* ventilating in detail Le Sage's system.[47] It clearly intrigued Laplace, for it was a subject of immediate pertinence to his research program with Lavoisier.[48] Most likely the suggestion originally made by Daniel Bernoulli about the corpuscular nature of heat particles that showed up in the beginning of their "Memoir on Heat" was prompted by Deluc's printed discussion.[49] Pressing his notions in person, he was also the most stimulating and compelling of informants for Laplace. Deluc, for his part, indicated that Laplace was his favorite interlocutor, whereas Lavoisier was "more eager for renown and ready to purloin ideas from others . . . He is jealous of them, longing to have discovered everything himself."[50] By contrast, Laplace was one of the few willing to entertain new ideas, even as he remained openly skeptical of them.

The possibility that some version of Newton's laws extended to the macroscopic domain was a cherished idea our Parisian scientist had always entertained, though somewhat imprecisely. In his earliest comments about determinism, Laplace had used a phrasing so general as to apply the laws to all things (*êtres*), not merely to celestial phenomena. In late 1783 when he was in the throes of his collaborative experiments, he declared that these recent experiments gave him hope that one could ascertain the laws of affinities to balance the repulsive force of heat just as accurately as astronomers had determined the gravitational laws. In a fit of unbridled speculation, he put forth a grandiose scheme that reveals why he remained so absorbed by the heat experiments: "In bodies of a small size, the attractive force of matter is negligible. It reappears in its parts in an infinity of forms and with a vitality so prodigious that it is difficult to believe it is the same force that makes heavenly bodies gravitate around each other. The hardness of bodies, their crystallization, the refraction and diffraction of light, the raising or lowering of the level of fluids in capillary tubes, and generally all the chemical combinations are the result of attractive forces whose laws have not yet been determined." It was as if Laplace were contemplating a research program in terrestrial physics of which the experimental work on heat theory was only a beginning. As we will see, his tenacious character made him return to this program a dozen years later. From Laplace's viewpoint, then, the collaborative enterprise with Lavoisier was part of his overall determination to follow a path vaguely envisaged early in his youth. For Laplace, its importance to the developing chemical reform so crucial to Lavoisier was probably of secondary concern.[51]

## Theories of Heat

Nevertheless, the differing motives held by the two close associates Laplace and Lavoisier did not in any way diminish either the solidity of their findings or its philosophical import. On the contrary; the classic memoir that is often taken as the origin of modern measurement techniques for heat—whatever its true nature—is also one of the earliest scientific memoirs to explicitly champion a positivist or instrumentalist approach. In the first section of their paper, they review briefly both material and dynamic theories of the nature of heat. They then assert: "We will not decide between these two hypotheses . . . Given our ignorance of the nature of heat, we can only observe its effects, chief of which consist in expanding substances, turning them into fluids and converting them into vapors." The ensuing memoir, which has been analyzed extensively by others, is far more than a simple recounting of carefully produced experimental data, extensive and detailed as they are.[52] It stands as a remarkable example of the use of a new instrument designed to answer precise questions raised in an explicitly theoretical context. The published paper reproduces what must have been intense discussions between the collaborators about fundamental issues in physics and chemistry. Every section begins with general considerations that are translated into real manipulations of nature, carefully calibrated to yield quantified results. Close attention is paid both to the abstract context and to the apparatus used to resolve issues raised about heat. Throughout this paper, Laplace and Lavoisier nonetheless reiterate their claim of theoretical agnosticism. Whatever one may surmise about their hidden personal allegiances, the calculated tone of the paper fit well with Laplace's diffidence about speculation that was not grounded on empirical information.[53]

Historians have often tried to determine which individual contributed the major ideas spelled out in this enterprise. Laplace has been credited at various times with favoring the dynamic theory of heat, with setting forth the hypotheses expressed in typical mathematical language, and with conceiving the innovative idea behind the ice calorimeter.[54] These may well all have initially been his proposals, but in a truly collaborative work between equal partners, assigning personal credit is a pointless effort. As indicated earlier, both Laplace and Lavoisier stepped into the venture to illuminate his own concerns, but found the work itself so engrossing as to lose his separate identity. Acting as professional natural philosophers, the scientists seemed to care more about the search for laws than the personal ownership of ideas. For Laplace, this must have been a more exhilarating and genuine adventure than had been his equally important epistolary exchanges with Lagrange.

The collaboration did not end with their work on calorimetry. Laplace also participated in the other classic set of experiments with Lavoisier on the synthesis and decomposition of water, which all historians and many contemporaries took as the empirical capstone of the chemical revolution. It is clear that their experiments performed at the Arsenal laboratory on 24 June 1783 were inspired by results first secured across the Channel, reported to the Académie the previous month, and carried out at about the same time on a larger and more precise scale by Monge.[55] Once again Laplace benefited from Lavoisier's ability to pay for expensive laboratory equipment, though in this instance Laplace complained that the experimental data was insufficiently accurate and urged that the experiments be performed again. They were in fact carried out the following year on a larger scale in the presence of a multitude of academic witnesses.[56] By that time, Laplace had been replaced as Lavoisier's collaborator by Monge's student Jean Baptiste Meusnier de la Place, a talented officer not related by family ties to Pierre Simon Laplace. To explain the substitution, one might wonder if Laplace and Lavoisier had a falling out. But Laplace continued to lend his presence and support to Lavoisier's exciting chemical reform, assisting in the fashioning of a new nomenclature, rebutting Richard Kirwan's published protests against it, and even offering an important suggestion in September 1783 for the source of hydrogen produced when hot steam passes over iron filings.[57] What seems more likely is that their experimental activities, which required so much time and attention, were designed to follow Lavoisier's exciting research program rather than elaborate Laplace's expectations for better understanding the gravitational forces on earth.

The two academicians continued to enjoy a close personal and professional association, but one different from that which they had from 1777 to 1784. In one important instance, Lavoisier became Laplace's confidant and co-conspirator. The year 1783 was a sad one for the scientific establishment. D'Alembert, Euler, and Bézout passed away, leaving a major void.[58] Lagrange and Laplace were already doing all they could to keep the flame of innovative work in the mathematical sciences burning. But Bézout's demise also set off a flurry of activity to determine who would succeed him in his lucrative post as the examiner of students in both the artillery and naval schools. Given Laplace's meager earnings, it was an opportunity he could not afford to miss.[59] To increase his chances of stepping into Bézout's post, he enlisted the aid of Lavoisier, who was well connected in ruling circles, and recruited other influential friends like the amateur astronomer and prominent jurist Bochart de Saron and the anatomist and court physician Dr. Joseph Marie de Lassone. In a detailed letter, Laplace reported on his strategy to capture the vacancy. It was to be the last time he felt in serious competition with colleagues in the Académie who also coveted the post.[60]

Intrigues and solicitations abounded. The head of the artillery corps, Jean Baptiste Vaquette de Gribeauval, was courted by Bossut, who was probably backing his friend Monge as a candidate, in return for which

Bossut's textbook in elementary mathematics—used for the engineering school at Mézières—would probably have been substituted for Bézout's textbook. But on 11 October 1783, Gribeauval recommended Laplace to the minister of war as examiner for the artillery.[61] An arrangement with the family to keep Bézout's text in force, as well as a reinforcement of the artillery corps' independence from the engineering corps, was probably an important consideration. With respect to the naval schools, Monge prevailed and was appointed to succeed Bézout, thereby dividing in two the appointment he had held. But with the benefit of sound advice or clever maneuvering, or both, Laplace managed to create a new position for himself as examiner of the naval engineering school in Paris, much to his competitors' chagrin.[62] In all this, Laplace's suspicions that his colleague Legendre was his secret competitor were not borne out.

The result of all these maneuvers was to more than triple Laplace's income without requiring an unreasonable expenditure of time for these mundane examinations. The artillery post brought in four thousand livres annually plus travel expenses; the naval engineering post with exams to be held in Paris was indemnified at twelve hundred livres.[63] Little did Laplace know that the routine examination of a cadet named Napoleon Bonaparte at the artillery school in Brienne in 1785 would have a major effect on his later life.[64] Laplace ranked him forty-second out of fifty-eight students, which was sufficient to provide Bonaparte with a commission in the artillery and launch him on his prodigious career.[65]

This fortunate turn of events, providing Laplace new responsibilities, gave him an additional reason to extricate himself from obligations he had with Lavoisier. But he did not abandon his hope that terrestrial physics could legitimately be linked to celestial mechanics with some mathematical precision. Not only did the experimental establishment of Coulomb's law in 1784 reinforce his faith in the inverse square law, but also a new aspect in terrestrial physics seemed promising. His new academic colleague Haüy had presented a mathematical theory for the nature of crystals that Laplace had reviewed favorably with Louis Jean-Marie Daubenton.[66] A few years later, the same Haüy presented a much-applauded *Exposition raisonnée de la théorie de l'électricité et du magnétisme d'après les principes de M. Aepinus* (Analysis of the Theory of Electricity and Magnetism According to the Principles of Aepinus), which featured the new mathematical laws derived from direct evidence. Laplace was a member of the committee praising this work on 21 July

1787. The terrestrial program, though far from realized, seemed well on its way.

By venturing in this new direction, Laplace greatly enhanced his reputation in France and abroad as an accomplished professional scientist who had mastered all the mathematical and physical sciences of his age. He could well claim the mantle left behind by Euler, d'Alembert, and Bézout as one of the Académie's leading innovators.

# 7

# Revolutionary Tumult

The year 1788 was a fateful one for Laplace. He was making headway with both his astronomical and his physical research, even though these programs remained incomplete. Colleagues at the Académie des Sciences had recognized his keen talents by continuing to appoint him to increasingly important review panels. On average, he sat on ten committees a year, choosing to serve as author for about half the reports. The topics he was assigned confirm his command of all the disciplines in the mathematical and physical sciences, but in addition his new post as examiner of the artillery won him a seat to judge projects submitted by naval, artillery, and engineering officers.[1] The secretary of the Académie, Condorcet, also called on him to review the famous population estimates of Jean Baptiste François de La Michodière and various projects relating to life insurance and annuities, presumably because a knowledge of probability theory was essential.[2] In addition, Laplace was selected to judge the many projects to reform hospitals in Paris and Bordeaux, most of which were chaired by his friend the astronomer Bailly, future mayor of Paris and victim of the Jacobin Terror, or by the surgeon Jacques René Tenon.[3] Clearly he was being recognized for his broad range of competencies. He seemed to be both at the peak of his powers and enjoying his reputation.

On a personal level, Laplace's financial position had improved to a suitable level. In addition to the pensions he collected from his lapsed professorship at the Ecole Militaire, from the duke of Orléans, and the small stipend from the Académie, Laplace was earning a generous salary as examiner for students at the artillery corps and naval engineering school. He was now a fully accepted figure in gentrified social circles, living closer to the center of Paris—on rue Mazarine in a house belonging to the Collège Mazarin, across the Seine from the Cour Carée of the Louvre, where the Académie convened.[4] He frequented respectable circles among the secularized elites of Paris and established close ties with colleagues of his generation, chief among them Lavoisier, Berthollet, and his own coterie of astronomers. Apparently he did not venture into the fashionable salons or the unofficial associations concerned with freemasonry or reform politics.

The year was filled with political turmoil and portentous cultural changes. France's financial crisis had led to the calling of the Assembly of Notables, as well as a formal request by Louis XVI for spelling out grievances and encouraging reform plans. The various municipal projects considered by the Académie to transfer cemeteries and slaughterhouses to the outskirts of Paris, to improve ventilation in prisons, and to transform the Paris hospital system, which did not come to pass until after the Revolution, were part of a reform spirit spreading throughout the land.[5] Calls for revamping regulatory powers of the government, coinage, the system of weights and measures, the calendar, and educational practices were sounded throughout France, particularly as censorship of pamphlet literature was lifted.

The momentum to solve municipal and national problems with a rational approach of the sort long practiced in the Académie made authorities turn increasingly to scientists for dispassionate evaluations and the drafting of new proposals. For several years before the taking of the Bastille in July 1789, academicians had been called on as consultants. The dramatic changes in governance only accelerated the process, drawing more scientists into the political process. A good number of Laplace's colleagues even ran for election to the various new deliberative bodies being established in the early years of the Revolution. Not so for Laplace. Though he was well aware of this societal ferment, it was not current events that first transformed his habits. Instead, family matters set off a chain of events that, combined with the coming of the revolutionary era, precipitated tremendous change in his life.

## Marriage

A bachelor until age thirty-nine, Laplace suddenly married on 15 March 1788. We have little earlier intimation of his interest in women.[6] For his spouse, he chose a pretty eighteen-and-a-half-year-old girl from a good family in Besançon, Marie Anne Charlotte Courty de Romange. She was obviously in good health, ultimately surviving her husband by some thirty-five years.[7] Precisely how the union was arranged is not known, but there is good reason to think that Berthollet and the Orléans circle acted as intermediaries. The bride's father, Jean-Baptiste Joseph Courty, came from a successful family of iron masters in Franche-Comté who served as workers, then managers, and finally owners of local mines.[8] After his own father's death, Jean Courty accumulated an ample fortune, bought a title as "Secrétaire du Roi" in 1767 for 66,000 livres, and was eventually ennobled on the eve of the Revolution, having added to his family name that of the estate he purchased at Romange. More relevant for our purposes, in 1777 he had acquired a one-third interest in a Paris real estate enterprise that built a series of houses not far from the Palais Bourbon—an investment that more than doubled his fortune. (Today, a small street off the Boulevard Saint Germain still carries his name.) By 1780, he could boast of assets close to 3 million livres. The Orléans family was also heavily invested in real estate ventures, and it is likely that Courty and his wife, Marie Hélène Angélique Mollerat, befriended the Orléans social circles and supported their business schemes.[9] The Orléans family, for example, owned woodlands purchased by the Mollerats, who were also iron masters, and both families retained local mines east of Paris.[10] Another member of the Mollerat family had just arranged for a sizable loan of 350,000 livres to the duke.[11]

The match was a good one, since both parties were among those rising in the social hierarchy. The elder Pierre de Laplace had for years been the tax collector for the Orléans estates in Beaumont-en-Auge. His son could vaunt his personal success, including a coveted academic title and a solid reputation, a respectable income, and social connections through his service in the military establishment and with the Orléans family. Laplace was a close associate of Berthollet, physician to Madame de Montesson, wife of the duke, who presided over Orléans society. For their part, the Courty and Mollerat families tendered wealth and titles, and a charming offspring. The marriage contract, witnessed by the duke of Orléans himself, was registered by a notary near the Palais Royal.[12]

Pierre de Laplace, already aged and ill, had deputized a colleague of his son at the Académie, the surgeon Raphaël Bienvenu Sabatier, to represent his interests at the marriage ceremony.[13] Vicq d'Azyr, a renowned academician who was soon to serve as physician to Marie-Antoinette and was also a Norman colleague of Laplace, acted as witness, along with several high-ranking military officers. The Courty and Mollerat clans were present in full regalia, with noble titles and the itemizing of their multiple offices in the marriage contract. The contract detailed Laplace's sources of income totaling 9,900 livres per annum and specified a generous dowry amounting to over 100,000 livres for the young Marie Anne Charlotte.[14] The wedding was celebrated the same day at the Saint Sulpice church. This time, Berthollet stood as Laplace's closest personal witness.[15]

Pierre de Laplace, then close to seventy, had retired in Beaumont, living on his favorite farm at Le Mérisier. Shortly after his son's wedding, he turned seriously ill, and Pierre Simon rushed to Normandy to take over the family's affairs.[16] He stopped in Pont-L'Evêque to select an agent for collecting rents and handling the debts his father had incurred.[17] In addition, his older sister Marie Anne, widow of a barely literate local merchant, traded her share of the anticipated inheritance for an annuity.[18] A family man faced with the prospect of an inheritance, Pierre Simon was now required to take full control of complex business affairs, which included farming out numerous small properties and periodically disbursing a few substantial debts, including the unpaid loan to Lavoisier. Pierre de Laplace held on to life for several months after his son's return to Paris, and Pierre Simon did not make another trip to Beaumont to attend his father's funeral.[19] It seems that neither the father-son relationship, nor any sibling bond, had been a source of affection for Laplace. By gaining a new set of relations through his young wife, Laplace could revamp his family life. Several months after their marriage, the newlyweds moved into a Right Bank apartment on rue Louis-le-Grand, near the boulevards, where a member of the bride's family, Mollerat de Brechainville, already lived.[20] The next year a son was born.[21] Charles Emile Pierre Joseph was baptized three days after his birth on 5 April 1789 at the nearby Saint Roch church. Three years later, on 18 April 1792, Sophie Suzanne was christened at the same local parish.[22]

## The Provincial Retreat

Details about the Laplaces' family life are scarce, but one document indicates that the Courty family helped the newlyweds in their household. Laplace paid his father-in-law three thousand livres annually for purchasing food for Laplace and his new wife starting in mid-March 1788, and this sum later included food for a nanny after the birth of Emile the next year.[23] Given her youth, one can assume that Marie Anne and their son Emile stayed close to her family, and even lived with them from time to time in the country. Their eventual choice of Melun as a haven from the turbulence of Paris may well have preceded the onset of the Revolution, since her family owned property in the region.[24] The Laplaces kept their Paris apartment, paying rent until April 1793.[25] But they also rented a house on the island in the center of Melun starting in October 1792, when their second child was old enough to travel.[26] In July of 1793 they moved nearby to a picturesque home on the edge of the forest of Fontainebleau overlooking the Seine in the hamlet of le Mée, bought for 15,000 livres on 23 February 1793 with a loan from Jean Courty, which they eventually repaid.[27] Passports of the era show that both Laplace and his wife moved back and forth between their abodes in Paris and Melun throughout the difficult days of the Revolution, and even after the Republic was proclaimed in September 1792. Melun and le Mée seemed to be quiet refuges from Paris rather than secret hiding places.

## Public Service

In fact Laplace could not have hidden from public view even if he had wanted to. By the early 1790s, many of his academic colleagues had found their way into the administration, either as representatives of Parisian districts or in the executive branches of government. As they engaged in politics and the management of public affairs, their scientific output suffered. Laplace was more committed to his research than to civic service, in keeping with his self-image as a professional savant. He did not run for political office, nor volunteer his services. Nevertheless he did not shirk responsibilities he had assumed earlier. One might think that he survived the Revolution unscathed because he withdrew from his official duties. Nothing is further from the truth. He was paid by the state to examine artillery and naval engineering students, which he continued doing faithfully even during the Terror. After enemy troops invaded the North, the artillery cadets were moved to Châlons-sur-Marne, not far from Melun. He conducted his mathematical exams as usual in 1789 at Metz, and in 1792 and 1793 in Châlons.[28] The contingent of hopefuls was smaller, but just as talented. They included future renowned officers like the field marshals Duroc, duc de Frioul, and Marmont, duc de Raguse, as well as the generals Aubry, Bicquilley, Desvaux, Foy, and Griois, who were all in later years well acquainted with Laplace's talents. He examined naval engineering students in 1794, and continued to perform his task even as the Ecole Polytechnique began to supersede the system of military schools.[29] To perform all these activities, Laplace had to hold an official appointment from constituted authorities, and to report his findings to the appropriate minister. He neither avoided his responsibilities, nor balked at serving the successive administrations, be they under royal or republican auspices. Like a loyal bureaucrat, he followed the political pendulum whichever way it swung.

Laplace was not oblivious to the unfolding events, however. He paid his "patriotic contribution" of 750 livres prescribed by the law of 6 October 1789.[30] He or his paid substitute performed guard duty in his Paris district for the national guard.[31] He obtained the required passports and residence affidavits for himself and his wife.[32] He registered locally in Melun as a nonémigré.[33] Clearly he kept abreast of the adventures of his colleagues in the Académie, a few of whom took an active political role in Paris. Most of them survived the winds of change.[34] Some were arrested and a few lost their lives tragically, ultimately because of their engagement in civic activities that he had shunned. He may well have been present when his friend the astronomer Bailly was arrested by local authorities in Melun and sent to Paris under guard to stand trial. Eventually Bailly was sentenced and guillotined. It is thought that this disgraced ex-mayor of Paris was in Melun to take up residence in the house rented from Laplace's landlord and vacated when Laplace moved to le Mée.[35] The director of the observatory, Cassini IV, was also incarcerated for his outspoken, ultra-conservative conduct. Very likely Laplace knew of the fate of Louis Alexandre La Rochefoucauld d'Enville who was slain in 1792, Condorcet who was arrested and died in his cell, as well as Philippe Frédéric baron de Dietrich, Lavoisier, and Bochart de Saron, each of whom was guillotined. Fortunately for him, Laplace was insufficiently notorious to be targeted for arrest in Paris. Because he was not conspicuously involved in political activity, there was no basis for protracted suspicion. At one point, on 18 September 1793, he was taken into custody by a local militia near Melun, but he was released quickly after neighbors testified that he was an authorized resident.[36] Only at the height of the Terror was he removed from the governmental commission to establish the new system of uniform weights and measures, along with a handful of other academicians, because the scholars were not among "those worthy of trust by virtue of their republican virtues and hatred for kings."[37] At about the same time he was also temporarily excused as examiner for the artillery, although only a short while later he was hired in the same role at the Ecole Polytechnique.[38] If Laplace held strong political views, he chose to keep them to himself. That was the key to his survival through the most difficult days of the Revolution. He remained true to his cautious character and circumspect behavior, persevering in his professional calling and often turning to his foreign colleagues for moral support.

The Revolution took its toll on the scientific projects of Laplace and his colleagues. Laplace maintained his detailed pursuit of a solution for the satellites of Jupiter, indirectly exchanging views with the astronomer William Herschel, who was then living in England, through his friends Deluc and Charles Blagden, secretary of the Royal Society. He kept up with his astronomical partners Delambre and Oriani in Milan who sent him data and analyzed his findings. But in the letters to his foreign colleagues, Laplace disclosed the difficulties the new political environment posed. To his old friend Deluc, now a lecturer to the Queen of England, he wrote on 7 November 1789:

> I was on the verge of pursuing some tranquillity in England which we were threatened to lose in France . . . I congratulate you upon living in a country possessing perhaps the most perfect constitution one can imagine . . . I trust that the changes we will introduce will be as worthwhile for the country as partisans of the Revolution expect.[39]

And then again on 5 July 1790:

> In France all heads are currently turned toward public affairs. I see with pain that science is suffering from this diversion. I state no opinion about the sweeping changes in our constitution; we are in the midst of a great experiment and only the future will tell if our legislators will have succeeded. I congratulate you on living under a regime that has for a long time been stable, and that seems one of the best results of human intellect. It is at about the same time at the end of last century that the true foundations of the system of the world and the

social system were laid down. We fought the first of these and finally adopted it; perhaps it will be the same for the social system.⁴⁰

He sent the same message in a 10 July 1790 letter to Oriani in Milan:

> You are fortunate to live in a peaceful country without distractions from the study of science. Here all eyes are turned to public affairs and, for some time now, science will suffer from this diversion. Up until now my only role in public affairs has been to harbor the wish for the well-being of my country. May the outcome of the great changes that government is experiencing compensate us for the sacrifices we are forced to make in this period of crisis.⁴¹

As the Revolution progressed, Laplace wisely corresponded less with foreigners and abstained from articulating such explicit political opinions. This agnostic posture he assumed matched in some ways his unwillingness to take a stand on metaphysical matters regarding the nature of heat. He was more than perfectly neutral politically, however; he embraced some of the revolutionary vocabulary and quickly learned how to present himself in a favorable light. For example, when asking for a reduction of special taxes imposed by the Department of Seine-et-Marne, he explained his need to preserve his capital for the education of his family and underlined his service to the nation through his official appointments.⁴²

In the midst of all these substantial changes, which completely unsettled his personal habits, Laplace remained on the surface calm and unruffled. As a self-assured, middle-aged man with a solid professional life already behind him, he held on to his persona. Indeed, Laplace expected to continue exercising his many talents in the framework of his venerable Académie. But that institution had been gradually changing to meet new societal demands. Slowly it metamorphosed from an assembly of disinterested natural philosophers into a symbol of scientific authority and a major governmental consultant.⁴³ Similarly, although he continued to pursue his clear scientific research goals, Laplace began to assume a new role as a public servant. To his foreign colleagues, he said that he regretted this modification from the path he had laid out for himself, but excused it by referring to the more immediate and pressing needs of the times. Privately, however, he embraced this new dimension in his life.

## New Weights and Measures

The Académie was entrusted by the legislature to develop a uniform code of weights and measures for the whole nation, and by extension to help rationalize its calendar, currency, cadastre, and census procedures.[44] Given his expertise in the mathematical, astronomical, and physical sciences, Laplace was immediately enlisted to help in all these projects. He and his colleagues understood all too well that it was in their common interest to assist in the execution of enlightened policies adopted by early revolutionary governments. Moreover, these reform initiatives offered a signal opportunity for the advancement of various projects the scientists had initiated. When Laplace was first consulted by Talleyrand, who introduced a bill in the National Assembly for a uniform system of weights and measures, the plan was to base the unit of length on a multiple of the length of a pendulum at forty-five degrees latitude beating a second of time. As a member of the Académie's standing committee to execute the plan, Laplace favored determining the unit from a fraction of the length of the meridian measured by survey from Dunkirk to Barcelona. The ideological rhetoric of a neutral value taken from nature, hence bearing universalist value, coincided well with his desire to confirm geodetic measurements made earlier in the century in order to obtain a more accurate value for the shape of the earth.[45] In addition, it became the occasion to test the merits of Borda's quarter circle, thought to be superior to English measuring instruments.

The astronomers Méchain and Laplace's close collaborator Delambre were assigned the task of running the geodetic survey from the English Channel to the Spanish Mediterranean, which demanded herculean efforts to accomplish given the state of disruption in the countryside during the Terror and wartime. Laplace kept a running correspondence with Delambre about each phase of the operation, offering encouragement, advice, and direction at each step. In addition, Laplace participated in the group working in Paris around Lavoisier and Haüy to establish a unit of weight related to the standard of length.[46] The academic commission worked through all the considerable logistical, fiscal, administrative and technical problems for over five years, with Laplace's specific role impossible to separate from that of the committee as a whole. As indicated earlier, Laplace was removed from the board shortly after Lavoisier was arrested, but once the political turmoil had subsided, he was brought back on the commission, where he worked with redoubled effort.[47] After the Terror, Laplace assumed the role of spokesman for the project before the legislature and maintained contact with policy makers in the government.[48] By seizing this opportunity, he transformed himself from a professional scientist to a public figure. Stepping into the shoes left by the now-deceased Condorcet, he expanded his activity as a consultant to publicize science's most auspicious contribution to the Revolution. In public, Laplace rejoiced over the new metric system that for him sealed the triumph of rational reform in the midst of unimaginable horror.[49]

Among the many significant innovations that Laplace supported was the replacement of fractions of units by decimals, not only for weights and measures, but also for currency, time, and angle measurements. The day was to be divided into ten hours, each hour into one hundred minutes, and each minute into one hundred seconds. The right angle was to be divided into one hundred degrees; and each degree into one hundred seconds.[50] For the new republican calendar, which began the year at the astronomically meaningful autumnal equinox, months were to be divided into three ten-day "weeks" called "décades," again revealing a fixation with a number system based on ten. His fascination with numbers extended to the calculations initiated by Gaspard de Prony as head of the national land survey (the cadastre) to obtain interpolated values for trigonometric and logarithmic tables. Two positive reports authored by Laplace to the Académie were submitted on 12 May and 11 July 1792.[51] Laplace expressed his concerns both privately and publicly on a variety of issues, including the choice of names for new units for length, volume, weight, and currency.[52]

It bears notice that Laplace's transformation into a public servant began before the Terror, but in an unobtrusive manner. Laplace had served the administration starting in 1792 as one of the academic members of the Bureau de Consultation des Arts et Métiers, a panel set up in September 1791 to judge inventions and recommend subsidies to their creators. The records of this working committee show that he executed his tasks conscientiously, even serving his term as president of the Bureau.[53] He sat on this panel not only with fellow academicians, but also with a bevy of artisans, some of whom were actively pressuring the legislature to reduce the role or entirely eliminate the Académie in the new republican regime. Like Lavoisier, Laplace learned to curb his feelings of superiority over less-educated craftsmen, whom he had rather belittled prior to the onset of the Revolution.⁵⁴ Clearly he had learned how to exercise restraint in a politically charged setting.

In handling all these matters, which were of great concern to the legislature and its committees, Laplace acted constructively and conscientiously. While he must have been aware of the ideological import of each of these activities, he was careful to comment solely on scientific and operational features. He functioned more like a technocrat than a politically committed creature. This stance may well have been what saved him from the most extreme ire and retribution of some revolutionary actors.

As he was shuffling back and forth from his official duties in Paris and his family near Melun, and before the Revolution had fully passed through its most violent phase, Laplace decided to join in the revolutionary fray by becoming the public spokesman for his profession. For over a quarter-century, that role had been brilliantly performed by the Académie's "perpetual" secretary, Condorcet, but the Revolution had brought about his tragic end, leaving behind his moving testament, the *Esquisse d'un tableau historique du progrès de l'esprit humain* (Sketch for a Historical Picture of the Progress of the Human Mind). Within the governing circles, only the chemists Louis Bernard Guyton de Morveau and Antoine François Fourcroy; the erstwhile minister of the Navy, Monge; and the manager of the military enterprise, Lazare Carnot, had meaningful input during the Terror. They fell out of political favor as Maximilien Robespierre fell from power.⁵⁵ Moving into this vacuum, Laplace saw a chance to promote his professional interests by pushing himself forward. He had watched from the sidelines, and now entered the dangerous game in the name of restoring stability. He ventured into two domains that had not been a central part of his outlook under the Old Regime: publishing for and teaching to a general audience. He composed his first and most popular treatise in astronomy, the *Exposition du système du monde* (Account of the System of the World), eventually to appear in five editions during his lifetime, and served as one of two professors of mathematics at the Ecole Normale.

## The Ecole Normale

It is not clear when Laplace first began to consider seriously composing his elementary text on astronomy, which was eventually published in

February 1796. He offered it as the complement to his cycle of lectures on mathematics delivered in early 1795 at the new and short-lived Ecole Normale. Together with the subsequent *Essai philosophique,* they constitute his most important and influential foray into popular scientific education. Moreover, they signaled a critical turn in the nation's attitude toward science as well as in his personal career. France was just emerging from the Terror, and the National Convention was attempting to restructure the nation's cultural life, which had been severely interrupted and transformed during the Jacobin era.[56] Pleased with the prospect of a restored political and social order, Laplace embraced the changes wholeheartedly, and had a hand in shaping some of them. As we will see, he was instrumental in founding the new Bureau des Longitudes and the Institut National, and he began to frequent political and intellectual notables who were setting the tone of the new era. Henceforth, without abandoning the scientific arena in which he had made his mark, Laplace would be fully engaged in civic affairs, operating as a public figure to an extent he never had before.

The experience at the Ecole Normale was the crucial stepping stone in this changeover. This school, which lasted only a few months, was quite distinct from anything he had experienced as a young mathematics instructor at the Ecole Militaire when he had arrived in Paris. Since 1769 a host of reform proposals for educating the young had been unfurled, most of which assigned a leading role for science and mathematics in the proposed modernist curriculum.[57] During the early years of the Revolution a heated debate was carried on in the press, the legislature, and legislative committees to find an appropriate substitute for the instruction of young men, who had previously been educated by the religious establishment.[58] At first, a special legislative committee in charge of public instruction and the famous Committee of Public Safety inaugurated several "revolutionary courses" to meet the immediate needs of the war effort, and created the Ecole Centrale des Travaux Publiques, which later turned into the Ecole Polytechnique. They also launched the concept of an "école normale," intended to mold a new cadre of teachers who would spread the republican gospel throughout the land and restore a sound basic education.[59] For that purpose, each geographic district was asked to select a proportionate number of its most promising citizens and, at government expense, send them to Paris to learn the elements of all essential knowledge. It was expected that they would return to their precincts inspired by their experience in Paris and spread its teachings around the nation. A bill approving this new institution was enacted by the National Convention on 30 October 1794.[60] The same day a list of professors, not including Laplace, was selected.

Laplace first officially heard about the Ecole Normale in his country residence, located in the district of Melun, which had selected him to be one of its "students."[61] He turned down the position on the grounds that he had a conflicting obligation as examiner of the naval engineering school.[62] A month and a half later his name had been added to the prestigious roster of instructors.[63] This was an honor he could not refuse. He was being asked to share with Monge and Lagrange the teaching of all elementary mathematics, alongside other prominent figures, including his former academic colleagues Haüy, Berthollet, Daubenton, and André Thouin, who had been called on to teach their disciplines. Given the fragile role countenanced for accomplished scientists during the Terror, this was an opportunity to be seized eagerly.[64] The political thaw following Robespierre's demise was a time for the public regeneration of the scientific enterprise. The nation's legislature was reaching out for experts' assistance, convinced as they were that the "Republic [definitely] stood in need of scientists."[65] As France's leading native mathematical scientist, Laplace was challenged and keen to show off his calling.

Laplace gave the inaugural lecture of the Ecole on 20 January 1795 in the amphitheater of the Jardin des Plantes to a crowd of some seven hundred mature students.[66] The event received conspicuous public notice.[67] Even though the school would not last more than four months, it was an intense experience for learners and teachers alike. They both had to endure the bitterly cold winter and many of the students lacked adequate food, clothing, and accommodations. The amphitheater was not well suited for orators with small, high-pitched voices, and no audiovisual aids could be used. It was nonetheless a truly revolutionary encounter, with the leading savants of the day offering a synopsis of their knowledge to a vast gallery, keeping to a strict time limit, and the students being given the chance to ask questions. All the sessions were stenographically recorded, and shortly thereafter published, presumably because of the definitive character of the material presented and the wish that it be taught throughout the land.[68]

The lectures were also revolutionary from a pedagogical viewpoint. Instead of offering an encyclopedic panorama or even a systematic survey of each discipline, the lecturers aimed to illuminate the principles behind their subject. Contrary to the mathematical traditions that Laplace had encountered when he studied at the Benedictine school in Beaumont or at the University of Caen, and the tradition established by the textbooks of Camus, Bossut, and Bézout used at the Ecole Militaire, he laid down a much more general and metaphysical foundation for the subject. As Laplace indicated, he intended "to present the most important discoveries that have been made in the sciences and to develop their principles, to show the subtle and happy circumstances that led to their birth, to display the most direct route leading to them, to indicate where the best sources to find details about them are, to show what is left to be done and the procedure for making new discoveries; such is the object of the Ecole Normale, and it is from this perspective that mathematics will be considered."[69]

It was neither the techniques of mathematical manipulation nor its applications for solving practical problems that was proffered, but the underlying elements of the discipline as understood by its masters.[70] In the first few lectures, instead of offering standard mathematical equations, Laplace tried to explain concepts using the common language of a *géomètre philosophe,* hoping thereby not to turn away those auditors unfamiliar with its symbolic language. He used no equations; instead he offered a literary rendition of an abstract art, in a manner not unlike the way Aristoteleans had once taught logic. Though he sprinkled his lectures with references to famous treatises, he had no models in pedagogy to fall back on. Laplace turned instead to actual mathematical memoirs by Descartes, Newton, Euler, Lagrange, and others. One marvels at his consistent treatment of all mathematics in an analytical mode, explicitly turning the concrete geometric issues into algebraic abstractions.[71] It was a truly creative spectacle, which even included a new discussion of the theory of algebraic equations (a topic he had never broached in his publications) as well as an early proof of the fundamental theorem of algebra—thereby proving that he had made this major accomplishment several years before the mathematicians of the next generation, namely Carl Friedrich Gauss, Jean Robert Argand, and Augustin Louis Cauchy, who are usually credited with it.[72] One would like to think that the boasts by Delambre, Sylvestre François Lacroix, and Biot, who claimed that these lectures "changed the face of public education," are essentially correct.[73] These lessons at the very least both set the pedagogical stage for manuals that were later prepared for secondary schools and helped determine the proper curriculum for students at the Ecole Polytechnique. The immediate success of this venture inspired Laplace to continue composing in the same vein.

Laplace's ninth lecture broke with the theoretical approach by explaining the new metric and decimal system. All the other topics were meant to be of a fundamental, timeless nature. Perhaps Laplace expected the reform of weights and measures to be of a similar character. The last lecture was devoted to probability theory, and would eventually be totally rewritten seventeen years later when he prepared his *Théorie analytique des probabilités* (Analytic Theory of Probability). This tenth and final lecture was hurriedly assembled and is the least satisfactory, following the earlier example of expressing mathematical concepts in words rather than equations, which Laplace thought would ease the narration of his sophisticated probabilistic notions. It was at the beginning of this last lecture that he announced he was preparing a work on the discoveries of the "système du monde," meant to survey mechanics and astronomy following the same approach as the lectures.[74]

## The *Exposition du Système du Monde*

While developing his lectures for the Ecole Normale, Laplace prepared his other project for popular consumption. Over the course of its long life, the *Exposition* became a classic text that was amended in several editions, was translated into German in 1797 and English in 1809, and appeared also in Russian and Chinese.[75] It is clearly the most elegant and readable work Laplace ever composed, and its literary merits were sufficiently recognized to have it assigned for French language courses at the Ecole Polytechnique and awarded as a student prize during the entire nineteenth century.[76] One can still read it with profit today despite its formal, stilted manner. Unlike the lessons at the Ecole Normale that were composed in a rush, the *Exposition* is more deliberately didactic and often verges on sermonizing. It was meant to not only impart the latest intelligence about astronomy, but also to display a philosophy of scientific progress reminiscent of the recently departed Condorcet's *Esquisse d'un tableau historique des progrès de l'esprit humain*, which appeared in 1795 just as Laplace was preparing to step into Condorcet's shoes as a popularizer of scientific knowledge. Was it not likely that

Laplace consciously measured himself against his former colleague and rival, who, despite his tragic end, had left such a deep impression on French culture? Such considerations would go far in explaining Laplace's eventual entrance into the Académie Française after the Revolution. This honor was tendered as a public reward for presenting science in a palatable fashion to the literate public, just as Bernard Fontenelle, Buffon, Bailly, Condorcet, and Georges Cuvier had done.

For the most part the themes that Laplace elaborates in the *Exposition* are verbal renditions of his astronomical research, something like a popular version of the *Mécanique*. Indeed, at the beginning of the second volume he admits: "The most abstruse mathematics has been essential to establish these various theories. I have assembled them in a treatise on celestial mechanics which I intend to publish. Here I will limit myself to presenting the principal results of that work, indicating the path geometers have followed, and trying to explain their rationale as much as one can without the help of [mathematical] analysis."[77]

What is new here is the form given to the popular treatise. He was either unwilling or unable to simplify and dramatize his métier. The *Exposition* shares little with Fontenelle's literary masterpiece the *Entretiens sur la pluralité des mondes* (Discourse on the Plurality of Worlds), nor is it targeted at persuasion, like Voltaire's *Eléments de la philosophie de M. Newton* (Principles of the Newtonian Philosophy). It does not at all resemble the religiously inspired moralizing treatises of the abbé Antoine Pluche, or the less well-known lectures of Pierre Le Clerc, *L'astronomie mise à la portée de tout le monde* (Astronomy for Every Man). Nor is it a manual for cosmographers or amateur astronomers of the sort that Lalande and the abbé Jacques François Dicquemare had composed before the Revolution. The new work is an uncompromising display of what Laplace characterized as "the greatest monument to the human spirit, the noblest mark of his intelligence."[78] He takes the reader by the hand from the appearances of motions of heavenly bodies to their "real" motions; from those to the abstract laws of physics; and from there to the law of gravitation, no longer presented as a conjecture, but now taken as a veritable cause of motion, a "grand principle of nature." The work continues with a revealing synopsis of the history of astronomy, drawing heavily on the tomes previously authored by Jean-Etienne Montucla and Bailly, and is aimed at showing the ideal way by which mankind has moved from observing the false appearances of phenomena to understanding their true nature.[79] Themes already familiar to him and to Enlightenment philosophy are in this book etched in stone and offered as guides to the future: mankind has moved over time from brute gazing at phenomena, in which reality is obscured by prejudice and superstition, to the more civilized use of critical empiricism, which is embraced by modern science. Just as was true for Condorcet and other commentators, the turning point of history was the era of Galileo, which ushered in the scientific revolution. Since then, progressive scientists had moved steadily from empirical compilations to mathematical laws, and thence from laws to causes. In the process, conjectures had been instrumental, but only when based on—and confirmed by—evidence.

## The Nebular Hypothesis

Following his historical account, Laplace ends with a short chapter entitled "Considérations sur le système du monde, et sur les progrès futurs de l'astronomie" (Considerations on the System of the World and on the Future Progress of Astronomy), which conveys some of his most compelling conjectures about the origins of the solar system. In the original edition they constitute a natural progression from the past to the future, in the same manner as Condorcet had ended his *Esquisse* with a section displaying his vision for the future of mankind. In some later editions, Laplace moved this portion to the notes, much as Newton had relegated his speculations to the famous "Queries" appended to the *Opticks; or, A Treatise of the Reflections, Refractions, Inflexions and Colours of Light*. He explicitly labeled these notions as tentative beliefs that should be considered critically, with great caution.[80] By the third edition of 1812, these suggestions, which had become the most novel part of his popular treatise, had become his "theory of the origins of the solar system." Throughout the nineteenth century, they were referred to incorrectly as the Kant-Laplace nebular hypothesis, providing for a century the stimulus for cosmological research that would surely have pleased Laplace.[81] In 1796, however, he could not have imagined how fruitful these ideas would become.[82]

In many ways Laplace's views were not totally original. Kant had proposed analogous notions in 1755 in his *Allgemeines Naturgeschichte und Theorie des Himmels* (General History of Nature and Theory of the Heavens). But there is no indication that Laplace or anyone in his entourage knew about them. The corpus of Kant's philosophical work was introduced in France around 1801 by Charles de Villers, who barely mentioned this early treatise. Even if Laplace had known about it, the two men approached the issue from fundamentally different contexts. Kant was concerned with the way the universe developed from its creation—that is, as the title suggests, with the "history" of nature.[83] The effort was directed to postulate a process leading from its primitive origins to the present. Laplace, however, was from the start concerned with showing the fruitfulness of a probabilistic approach to natural phenomena, moving from the present back to the past. It is revealing that he first broached the nebular hypothesis in his last lecture at the Ecole Normale devoted to probability, as a documented example of the power of mathematical inference—an issue that had been on his mind since 1772. The way he introduced this original concept in his lectures is particularly indicative. It was an integral part of his assault on first causes:

> Phenomena often appear to arise from a recurring cause even though they depend on irregular, variable or unknown causes, which we label with the word 'chance' [hasard]. It is up to analysis by probability to determine to what extent a recurrent cause is probable by virtue of the phenomena, and to alert philosophers about it as a topic worthy of research. The solar system unquestionably offers a cause of this type.

This remark was followed by a full-blown exposition of the probabilistic reasons to believe that the physical disposition of the solar system implies that it stemmed from a single physical cause. The wording of his summation is equally significant: "This cause . . . can only have been an immense fluid disposed like an atmosphere around the sun . . . that extended beyond the orbs of all the planets and which gradually condensed to its current limits."[84]

These remarkably terse but tentative utterances led to unanswered questions about the cause of the rotary motion and other speculations about the manner in which this fluid atmosphere contracted. All are repeatedly labeled "conjectures." As if he were running out of time, Laplace brushes them all aside with a phrase revealing his true intent in offering these conjectures at the Ecole Normale: "Whatever the origins of the planetary system, it is certain that its constituents are organized so as to enjoy the greatest stability as long as external causes do not intervene . . . It seems as if nature has disposed everything in the heavens to insure the permanence of the system by modes similar to those which it seems to follow to insure the conservation of individuals and species on earth."[85]

Laplace broached this subject to advance his original philosophical goal of fully understanding the fixed laws of physical astronomy that insure stability and permanence. Like so many classical natural philosophers, his deepest wish was to reduce apparent change to essential reality, as he envisioned it. The evolutionary and developmental notions implicit in Kant's approach were not part of his mindset.[86]

Laplace's probabilistic ideas were far from original. He had opened an earlier paper of 1776 on the mean inclination of the orbits of comets by noting that "one of the most extraordinary phenomena of the system of the world is the motion of planets and their satellites in the same direction, and approximately in the same plane."[87] Immediately he calculated the improbability that this motion was due to chance and asserted confidently that "it is absurd to doubt the existence of a natural cause" to account for it. Laplace reminded his readers that the probability issue had been treated adequately by Daniel Bernoulli in his winning essay of 1734, which was a response to a prize question set by the Académie years earlier.[88] Bailly reiterated the idea in 1785 in the last chapter of his *Histoire de l'astronomie moderne*.[89] While the problem posed by this "extraordinary" phenomenon was commonplace, its solution was not forthcoming. In 1776, Laplace admitted to having pondered about it without coming to a satisfactory answer.

In the 1749 *Histoire naturelle,* Buffon had tendered an explanation, that the configuration of the solar system was a consequence of a comet obliquely striking or passing close to the sun, stripping off fragments of the hot sun to spin them into orbits, which, through the action of universal gravitation, had eventually settled into an orbit and cooled off to become planets and their satellites.[90] While it may have been qualitatively attractive, offering also an explanation of the original molten character of planets, the notion was dismissed after his death by mathematical astronomers—for example, by Dionis du Séjour in several paragraphs of his 1789 *Traité analytique des mouvements apparents des corps célestes* (Analytic Treatise of the Apparent Motions of Celestial Bodies).[91] However "ingenious" this hypothesis, it would necessarily have yielded planetary orbits of a highly eccentric shape. Planets generated by Buffon's proposal would all have to pass close to the sun, which they obviously do not. Dionis du Séjour, like Bailly and Laplace before him, admitted that the solar system depended on "particular causes that are unknown to us."[92]

Taking a cue from Dionis du Séjour and Pingré, Laplace introduced two important new features to this set of observations. One was in his systematic catalog of five sets of phenomena that pointed to an antecedent cause. To Daniel Bernoulli's observation about the motion of planets in the same direction and nearly the same plane, he added the similar patterns for satellites, the rotation of the sun, the small eccentricities of planetary and satellitic orbits, and the large eccentricities of comets coupled with their seemingly uncoordinated planar movement.[93] He was fascinated with the notion that comets were extra-solar phenomena, behaving in a quite unruly manner, whereas the other objects in the solar system conducted themselves in a much more disciplined fashion.

The other new idea was a suggested mechanism for this set of regularities. It too was brought to his attention by others' work; in this case, Bailly's historical treatise and Dionis du Séjour's publications.[94] A series of proposals for the formation of Jupiter's ring were on the table, and provided by analogy a possible mechanism. Dionis du Séjour had refused to pronounce whether Jupiter's ring was "a residue from a once-larger globe; . . . or if it broke off from Saturn owing to an excess of centrifugal force."[95] Earlier astronomers had added other suggestive ideas: Jean Jacques Dortous de Mairan and Jacques Cassini, for example, had commented on the presumed atmosphere surrounding Saturn that might have been responsible for the configuration of its satellites, a topic that most likely had intrigued Laplace while writing about satellites in the 1790s.

The idea of a solar atmosphere intrigued Laplace so much that he developed notions about its heat content and luminosity, and began to make use of analogies taken from research on the extra-solar world that was actively being conducted across the Channel. Laplace was fairly well informed by Deluc, Blagden, and Herschel himself about advances reported to the Royal Society. John Michell, for example, was a name known to him, even though he systematically misspelled it in his publications.

The initial problem Laplace toyed with was the probable consequence of atmospheric contraction. In the first two editions of the *Exposition*, Laplace reminded his readers of the transient brilliance of the "new" star of 1572 studied by Tycho Brahe, attributing the luminous variation to the likely change of size of its "atmosphere." This led our astronomical speculator into a curious digression that postulated the phenomenon we now know as "black holes."[96] Laplace indicated that the large size of a star (he arbitrarily says 250 times the sun's diameter) would absorb the surrounding light rays, rendering the star invisible.[97] He suggested that the study of the variation of light and color in stars was a central agenda for the future of sidereal astronomy. In turn this provoked a paragraph on heavenly nebulosities of the type observed in France by Messier. Clearly he had also read Herschel's findings, and knew about Michell's conjectures, either directly or via Priestley.[98]

A major change in Laplace's treatment of his nebular hypothesis occurred following his English correspondent William Herschel's pioneering paper on the evolution of nebulae, read before the Royal Society on 20 June 1811.[99] On the basis of new observations, Herschel established that certain stars pass through several stages of nebular condensation as a result of gravitational action. Elated by this fortuitous confirmation by analogy of his own views of the transformations of the solar atmosphere, Laplace immediately had a synopsis of Herschel's accomplishments broadcast in the 7 July 1812 edition of the official government newspaper, the *Moniteur*.[100] The gist of the paper had been transmitted to Laplace by his friend Blagden in a letter that Laplace triumphantly read to his astronomical colleagues at the Bureau des Longitudes on 10 June 1812.[101] Now his admittedly precarious conjectures of 1795 seemed to be confirmed with a set of independent observations. In Laplace's mind, these conjectures were transformed into a believable theory, one that he began to treat with much fondness despite the many objections to it.[102] The nebular theory could account for some of the most striking phenomena, but clearly not all.

Confident as he was in his own intuition, Laplace made important emendations to the fourth edition of the *Exposition*, leaving out all the conjectures about the solar atmosphere and "black holes," and replacing them with a tribute to Herschel's powerful instrument. His scientific imagination, which had been given exposure only because he had embraced a new popular medium, was apparently totally justified. Laplace's deep commitment as a professional scientist to the authority of empirical evidence was suddenly, and stunningly, fused with his newfound venture.

As the most hazardous moments of the Revolution passed, Laplace resolutely donned his new personality as a public figure. With his successful podium appearance at the Ecole Normale and the highly prized *Exposition du système du monde,* he stepped into the public limelight, never to retreat solely to his previous vocation as an ivory-tower academician. He chose to forge his own persona, guided by the example of his former colleague, Condorcet, but he wisely chose to eschew the political engagement of the Girondin politician. The themes he addressed were much the same and centered on the triumph of rationality through science. Like Condorcet, Laplace was willing to take advantage of the utilitarian byproducts of his métier, but he always returned to the assertion that the pursuit of science was the noblest of human activities. The exact sciences that he so publicly commanded were for him the modern substitute for the religious pabulum he had abandoned in his youth. He even went so far as to cast public aspersions on great scientists who had inadvertently slipped back into "superstition." In his first lecture at the Ecole Normale, for example, which was pointedly reported in the *Moniteur,* he made fun of Leibniz's use of the binary system of numeration to assert the existence of God.[103] In the next sentence, he chided Newton for straying from science in dealing with the Apocalypse. Other examples of the misguided judgments of great scientists were used in the *Exposition* to drive home to popular audiences the value of adopting a strictly rationalist and empirical stance if progress was to be sustained.

Of all the major scientists who survived the excesses of the Revolution, Laplace, Monge, and the chemist Fourcroy were the only ones who participated actively in the reconstruction of France's scientific life after the Terror. They operated not only as experts in their field and popularizers of their respective disciplines, but also, and more centrally, as political actors. It was Laplace's initiation into politics that ultimately catapulted him to high status in the Napoleonic era. He followed the natural sequence from professional to popular science, and thence to public affairs. But unlike Monge and Fourcroy, he did not reduce his professional contributions to science in order to make room for politics.

# 8

# The Politics of Science

Perhaps it was Laplace's Norman upbringing that had kept him politically neutral early in life. As a young man, he instinctively held back from making quick, emotional commitments regarding any topic he did not fully understand. Perhaps it was also part of his cautious nature to refrain from taking firm stands on political issues that did not touch him personally. Before abandoning the Church, he had made sure a new career in mathematics was open to him, and then resolutely took a stance as a modernist worker in science. Once in Paris, Laplace had the opportunity to join the encyclopedic circle around d'Alembert; the freemason associates of Lalande; or the freethinking circles around the Baron d'Holbach. But instead he opted to develop himself into an accomplished scientific expert, shunning the salons and the Enlightenment fraternity all around him. When the Revolution came, he stayed away from clubs and political parties, confining his contributions to his earned reputation as a competent academician. He remained somewhat aloof, and even as a bachelor boasted few close friends. As far as we know, these were principally academic colleagues, namely Lavoisier and Berthollet. He even stayed out of most political squabbles in the Académie, by speaking out forcefully only on scientific matters, and in his official capacity.[1]

The few reports we have about Laplace's personality show him to be a man of strong scientific opinions, willing to exchange views with informed associates, but not given to cultivating warm relationships with visitors or would-be scientists.[2] He managed to find a soft spot for talented mathematicians, particularly if they could be enlisted for his own scientific program.[3] Early in his career he developed a habit of encouraging younger scientists to help them start their own careers. Delambre,

Oriani, and Lacroix were among his best-known junior picks, each in need of his patronage. He struck up good relations with them, but without dropping his air of superiority.

As mentioned earlier, on the eve of the Revolution, Laplace made his move up the social ladder by marrying into a recently ennobled family of middling fortune. Members of the Courty family were not particularly engaged in the political disputes of the Revolution; they were concerned more with their assets than with the rights of man. Major political issues first intruded when the Académie tried to change its structure to be aligned with the current political mood. Laplace was named to the committee to draft a new "constitution" for the learned society. There he witnessed the rancorous divisions that pitted colleagues against one another, but he did not declare himself on any side.[4] He managed to work equally well with the ultra-conservative director of the observatory Cassini IV, the moderate reformers Condorcet and Lavoisier, and with his more activist colleagues Monge, Guyton de Morveau, and Fourcroy. He even learned to cooperate with artisans and inventors of the Bureau de Consultation, whom he had once treated with some disdain. The repercussions of the Revolution affected Laplace most personally once the Académie was shut down in August 1793. Bailly was arrested in Melun in September, Lavoisier in late November, and Laplace's former patron Bochart de Saron shortly thereafter in December.[5] Laplace was purged from the Committee on Weights and Measures in December. His tidy world as a new father and an accomplished academician was rapidly breaking up, and his financial security was threatened. He also understood the need to protect his family from the sporadic violence of the capital. With the advent of the Terror, he began not only to lose other esteemed colleagues—Old Regime political figures like Condorcet, Bailly, Lavoisier, Vicq d'Azyr, and Bochart de Saron—but to learn of others arbitrarily jailed, under suspicion, or even temporarily arrested, like Delambre on his official mission to measure the meridian. At first, he tried to steer a middle course as an uncommitted technocrat, ready to do the government's bidding. Whatever private views he harbored about politics, he kept them separate from his professional activities.[6]

## The Bureau des Longitudes

Faced with these major changes that effected his customary outlook, Laplace began to construct a new role for himself. The opportunity to serve his calling by stepping in to assume roles once performed by departed colleagues coincided with his self-confidence, which had been reinforced by the public recognition he received at the Ecole Normale. The likelihood that this new direction would also provide new revenues must have figured into his decision. It must be added, however, that he never solicited the government for more support or positions.[7] All these occasions came after Robespierre fell from power in July 1795.

One major opportunity presented itself just as he was being named to the Ecole Normale. The timing makes one wonder if there was not a connection between the two events. The people's representative Joseph Lakanal, who was entrusted by the National Convention to draw up plans for higher education, called on Laplace for advice about astronomy in December 1794.[8] Aside from the aging Lalande, Laplace was the astronomer with the strongest and most neutral standing, untainted by partisan politics. He quickly seized the opportunity, writing to the legislator with a stirring manifesto endorsing astronomy. Lakanal was so impressed that he copied it verbatim when offering his plan to the Committee on Public Instruction on 10 April 1795.[9]

Lakanal had called for a central commission of astronomers in Paris to coordinate the activities of ten observatories that he urged the committee to support. By the time the bill reached the National Convention on 26 June, this commission had taken on the British-inspired name of Bureau des Longitudes, and was charged with both supervision of the two Paris observatories and correspondence with provincial and foreign institutions. The Bureau was asked to issue the French version of the *Nautical Almanac,* the *Connaissance des temps,* and to offer annually an elementary course on astronomy for the public. Historians looking at this foundation in retrospect may rightfully conclude that the new Bureau merely assumed tasks that had been carried on before the Revolution—the only difference was that they were placed under a single administration.[10] But at the time, the Bureau was acclaimed as a significant innovation symbolizing the Republican government's resolve to support the scientific enterprise. It was taken as another sign of the rebirth of official science.

As an unofficial architect of this new institution, Laplace was quite naturally appointed a founding member, with a respectable government salary of eight thousand francs annually. He shared the mathematics section with the elderly Lagrange, but took to his appointment much more energetically. The minutes of the meetings show him to have been from the very beginning an activist member of the governing board.[11] He pushed for new mathematical tables for logarithms and trigonometric functions to serve the decimal system he had helped to create; he worked out an arrangement with Prony's cadastre office to have some of the astronomical tables prepared for the almanac; and he argued for the completion and establishment of the metric system.

Whenever there was need to negotiate with government agencies for payment of legitimate activities assigned to members of the Bureau, he was delegated to act on its behalf, giving him ample opportunity to interact with various legislators, ministers, and their cabinets. He quickly became the spokesman for all matters related to astronomy, and even the other sciences. He initiated discussions about the republican calendar, the expenditure of funds for new instruments, and the collection of data on tides. As the most dynamic voice in the Bureau, he was repeatedly elected its president, with his friends and former collaborators Lalande and Delambre often serving as administrator or secretary. Laplace was instrumental in promoting his disciples to vacant posts, and then using their services for his own ends.[12] His presence at the meetings loomed large, enabling Laplace to control the scientific discussion, during which he often pushed aside other perspectives. At times, it was as if he considered the new institution to be a laboratory for the application and verification of his theories circulated in the *Mécanique*.[13] Several years later, when writing an account of this period, Delambre privately blamed Laplace for monopolizing the Bureau's research program.

> One should never place a mathematician at the helm of an observatory, for he will neglect all observations except those needed to test his formulae . . . There would have been no difficulty if all the mathematicians had had Lagrange's character. He had his views, but never imposed them. But the other mathematician was well known for wanting to supervise everything. As Monge said, "He is a man full of ideas and governs well, but likes it too much and lets everyone know it." More than anyone I know what magnificent things he has done for astronomy with his theory of planets and satellites, but he should have restrained himself or carried on differently with respect to the Observatory whose sole direction he assumed. He can be condemned for not letting the Bureau des Longitudes make any stellar observations or for failing to make a single star catalogue in twenty years' time![14]

However one may evaluate this criticism, it is clear that Laplace was considered by colleagues and the public alike to be the leading theoretical astronomer of the era, and that this new government-supported institution was the administrative locus for research.[15]

## The Institut National

Six months after the creation of the Bureau, the government established the Institut National des Sciences et des Arts, in which Laplace also played a major role.[16] The Institut not only resuscitated the defunct Académie des Sciences, but also organized the entire academic system under the double banner of the French flag and science. A symbolic poem by Jean François Collin d'Harleville entitled "The Great Family Reunited" was read at the inauguration of the Institut, which was attended with great fanfare by the leaders of government.[17] Its creation marked the full restoration of the intellectual life of France. Significantly, science was now in the vanguard. The Institut National was subdivided into three "Classes," the first of which was entirely devoted to natural sciences.

Laplace and Lagrange were named to the mathematics section of the Institut's First Class and given the task of selecting other academicians. Just as they did at the Bureau des Longitudes, he and Lagrange pulled all the personnel strings, decisively naming which fellow scientists should once more bear the title of academician. Their original choices were unimpeachable, based largely on seniority and prowess as scientists. Laplace and Lagrange were publicly acknowledged as the supreme judges of scientific merit, and were considered by younger workers as natural sources of patronage. While Lagrange handled this standing with a light touch, Laplace thoroughly relished the authority it gave him for the rest of his career. He also collected a salary of 1,500 francs annually, adding to his prosperity and financial security. In the short span since the start of the Revolution, Laplace had raised himself from a relatively modest position as an academician to that of a major player in France's intellectual life.

As a founding member of the Institut, Laplace was delegated to present the new body's first annual report to the legislature at the end of the Republican year four. It was his first political speech, filled with his usual combination of succinct detail and precision, but also dressed up with appropriate politic declarations. He had already successfully tried this kind of rhetoric in the closing lecture of the Ecole Normale, the *Exposition,* and with Lakanal. Speaking before the legislature, he indicated how advantageous was the liaison between the political power structure and the enlightened community of savants. He ended his address by reiterating the value of learning for the political health of the nation:

> This account would be incomplete if we did not refer to our efforts to propagate the eternal principles of justice and equality that are at the root of the French constitution . . . We declare that it has no more sincere champions than savants and artists. Nature, which is the object of their constant concern, reveals at every turn the rights and dignity of man . . . they are passionately stimulated by all that is grand and orderly, so that equally far from servitude and anarchy, everything ties them to a government that steers a middle course between these two extremes, and whose existence is intimately linked to the progress of science and the arts, without which there is no durable liberty, no happiness.[18]

Laplace engineered another event that secured his reputation both nationally and internationally. Research for the new metric system, which had been decreed in 1791, was finally to be completed in 1798. It had been science's trademark to justify its commitment to public service. Now Laplace saw an opportunity to celebrate science on an international level by calling together a congress of scientists, ostensibly to verify the metric system's validity. On 24 January 1798 he proposed that the Institut invite distinguished foreign scientists to Paris for this purpose. In the discussion that ensued, he was backed by a newly elected member of the Institut, General Bonaparte. Laplace then worked with the foreign minister Talleyrand, who had originally introduced the bill in the 1791 National Assembly, to convene representatives from all the French-dominated territories and to put a seal of approval on the weights and measure reform. In his letter to Delambre, Laplace acknowledged that the motive for this meeting was principally political. He said, "You must know that all this is just a formality so that they can claim the system as their own, to nullify the feeling of national jealousy, and to give them a reason to adopt the new measures."[19] Many of his astronomical acquaintances were called to Paris, including Oriani and Gregorio Fontana from

Italy and Van Swinden from the Lowlands. Laplace basked in the limelight of this first international congress of scientists.[20]

## General Bonaparte

All these activities put Laplace in close touch with the rising star of the national scene, General Bonaparte. Laplace had helped him over a mathematical hurdle at the start of his career, when he examined the student at the Ecole Militaire, recommending that he be given a commission as a lieutenant in the artillery in 1785. The young officer continued to work with the mathematical sciences at Auxonne, aiding professor Jean Louis Lombard to verify his ballistic tables.[21] Bonaparte fed his sense of personal curiosity in a lull during his military career in mid-1795, when he followed a set of unspecified scientific lectures and visited other cultural sites in Paris.[22] The lure of science was part of his inner being. When the general conquered Piedmont for France in 1796, he penned an open letter to Laplace's friend, the astronomer Oriani, that could only have been written by a politically astute official familiar with science: "The sciences that honor the human spirit and the arts that grace life and transmit great deeds to posterity must be especially honored by liberal governments. All men of genius . . . are French, whatever their country of origin . . . All who desire to move there will be welcomed with distinction by our government. The French people value the acquisition of a learned mathematician or a renowned painter . . . even more than the richest and most populous city."[23] This remarkable assertion by a conquering general was reproduced in the *Moniteur* and drew immediate praise in France.[24] The shrewd general understood that durable conquests must include cultural hegemony as well as military domination. He shared in what was already France's self-appointed mission to amass the products of high culture—human or material—wherever they might be found. To this effect, Laplace's friends Monge and Berthollet accompanied the general on his Italian campaigns to sort through objects of art and science to be "preserved" for posterity by being shipped to Paris.[25] The close association between power and culture that was being forged was immediately grasped by Laplace as an additional opportunity for renewing France's scientific life.

Laplace and Bonaparte met again when the general returned triumphantly from his Italian victories in 1797. Shortly thereafter, he launched an audacious expedition to Egypt, taking with him a host of scientists (including once more Monge and Berthollet), students at the Ecole Polytechnique, and their instruments—with the express purpose of creating a research outpost in the Middle East.[26] Bonaparte gathered his scientific cadre around him and created the Institut d'Egypte in the likeness of the Parisian Institut. This time the general offered more than rhetoric. All kinds of scientific observations were made during this exotic venture. By then, too, Bonaparte was more than just a supporter of the French Institut National des Sciences et des Arts; he was an elected member. The election had come in December 1797 between Bonaparte's missions to Italy and Egypt.

Given the high state of excitement about the young officer, it takes little imagination to picture how professionals could have been seduced by Bonaparte's personality. Laplace was reported in the public papers to have marveled at the general's interest and knowledge of science.[27] For Laplace it was flattering to be linked to a star of such national prominence. Others imagined a close relationship.[28] Moreover the fresh accounts he received from Bonaparte's scientific companions Berthollet and Monge added to his favorable disposition. Most of all, it seemed politic to recruit Bonaparte into the family of academicians. One must recall that elections to the Institut were voted on by all its members, including many who had no special knowledge of the sciences, and none of the alternate candidates carried his name recognition.[29] Obviously the election was understood to be a reward for Bonaparte's constructive support of science and art on his Italian campaigns, and his active promotion of French culture in the conquered ("liberated") territories.[30]

Laplace's new association with the general was immediately exploited for astronomical research. On several occasions before he left for Egypt, the Bureau des Longitudes received a promise from the general to obtain ten thousand francs from his military budget for badly needed astronomical instruments for the observatory.[31] The promise was never honored, but it reinforces the extent to which all parties were prepared to exploit one another.[32]

## Minister of the Interior

Bonaparte returned from Egypt with clear ambitions, anxious to restore political stability to the shaky Directorate government. His brother

Lucien helped to engineer the successful coup d'état of 18 brumaire (9 November 1799), which catapulted the returning general to the central post as first consul. This major event signaled the end of legislative bickering that promoted a sense of national insecurity. As the first consul, the young Corsican hero took the reins of the executive council to establish decisive policies. As far as we know, Bonaparte's scientific entourage was not directly involved in the plot to take over the government. Yet it is likely that Laplace, like many of his colleagues at the Institut, welcomed the emergence of a scientifically aware strongman to guide the nation. Scientists all yearned for the restoration of stability in public affairs. As always, Laplace stood ready to do his government's bidding. To his complete surprise, he was named by the consuls as minister of the interior three days after the coup.[33] He was being asked to join ministers Joseph Fouché (police), Martin-Michel-Charles Gaudin (finances), Charles Frédéric Reinhard (foreign affairs), Louis-Alexandre Berthier (army) and Marc Antoine Bourdon (navy), appointed the previous day, to run the country with the three consuls.[34]

We know in retrospect that Bonaparte was in charge of appointments. For the post of interior minister, he had his eye on his brother Lucien, who had been a key figure in the takeover. But for the sake of appearances and to still cries of nepotism, he wanted a member of the Institut as part of his cabinet. His first choice was Bernard-Germain-Etienne Lacepède, who declined this distinction, but later accepted the job of heading up the Legion of Honor.[35] Laplace was a sensible alternate choice, given his knowledge of scientific institutions and his keen interest in the development of public education. He had been serving effectively for over four years as a prominent member of the board to select students for the Ecole Polytechnique, and more significantly had assisted in the selection of instructors for the new écoles centrales in Paris and in his home Department of Seine-et-Marne.[36]

The Ministry of the Interior was charged with the administration of all the Departments and set the rules for all civilian affairs, especially transportation, commerce, industry, welfare, education, and the appointment of personnel.[37] From his predecessor Nicolas Marie Quinette, Laplace inherited a well-structured bureaucracy run by a capable group of civil servants. They included the former abbé François Joseph Michel Noël, once minister to The Hague and to Venice, sometime co-editor to the *Magasin Encyclopédique,* and in charge of prisons, hospitals, and welfare;

as well as the former abbé Jean Baptiste Dumouchel, formerly rector of the defunct University of Paris and in charge of education. Noël was soon to be named prefect of the Haut-Rhin. Christophe Dieudonné, also a former priest, headed the ministry's first division; he was destined to sit in the legislature before also becoming a Departmental prefect. These "secretaries" were devoted and informed public officials who understood the administration and knew how to pull the right strings to make policies work. Laplace's role consisted of hearing their proposals at biweekly meetings, reporting his activities to the Council of Ministers, where overall policies were set, and signing directives prepared for him. Since he remained in office for only six weeks, it is difficult to judge his competence or to isolate which items he initiated. Several years later the exiled emperor graded him rather severely, claiming that though "a mathematician of first rank, Laplace never proved himself a more than mediocre administrator. From the very first the Consuls realized they had made a mistake. Laplace never grasped issues from a proper perspective; he sought quibbles and saw problems everywhere, bringing the spirit of infinitesimals into the realm of governance."[38] In fairness, the defeated emperor who dictated this assessment in Saint Helena was still smarting from what he felt was Laplace's disloyal conduct in 1814 when the Senate removed Napoleon as head of state, and in 1815 when Laplace failed to rally to his cause during the Hundred Days.

There are a few notable actions we may nonetheless credit to Laplace. On one of his first days on the job, he increased the national pension awarded earlier to the widow of his former colleague Bailly.[39] He also inquired whether the popular physicist and showman, Jacques Alexandre César Charles, should continue to be rewarded with a pension for his part in the invention of balloons.[40] He ordered the establishment of a welfare board for the capital, which later won him the presidency of the private association called the Société Maternelle.[41] He assisted the Bureau des Longitudes in speeding the publication of various mathematical tables.[42] In a letter to the Dutch physicist Van Swinden, who had written the report that ratified the determination of the length of the meter, Laplace reassured him about the fate of the reform of weights and measures.[43] He accelerated the spread of the metric system throughout the land by recommending the use of Brisson's *Instructions sur les mesures et poids nouveaux comparés aux mesures et poids anciens* (Instructions for Comparing the New Weights and Measures to the Old).[44] He ordered an inquiry into the local military school, the Prytanée Français.[45] Above all, he signed a bill reorganizing the Ecole Polytechnique, a bill that had been held up in the previous administration's legislature.[46] Perhaps his greatest satisfaction came when he ratified the recommendation of the Ecole Polytechnique administration, which named his protégé Lacroix to assume the professorial post vacated by Lagrange's resignation. Laplace wrote him a personal letter to accompany the edict.[47] A few days later, his Norman colleague Jean Baptiste Labey was also appointed to teach mathematics at the Ecole Polytechnique.[48] In addition, Laplace was vigilant about demanding the prompt submission of records of births and deaths from the Departments, which were grist for his probabilistic research.[49] When his secretary Dieudonné was later promoted to a prefecture, he gave him advice toward the preparation of the important *Statistique du Département du Nord* (Statistics of the North Department), which he had promoted while minister.[50]

The record suggests that Laplace considered his position as a means to further his aim to turn France into the leading scientific nation of the world, now that political stability had returned. In one letter, Laplace refers to his office as "the Ministry of Science and the Arts," a sort of executive branch of the Institut National.[51] To his mind, there was no conflict of interest between serving the Institut as a member and making ministerial decisions about it. Attendance records show he continued to sit in the First Class even as he assumed his ministerial post. He also made an effort to attend meetings of the Bureau des Longitudes when time permitted.

In his subsequent career, Laplace rarely said much about his fleeting tenure as interior minister. He quickly recognized how tricky and contingent was the world of politics. On several occasions he made a distinction between political revolutions, which were unpredictable, and the revolutions he studied in the heavens.[52] With one of his former collaborators at the Ministry of the Interior, he was more expressive: "I do not predict in this domain [politics] any more than with the weather, for events in the moral world are as difficult to foresee as they are in our atmosphere; but I put my trust in the wisdom and foresight of our government. Even though many look upon those who govern with envy, I recognize the heavy responsibilities with which they are saddled. So rather than crave their lot, I am more likely to pity them. And when they govern well, I extend them my most fervent appreciation."[53]

## The Senate

Bonaparte's replacement of Laplace by Bonaparte's brother on Christmas Day 1799 was technically not a dismissal; Laplace was already slated for another, less sensitive political post in the newly constituted Senate. This deliberative body was meant to replace one of the two houses of the Directory, which had been a less-than-efficient arm of the government. Bonaparte packed this assembly with his subjects so that they would ratify the consul's decisions and bring about the political stability he so desired. Laplace was named to the Senate on 24 December 1799, along with dozens of his colleagues at the Institut.[54] The new organization had fewer lawyers and more administrators and men of learning, who would be more pliable. Laplace was not merely appointed as a senator; Bonaparte bestowed on him a leading role in its organization, successively naming him secretary in 1800, president, vice president, and finally chancellor of the Senate in 1803.[55]

This upward move in the political hierarchy changed Laplace's living situation dramatically. During the Directorate, he had continued to keep his main residence near Melun in le Mée, using an address at 1334 rue des Bons Enfants for his frequent stays in Paris.[56] One assumes he was offered more permanent quarters when he became minister. Now, the post of senator also required a more permanent presence in Paris, and by the end of 1800 he had moved, with his family, to 2 rue Christine, located between the Institut and the Senate.[57] The appointment came with all sorts of perquisites. There was a handsome salary to go along with the promotion. Bonaparte eventually endowed each of the senators with a major property, often taken from conquered territories.[58] And as vice-chancellor, Laplace could afford to rent one of the stately homes adjacent to the Jardin du Luxembourg on 1133 rue de Tournon.[59] The position came with horse and carriage, domestic servants, and an extravagant annual salary of 72,000 francs, which included his expenses for entertainment and ceremonial occasions. By contemporary standards it was a major sinecure. Laplace now lived like the privileged "notable" he had become.

We can observe this lifestyle change most concretely by comparing his wife's ledgers for the household before and after the move to Paris.[60] While living at le Mée, she personally kept accounts of cash on hand. There was regular income from her husband's salaries at the Institut, the

Lithograph by Alphonse François of Laplace in senatorial costume, from the author's collection

Bureau des Longitudes, the artillery corps, and from the properties he had inherited from his father. Before 1799, they lived a reasonable life, spending around seven hundred francs a month, quite within their means. Aside from new clothes for the stylish Madame Laplace and for their offspring, they spent normal amounts for lessons for the children, laundry, repairs to their lodgings, gardening, and food. The most "extravagant" expenses were for sugar and coffee.[61] By 1807, in contrast, the expenditures for food and the wages of their cook Demollien and his assistant were over a thousand francs a month, sometimes as much as 1,500 francs.[62]

The entire family was affected. As the son Emile approached adolescence, he was sent off to day school, where he began to prepare himself for the entrance examinations of the Ecole Polytechnique. And Madame Laplace, now in her early thirties with growing children, moved from the quiet countryside near her family to the bustle of the capital. Armed with a commanding social position as the wife of an important government official, she plunged into Parisian society. The couple's social circles included all the relatives of Bonaparte and the new governing elites who eventually formed the Napoleonic aristocracy. The Laplaces rapidly integrated themselves into this culture, thriving in the new atmosphere, which was replete with special art forms, fashions, and social graces.[63] Madame Laplace was well liked, and developed a close relationship with Bonaparte's sister Elisa, whom she would eventually serve as a lady-in-waiting in 1805. The esteemed savant and his charming younger wife cut a fine figure in the salons of Paris, and most likely enjoyed their sudden rise to social prominence.[64] While maintaining his scientific research and his standing as an expert advisor, Laplace also began to take on the bearing of a courtier. Bonaparte knew how to use his services, turning to him for technical advice on several known occasions and basking in the reflected light of his scientific prominence.[65]

Laplace also knew how to use his position for the benefit of science and its practitioners. He repeatedly offered his influence with the government to support the work of his astronomer collaborator Oriani at the Brera Observatory in Milan.[66] He also, with Bonaparte's support, joined in the more dramatic effort to have the geologist Déodat Dolomieu, his colleague, freed from a prison cell in Messina after his capture by English vessels.[67] A striking letter to Sir Joseph Banks, president of the Royal Society, was signed by Laplace on 16 May 1800 asking for safe conduct for Captain Nicolas Baudin's expedition to explore the coast of Australia through waters controlled by the British, and pleading for Dolomieu's release. Both were eventually granted with Banks's intercession. In the letter, Laplace and his colleagues appealed to the universal value of scientific research as transcending national rivalries: "It is especially in the midst of war . . . that friends of humanity should work on its behalf to extend the limits of our knowledge . . . [by launching] enterprises similar to those that have immortalized great navigators and famous scientists of our two nations . . . [The issuance of a safe passage] will renew the marks of our respect for science that our two nations have often furnished."[68]

## Government Support

Shortly after being named vice-chancellor of the Senate, Laplace secured promises from the First Consul for a three-thousand-franc grant to survey the perpendicular to the meridian, extending from Brest to Strasbourg, under the command of the military engineer Maurice Henry.[69] He also initiated in the same year a project to extend the meridian measure as far south as the Balearic Islands.[70] By 1806, Laplace had shepherded this undertaking through the government and had Biot and Arago sent on this expedition. He even urged the British scientists to join in this enterprise by connecting it to their geodetic computations from Dover to Yarmouth, but to no avail.[71] The scientific community recognized Laplace's ability to further its aims. In Baron von Zach's monthly journal, he is continually referred to as "Senator" or "Chancellor" Laplace. Laplace also received praise from von Zach for having the Institut award the Viennese astronomer Bürg a hefty prize subsidized by Bonaparte.[72]

Laplace was not always successful in his encouragement of scientific activities endorsed by the government. The most notable "failure" came with the less-than-full application of the metric-decimal system so dear to his heart. During his short tenure as minister of the interior, he had successfully engineered the decree of 17 frimaire an 8 (10 December 1799), which established the platinum standards for length and weight and mandated the use of the decimal system. As it turned out, however, decimalizing time proved to be insurmountably difficult. Various other features of the original system, too, were amended to make it easier for common people to compare their old measures with the new ones; some of these amendments involved changing the names of units. By 1804, other problems had surfaced. The issue came to Napoleon's attention, and his minister of the interior agreed that new regulations must be decreed to mollify those who were resisting the change to the metric system. Unfortunately for Laplace, who strongly opposed this relaxation of measures, the minister was at the time his scientific colleague and friend Jean-Antoine Chaptal, who had Napoleon's ear as both an effective administrator and a savant. Laplace penned a reproachful letter to Chaptal on 2 February 1804:

> Let me send you, dear friend, some observations on the changes in the system of weights and measures we talked about at the Tuileries. It distressed me to know you had [already] proposed them to the Consuls. I would have preferred it if you had first conferred with me, Berthollet, Delambre, and several other of our friends who have been particularly involved with the matter. I would be pained to see our beautiful metric system altered in its most essential aspects, with the exception of its nomenclature, which, as you know, was not of our doing. I think of it as the most perfect [system] that one can imagine. We should have expected it would encounter difficulties. But I ask you in the name of friendship not to become discouraged . . . Time will eventually remove them, and future generations will be grateful for your care and steadfastness.[73]

Finally, a decree of 12 February 1812, which Laplace was talked into supporting, gave authorities the permission to develop common standards (*mesures usuelles*) that approximated the official metric standards and that allowed subdividing these common standards using a modified duodecimal system. Laplace had written a brief for Napoleon against it, but was not sufficiently persuasive. A decade after his death, his son Emile Laplace chaired the commission that presented a new bill restoring Laplace's wishes.[74] Time was indeed on his side.

## The Société d'Arcueil

It was during this period of his greatest influence that Laplace began to assemble a "school" of disciples. As we noted earlier, he had begun to act as patron for younger or less-established scientists as early as 1785, when he included Delambre in his projects. With his meteoric rise in public prestige, Laplace chose to establish an informal scientific salon, carefully selecting its participants. In part this venture was developed to liberate him and his entourage from the formality that operated in official Parisian social and academic circles. He yearned to promote an informal club of budding scientists who could test their ideas freely and engage in vigorous debate without violating the established norms of behavior. To carry out this plan, he selected a country estate on the outskirts of Paris where he held a "seminar" every Sunday afternoon. The group came to be known as the Société d'Arcueil, named after the village chosen as a haven from the whirlwind of Paris.

Actually it was Madame Laplace who chose the site. Biot, a central participant in this Société, recalled that she negotiated the purchase of the country house in 1806 without having shown it to her husband.[75] The home was modest, much like their abode in le Mée, and it had the winning feature of abutting a property owned by Laplace's close friend Berthollet. The Laplace and Berthollet families had over the years become quite close. Madame Berthollet befriended Madame Laplace, sending her chatty letters while her husband was away on missions with Bonaparte in Italy and Egypt.[76] Each had a son, and Emile Laplace and Amédée Berthollet both had a strong interest in science, no doubt cultivated by their fathers, who worked together at the Institut and the Senate.[77] Moreover, Laplace and Berthollet admired each other's endeavors. Berthollet was engaged in a type of work that Laplace had hailed many years earlier as the most promising activity for the establishment of chemistry as a proper quantitative science.[78] Berthollet had discussed this goal during his Ecole Normale lectures, praising Laplace for his contributions.[79] And in 1803, he published an *Essai de statique chimique* (Essay on Chemical Statics), an expanded version of his "Recherches sur les lois de l'affinité" (1800; Research on the Laws of Affinities), once more thanking Laplace for his inspiration.[80]

Berthollet had tried to recapture the exciting atmosphere of the pre-Revolutionary gatherings held by Lavoisier at the Arsenal by building a chemical laboratory of his own at Arcueil, where he hoped to assemble co-workers. In the 1780s Lavoisier and his colleagues, including quite prominently Berthollet, Laplace, and Monge, had speculated about the laws of affinities that were yet to be tested.[81] As professor of chemistry at the Ecole Polytechique, Berthollet had recruited his best students to assist him in his undertaking, notably Joseph Louis Gay-Lussac and Louis-Jacques Thénard.[82] Now that Laplace had become his neighbor, they were able to develop an even more productive assembly. At first on Thursdays, then later every other week on Sundays during the warmer months, they would invite scientists, young and old, to share a meal and carry on scientific discussions in an informal setting. The secretary of the Royal Society of London, Charles Blagden, was a frequent visitor to Arcueil, bringing news of research in Britain. Alexander von Humboldt, when not traveling in Central America, was also a member, brought in no doubt by his close association with the physical chemist Joseph Louis Gay-Lussac. Chaptal was also closely associated with the group, as well as a promising young mathematician named Augustin Louis Cauchy, whose father owned the property adjacent to those owned by Berthollet and Laplace. Several recent graduates of the Ecole Polytechnique formed the core of the assembly, including scientists who are now household names in history—Arago, Pierre-Louis Dulong, and Etienne Louis Malus—as well as lesser stars in the hall of fame, including Thénard, Hippolyte Victor Descotils, and the sons of Berthollet and Laplace, Amédée and Emile. It was a brilliant assemblage.

The younger generation of scientists, mostly trained at the Ecole Polytechnique, considered these informal sessions to be a valuable postgraduate experience. Not only could they present the results of their new findings and engage in free-spirited discussions, but they also could get feedback from the two wise men of the Institut and their guests. Detailed records of their assemblies do not exist, preventing us from knowing if experiments reported there were carried out on the premises in the same way they had been at Lavoisier's Arsenal laboratory. As was the case at the Arsenal meetings, though, the agenda was set by the hosts. It was in this way that Laplace could focus attention on his own research program, much as he did at the Bureau des Longitudes. The Laplacean school of physics, as it has been called, could flourish under his gaze. In the end, the circle served both the conveners and their new followers.

The name Société d'Arcueil was immortalized when the participants of the gatherings published the first two volumes bearing the fruits of their labors. The preface to these proceedings ends with a reference to Berthollet and Laplace's personal friend, Napoleon, who wanted to be known as a patron of science:

Drawing by Bosio of Laplace in a garden, from the David Smith Collection, Columbia University

> The progress of physics is of great interest because it has as its goal to ascend to the true causes of phenomena, to identify the forces of nature, and to indicate their application to the ingenuity of man. May the zeal of the society win the approval of the august chief of our government! May the peace, which has for a long time been in the heart of this triumphant hero, permit his genius to spread a fruitful influence on the arts and sciences which alone could have established his fame, if the destiny of this world had not been entrusted to him![83]

Once more, politics and science were coupled in a manner that was advantageous for Laplace.

# 9

# Celestial Mechanics

Politics and social notoriety did not totally crowd out Laplace's professional ambitions. For more than a dozen years he had wanted to write an exhaustive monograph about the noblest of human accomplishments, astronomy. Already he had made an indelible mark as a leading contributor, solving some of the most vexing problems of his day. Not only had he shown mathematically how two distant planets mutually attracted each other to preserve the stability of the solar system; he had also recently proposed a suggestive theory for its origins in the popular *Exposition*. The inverse square formula for universal gravitation worked so well that he assumed it was a demonstrated law of nature, to be accepted beyond a shadow of doubt. To write a single treatise that would incorporate the labors of a handful of mathematical astronomers who had followed in the footsteps of Isaac Newton was his fondest desire, one to which his entire career had been devoted. Writing this monograph was to be the crowning achievement of his life's efforts.

A relentless and determined scientist, Laplace achieved his goal by publishing a monumental work over a period of six years. The first two volumes of the nearly 1,500-page *Traité de mécanique céleste* (Treatise on Celestial Mechanics)—or *Mécanique*—those dealing with the theory behind the mechanism of the solar system, came out in September 1799, before he was fully engaged in his political career. The next two volumes, which addressed the application of his theory to positional data collected by observers, followed in 1802 and 1805. Twenty years later, he added a fifth tome covering over four hundred pages, which he considered as a supplement, bringing his readers up to date with his latest research. This fifth volume was also of particular significance to mathematical physicists; it contained new studies that were both a departure and an extension of his original intent.

## Planning the *Mécanique Céleste*

Laplace had first indicated he was working on a major "treatise on physical astronomy" to his Genevan friend Deluc as early as March 1785, without hazarding a guess when it would be completed.[1] His privately printed *Théorie du mouvement et de la figure elliptique des planetes*, which appeared in 1784, had been a mere teaser for what he had in mind. At the Collège Royal, his colleague, the academician Cousin, had offered a series of lectures on this topic starting in 1767, which eventually appeared in 1787 as a substantial monograph entitled *Introduction à l'étude de l'astronomie physique*, preparing the ground for Laplace's more ambitious project.[2] The following year his old friend and new colleague in Paris, Lagrange, published his milestone treatise, the *Méchanique analitique*, which far surpassed any other contributions to theoretical mechanics since Euler and d'Alembert had penned their works. Laplace was impressed, and it was this work that prodded him to develop his expressed desire into a carefully crafted major work. From Lagrange, Laplace adopted the word "méchanique," which he substituted for the term "astronomie physique." If ever there was a contemporary scientist worth emulating, it was Lagrange, whose accomplishments were of the very high level to which Laplace aspired. The praise lavished on this Piedmontese star of the scientific heaven in Paris, as penned by Legendre and co-signed by Laplace in their report for the Académie des Sciences, was well deserved.[3] It was a testimony of genuine appreciation, which probably caused a twinge of jealousy for Laplace.

In July 1789, Laplace still referred to a large work on "physical astronomy" as his main preoccupation. Realistically, though, he indicated that the effort was of such magnitude that "it would take considerable time" to complete.[4] Not until 1797 did Laplace begin to speak of the "Mécanique céleste" as the title for his treatise.[5] A budding twenty-two-year-old Genevan astronomer named Jean Frédéric Théodore Maurice, who had come to Paris to seek guidance from Lagrange and Laplace, reported his conversation with his Parisian contacts to his father at home. Word quickly passed on to their elderly mentor, Le Sage, who was anxious for Laplace to embrace publicly his speculation about the mechanical cause of gravity, and who urged him by mail to abandon the term "mécanique" in favor of "dynamique," which would force Laplace to take a stand on this unproven hypothesis.[6] Like many others, Laplace had already dodged the matter, convinced that no experiments could test this ingenious concept. He diplomatically explained to Le Sage that his aim was merely to write of universal gravitation without considering its causes. In a few sentences, he expounded his "positivist" philosophy, couching it impersonally in the third person: "Other philosophers on the contrary admit their ignorance about the nature of matter, space, force, and extension, and worry little about first causes, seeing in attraction only a general phenomenon that, when subjected to rigorous calculations, yields a complete explanation of all celestial phenomena and a means of perfecting the tables and theories of the motion of heavenly objects. It is uniquely from this standpoint that I have considered attraction in my work."[7] Le Sage did not return to his demand in subsequent correspondence.

The publisher's contract with Duprat, signed in August 1798, specifies a work in four volumes, the first two of which are to be sold as a set, and printed with a maximum run of 1,500 copies for the set.[8] No royalties were specified. Printing of the first set began in October, and already Laplace was planning for the second. He enlisted his astronomy colleagues to supply him with the latest data on all the planets, from Mercury outward, in order to determine if perturbations due to cometary or other sidereal effects should amend his equations derived from theory. Sensitive to the need for reliable empirical data, he recruited astronomers and calculators doing work independently of each other. In Paris, he continued to call on his faithful collaborator Delambre, who was preparing more tables. A young astronomer sent by the Duchess of Saxe-Gotha, Johann Karl Burckhardt, was set to work for him on the Jupiter/Saturn calculations. And in Milan, Laplace asked Oriani to assist him with his data.[9] All complied with their senior patron's wishes, appreciating the magnitude of the enterprise and expecting that they would be recognized publicly as participants. Laplace pressed into service the young Maurice to assist him with the mathematical formulae in the first two volumes as a training exercise for the budding scientist.[10] Laplace also sought out Biot, a former star student at the Ecole Polytechnique now employed as a mathematics instructor in Beauvais, to read proofs for the first part. He mailed Biot the first batch in April 1799, asking him to prepare summaries of each chapter to be placed in the table of contents.[11] Ultimately Biot published an extended synopsis of the work under his own name, presumably with the approval of Laplace and with the same publisher.[12]

The third volume was in press in April 1802 when Laplace notified Oriani that it had been delayed by the tedious rechecking of details and calculations needed to bring lunar theory in line with the observational data. In his missive, Laplace commiserated with his latest astronomical assistant, Bouvard, who had also toiled for a year and a half to rectify a new formula for the Jupiter/Saturn interaction.[13] For these volumes, Laplace prevailed upon another graduate of Polytechnique of great promise, Siméon Denis Poisson, to read proofs and to verify the calculations.[14]

The fourth volume was finally ready in May 1805, and marked the completion of the original agreement with the publisher Duprat. Each section had been constructed as part of a general plan that Laplace had in mind years earlier, and that he had executed with the assistance of a small army of observers, calculators, and proofreaders who were either acknowledged in the text or benefited from Laplace's patronage. He orchestrated the monumental work as a collective enterprise of which he was clearly the originator, conductor, and executor, and for which he received immediate and extensive praise. Both the scientific community and the reading public acknowledged the magnitude of the enterprise, not merely in France but abroad as well. There was an immediate translation of the *Mécanique* in German by Burckhardt, who was temporarily residing in Paris.[15] The English translations, by John Toplis, Henry Harte, and Nathaniel Bowditch, did not appear until some dozen years later, most likely because that scientific audience was not sufficiently familiar with the Continental calculus notation.[16] Nonetheless, Laplace's work had been reviewed at length in the 1808 *Edinburgh Review,* receiving high praise for its synthetic and technical character.[17]

## Organizing the *Mécanique Céleste*

The *Mécanique* was a thoroughly mathematical treatise written for advanced specialists. Some complained that Laplace often short-circuited proofs and demonstrations that were difficult even for students well trained in mathematics to follow, and most of the English translators found it necessary to add explanatory notes to assist the reader.[18] The American translator Bowditch, for example, offered extended footnotes that make explicit the compact mathematical procedures used by Laplace. Successive editions of the *Exposition* were more tailored for the layman uninterested in and unable to follow these mathematical details, but Laplace's fundamental message in both works was the same: Newton's principle of gravitation was the basic true law of nature that fully governed the solar system.

Laplace must have struggled with the organization of the *Mécanique*. Although he divided the work between a theoretical discussion of positional astronomy and its applications to the "system of the world," as Newton had, he did not follow the Euclidean scheme of starting with definitions, axioms, and laws, thereafter deriving propositions from them. The intervening century had rendered this geometrical mode of presentation obsolete, in favor of an analytic presentation of the sort deployed by Euler, d'Alembert, and, above all, Lagrange. Laplace sketched a structure for the *Mécanique* in an undated draft that may have been conceived at the time he was working on the *Théorie du mouvement et de la figure elliptique des planetes*. He began with a familiar theme: "I am proposing in this work to show how all celestial phenomena derive from the principle of universal gravitation."[19] At this early stage of planning, he intended to have his work divided into eight sections, the first of which—a "succinct exposition of the system of the world as observation reveals it"—was eventually adopted for the *Exposition* but left out of the *Mécanique*. The next section was to establish general principles of equilibrium and motion "with the aid of these principles and observed phenomena."[20] This plan corresponds partially to book 1 of the *Mécanique*, which reads like what one might expect from a general treatise of mechanics. Nonetheless it is evident that Laplace wished to broach only those general issues needed to prepare the ground for solving problems particular to the solar system. Unlike Lagrange's *Méchanique analitique*, this work was designed for resolving issues likely to occur in astronomy, which suggests that Laplace was not at this point concerned with expanding its scope to include issues in classical terrestrial physics. In book 1 of the 1799 printed version, for example, he extended the general principles to deal with the equilibrium and motion of liquids because they would be needed later to attack the shape of the earth and tidal phenomena—but he left any extended discussion of empirical observations for volumes 3 and 4. On a few occasions, such as in chapter 7 (which addressed the motion of solid bodies), he deviated from this general pattern by considering the special case in which the solid body happens to be a planet rotating on its axis at the same time as it is moving around the sun.[21]

The next section in his draft statement corresponded to the heart of his convictions about the solar system. "In the third book, I turn to the law of universal gravitation and offer the differential equations for a system of bodies operating under its influence."[22] He followed this scheme in 1799 by devoting book 2 to the proposition that this law of nature is sanctioned by the evidence of astronomical phenomena. Referring readers to the details in his *Exposition*, he developed the now classical arguments about the elliptical motion of planets around the sun, derived Kepler's laws, and applied them to satellites—though he stopped with the earth's satellite, which required additional explanations to match observations. Here as earlier, he deviated from a strict adherence to a general theory, in favor of a theory devised for a special purpose. Notwithstanding, Laplace returned to his idée fixe, claiming that the Newtonian law operates with equal success for terrestrial as well as celestial matters. He reasserts his firm conviction that

> The same law applies on earth. It has been verified by very precise experiments with the pendulum. Discounting air resistance, all bodies would fall with equal speed to the center of the earth. Hence bodies on earth weigh in proportion to their masses, just as planets do with respect to the sun, and satellites with respect to their home planet. This conformity of nature with itself, on the earth and in the heavens, shows in the most striking manner that the gravitation manifest on earth is but a particular case of a general law for the universe.[23]

As one reads further in the *Mécanique,* one finds that Laplace's analysis becomes increasingly tied to astronomical problems he had resolved in the preceding twenty-five years. He touches on all the issues close to his research, from elliptical motions of planets and satellites, to cometary orbits, to the shape of the earth, to secular inequalities stemming from perturbations, to invariant planes, and so on. The *Mécanique* became the vehicle for a synthetic review of his quarter-century-long research experience, reorganized in a systematic and rational order. Along the way,

Laplace could not resist exclaiming how well his theoretical approach fit with empirical data recently uncovered—for example, in the much-heralded measurement of the meridian passing through Paris. He repeatedly underscores his delight at finding a tight fit between his formulation of theory and observed phenomena.

## Celestial Theory

The first two volumes of the *Mécanique* have usually been taken as a landmark in the history of astronomy, summarizing its theoretical bases in the eighteenth century, as well as establishing a model for future progress. Both these judgments have considerable warrant, but such generalizations serve to obscure facets of Laplace's actual role in ongoing developments. A closer analysis reveals that while he drew generously and appropriately on a handful of immediate predecessors and colleagues to fashion his synthesis, he shaped it to suit his distinctive mark as a scientist. Likewise, the individual stamp he placed on theoretical astronomy in 1799 left a mixed legacy that is not easily grasped in the midst of the excessive fame he has generally acquired. Like so many important scientists, he needs to be seen both as an integral part of tradition and as an innovator, some of whose approaches quickly became obsolete.

In 1787, the Paris academicians Bossut and Legendre expressed their gratification in claiming that progress since Newton's system had largely been accomplished in France. In their report for the Académie on Cousin's *Introduction à l'étude de l'astronomie physique,* they stated: "We cannot refrain from noting, along with Mr. Cousin, that the great progress made in physical astronomy is indebted principally to the Académie des Sciences, through its prize competitions. If France did not first lay the building blocks of the system of the world, it can take pride in having contributed more than any other nation to the elaboration of the system in all its aspects, and in establishing its truth in an undeniable manner."[24]

While Laplace agreed with this assessment, he was quite willing to acknowledge the theoretical contributions made by foreigners like Euler and Colin Maclaurin, and esteemed all the considerable observations made by British and Continental astronomers alike. As for Lagrange, whose main innovations were produced while living in Berlin, he was now clearly being counted as a Parisian colleague. Yet Laplace borrowed indiscriminately from all scientists, whose writings he commanded regardless of where they were published. Among his remaining papers, for instance, there is an accurate list of mathematical papers from the St. Petersburg Academy.[25] He was equally familiar with the *Memoirs* of the Turin and Berlin academies, and the *Transactions* of the Royal Society, even though he was not fluent in English. The twenty-two pages of references to other scientists in the index of his complete works testifies to his reliance on this literature, and the *Mécanique* shows he knew how to cull the best ideas from this rich panoply of Newtonian followers.[26] Proud of French accomplishments as he was, Laplace was eclectic and cosmopolitan in his outlook.

Laplace's earliest and most continuing debts were to his patron d'Alembert, whose writings he knew intimately, even while at times he disagreed with the master. In Laplace's earliest paper dealing with astronomy, he looked to d'Alembert's essay published in the year of his birth, the "Recherches sur la précession des équinoxes, et sur la nutation de l'axe de la terre, dans le système de Newton" (Research on the Precession of the Equinox and the Nutation of the Earth's Axis in the Newtonian System).[27] The introduction to this monograph was a masterpiece of descriptive and philosophical reflections on the issues facing the modern astronomer, and the views expressed are clearly echoed in Laplace's early papers.[28] By the time he penned the *Mécanique*, then, Laplace was focused on the technical details that had been brought to light by d'Alembert, but much improved by Euler's *Theoria motus corporum solidorum* (1765; Theory of the Motion of Solid Bodies). With respect to the nutation of the earth's axis and the precession of the equinoxes, Euler had also corrected and generalized d'Alembert's equations in a series of articles published in the 1750s in the *Memoirs* of the Berlin Academy.[29] What Laplace was able to add in the *Mécanique* was the concept of an invariant plane about which planetary orbits oscillate, taken as a frame of reference, that he linked mathematically to the sum of angular moments of the planets of the solar system. When combined with the conservation of areas swept by radii vectors, this technique yields a function that is equivalent to the principle of least action. In the space of three dense pages in chapter 5 of book 1, Laplace managed to generalize a problem that had dogged his predecessors for a half-century by introducing this new and fruitful concept.[30] By refocusing on this old and central problem in celestial mechanics, Laplace was also able to make new advances thanks to events that had transpired since the Revolution began. The precession of the equinoxes and the nutation of the earth's axis were dependent on the shape of the earth and linked with his ideas about explaining tidal motions, for which new data had been supplied since 1789. In turn, these discoveries offered the possibility of explaining better the anomalies of the moon's motion, which he had not earlier been able to resolve to his satisfaction. Because of Laplace's conviction that the law of universal gravitation was sufficient to account for all the known phenomena in the solar system, he was led to extend and amend his earlier conclusions written before 1789. All of these special problems had to be explored as he was preparing and composing the *Mécanique*.

## Using Astronomical Data

It was fortunate that Laplace had sponsored Delambre's rise to prominence in the years before the Revolution. Delambre had supplied much data to his mentor, and continued to collaborate with him by publishing tables of Jupiter and Saturn's motions based on the elements drawn from Laplace's new theories.[31] Their close professional bonds were cemented when Delambre was assigned the task of measuring the Dunkirk-Barcelona meridian in order to establish a unit of length taken from nature, a project that the Académie had cleverly instigated in 1790, and which began in earnest in 1792.[32] In spite of the personal difficulties faced by Delambre and his partner Méchain during the Terror, the two were able to obtain a more accurate measure of the shape of the earth, which was required by Laplace for his master work. Laplace stayed in contact with Delambre throughout the difficult period of the Revolution, and benefited from the improved data, which he exploited to modify his equations in book 3 of the *Mécanique*. In turn these equations were factored into the rest of the theory of the earth's "wobble," which was a fundamental factor in the complex calculations of the solar system. As secretary of the new Bureau des Longitudes, Delambre continued to supply his colleague with data, either those collected by him or by Laplace's newest collaborators, Bouvard, Burckhardt, and Bürg, all of whom were associated with the Bureau. Eventually, when Delambre became "perpetual secretary" of the renewed Academy of Sciences, he transformed his role from that of subordinate to that of equal partner. In his new capacity, he continued to assist Laplace, for example by cooperating with the convening of an international meeting to establish the metric system, and by requesting new tidal data from Brest to confirm the new Laplacian equations.

Increasingly, as each chapter of the *Mécanique* progressed, Laplace devoted more attention to empirical data—not because he was embracing empiricism as such, but because he recognized that his theory needed to be continually adjusted to match physical reality. Even in the solar system, the complexity of gravitational forces acting on each planet required slight modifications from the initial simple model derived from a general theory of motion in order to represent the true system of the world. The triumph of his version of physical astronomy consisted in showing that astronomy was no longer a mere science of observation, but one based on immutable laws whose astronomical elements were teased out in the master work. He said as much in a letter written to his friend Deluc at the height of the Revolution: "I am sending you a copy of my theory of the satellites of Jupiter . . . The new tables of Jupiter's satellites . . . prepared by Delambre are based on this theory . . . Finally [this branch of astronomy] has been removed from empiricism and founded upon the law of universal attraction, toward which I was aiming."[33] Laplace's *Mécanique* spelled the complete victory of theory over brute data, generalized in empirical formulae. It remained the driving force behind the two other volumes of the monumental work despite his immense concern with details of observation. The paradox of Laplace's approach to natural philosophy was his strong theoretical bent existing side by side with his devotion to the empirically recorded phenomena of nature. Convinced that one could never know with certainty the essence of nature, he was nonetheless deeply committed to the evidence provided by the senses. That evidence had to be measured with increasing precision, taking into account all the human and natural errors that observation entailed.

As Charles Gillispie has noted, Laplace began to turn his attention more and more to instrumental issues that he had never focused on before, though they had been on his mind for a long time.[34] One of the classic issues for observers was the effect of the earth's atmosphere on the refraction of light. Most observational astronomers since the sixteenth century had tried to take it into account. Laplace, however, treated the issue in a more general fashion, linking refractive indices in both terrestrial and celestial situations. To do so, he relied at first on the work on meteorology of Deluc, with whom he shared his renewed concern with this phenomenon. In February 1793, he wrote, "For some time now I have been doing research on [atmospheric] refraction because of the Academy's project on fixing a universal measure [of length]. It has led me to reread your *Modifications sur l'atmosphère,* which I consider one of the works that eminently graces physics."[35] Laplace, fully occupied by the task of calculating the height of a volcano in the Canary Islands, relied on Deluc's empirical equations to correct its value as well as determine the gravitational pull near the equator. All these small adjustments due to water vapors saturated in the atmosphere affected geodetic calculations, and became even more significant for observations in the heavens. He was still puzzled by all this in 1797 when he wrote to Delambre:

> The law of the variation of the density of air as a function of temperature, is it known exactly? At equal densities of the air, is heat not a factor that affects refraction? Is the hygrometric state of the air not a factor that alters it? Here are a set of questions for which we do not have satisfactory answers. Since we are dealing only with very small quantities, I think we need to be concerned with the minor equation for nutation and precession, which we have neglected to this point.[36]

Laplace sorted all this out in a few years, but held back his conclusions until the last book of volume 4 of the *Mécanique,* in which he developed an equation relating the refractive index as a function of temperature, pressure, and humidity.[37] By the time he published these results in 1805, he had been able to recruit Humboldt and Gay-Lussac to obtain samples of air high above sea level to test the variation of gas content at five to six hundred meters above the earth, and to calibrate thermometers used to establish temperatures at various altitudes. He also recruited the renowned mountaineer Louis François Ramond de Carbonnières to discover the proper empirical coefficient for the Laplacian formula relating hygrometer readings to the height of mountains and to atmospheric pressure.[38] By this point, he had abandoned using Deluc's formulas, which were insufficiently accurate for his purposes. In order to develop better ones, he was able to convince younger scientists like Gay-Lussac and Ramond to assist him, much as he had earlier enlisted Delambre, Bouvard, and Burckhardt to carry out the tedious observations and calculations he needed to plug into his theoretical apparatus. The overarching significance of his magnum opus was sufficient to excite younger men to participate in his gigantic enterprise. While all this was done before the establishment of the Société d'Arcueil, the pattern presaged his appeal to budding scientists, whom he enlisted in his cause and then supported in their careers.

## Contributions from Others

The *Mécanique* abounds with issues central to astronomers, many of which were stimulated by classical problems that others before Laplace had broached and that he found productive to reexamine. Laplace kept fully abreast of current research. Even at the height of the Revolution, when commerce with England was difficult, he continued to ask about the work of British astronomers through his contacts with Deluc or the secretary of the Royal Society, Charles Blagden. He would send offprints of his work to the prominent British astronomers Maskelyne and Herschel, and include sentences that a censor might have found disloyal if his letters were intercepted. In one written early in 1793, he laments: "I regret not being able to see you and several colleagues from the Royal Society. A trip to England has always been my ambition, but given my wife and two children who make my happiness, I do not know when this will be possible . . . I end this letter by congratulating you on the good fortune you have in being able to cultivate the sciences in a tranquil atmosphere, in the service of a Prince and Princess who appreciate you."[39] He was wise to interrupt his correspondence with France's enemy until four years later, but surely must have kept up with the *Philosophical Transactions*. The work of Herschel was of particular interest because this astronomer had made important observations of the satellites of Saturn and discovered two and possibly more satellites to his Georgian planet, Uranus. Laplace was later to use Herschel's observations of nebular formation to support the conjecture he had made in the *Exposition* about the origins of the solar system. The full use of data about satellites, including those around Jupiter, was explained in the lengthy and detailed mathematical book 8 of the *Mécanique*.[40] In the preface, Laplace admits the difficulty of developing a complete theory, given that the masses of the satellites are difficult to ascertain accurately, and acknowledges that further observations are required to verify his values. In it he also gives full credit to Delambre and Bouvard for their arduous calculations needed to carry out his program.

The issue that Laplace found most troubling, and to which he returned several times during his career without ever resolving it satisfactorily, was the secular equation of the earth's single satellite: the moon. D'Alembert, Euler, and Lagrange had each struggled to find the cause of this anomaly. Laplace's earliest supposition was that it could be accounted for either by assuming a finite speed for the propagation of the gravitational force or by the friction of trade winds and tidal phenomena slowing the earth's diurnal motion and speeding up the moon's. He dismissed these conjectures in favor of the oscillation of the eccentricity of the earth's orbit.[41] On 19 December 1787, he had presented an abbreviated version of this idea, elaborated and revised a year later.[42] Still not satisfied, he sought better data of eclipses recorded in earlier years, directing Bouvard to study tenth-century observations of Ibn Yunus recorded in a manuscript held at the University of Leiden.[43] For this purpose, and with the assistance of the physicist Van Swinden, who lived in the conquered Lowlands, the volume in question was lent to the Institut National and translated by the professor of Arabic at the Collège de France, Jean Baptiste Jacques Antoine Caussin de Perceval.[44] This intense collaboration yielded data that Laplace used to correct values of the moon's path that needed to be accounted for by theory.[45] Nonetheless, only a portion of the moon's secular acceleration was explained. Laplace's solution was vigorously challenged by Giovanni Plana and Francesco Carlini in 1820, and defended successfully by Laplace's disciple Théodore Damoiseau.[46] Another Laplacean recruit, Gustave de Pontécoulant, continued to argue for Laplace's approach, but the problem was not better treated until the brilliant work of the mathematician and philosopher Henri Poincaré, conducted at the end of the nineteenth century.[47] Laplace returned to the same issue in 1809, 1811, and 1820. Finally, as what Bowditch called "the last gift of Laplace to astronomy," he adopted a new formula, which was presented posthumously in 1827 and published in the *Connaissance des temps* for 1830.[48]

Laplace was more satisfied with his treatment of comets, a subject he had had in mind since his sharp dispute with Bošković in 1776. The problem of calculating cometary orbits seemed at first glance simpler, because perturbations caused by planetary attraction were the main factor to consider once the basic orbital elements had been established. Unfortunately, there were very few examples of comets that return periodically, so it was very difficult to establish their basic elliptical orbit. To Laplace's irritation, he could not treat comets in the same general fashion that he had used to consider planets and their satellites. Halley's comet of 1759 was one of the few available, for which he offered a sensible path to resolving its elements. Another comet, first observed by the academician Messier in 1770 and known as Lexell's comet, had been greatly perturbed by passing close to Jupiter, which modified its path. Laplace recognized that each case had to be approached individually, and he offered these two as examples. Using a method necessary for the calculation of each comet's orbit proposed by Lagrange in a prize essay of 1778, and frustrated by the difficulty of a complete analytical integration of these equations, he turned to numerical integrations. Laplace proposed a new technique using generating functions that permitted finding approximate values.[49]

It was characteristic of Laplace's approach to invent a mathematically reasonable—even if cumbersome—way to skirt the issue by proposing a method leading to a practical solution. Though inelegant, it satisfied his desire to solve the astronomical problem at hand. Without even attempting an explanation, he shied away from discussing the foundations of his technique in order to show an end result. Mathematics was, as he once said, an "intellectual tool," similar in its usefulness to observational instruments.[50] The oft-quoted observation that Lagrange was the superior mathematical theorist, whereas Laplace preferred the productive manipulation of these tools, fully applies.

Although Laplace seemed to dominate the entire field of solar astronomy, one important novelty escaped his grasp, somewhat to his annoyance. In the midst of the preparation of volumes three and four, a reputable observational astronomer and director of the Palermo Observatory, Giuseppe Piazzi, detected what was later recognized to be the first known asteroid, Ceres.[51] Normally, this celestial discovery should have been fodder for his formula to calculate cometary orbits. Piazzi, who had studied in Paris with Lalande, communicated the observations privately to Lalande, Oriani in Milan, and the Baron von Zach in Gotha-Seeberg. A young mathematician named Carl Friedrich Gauss, who befriended von Zach, calculated the orbit first, devising a method of his own using error theory and approximating values, and providing data that confirmed it was a planetary body occupying a space between the orbits of Mars and Jupiter. Another asteroid, Pallas, was discovered shortly thereafter. The entire category of asteroids is not treated by Laplace, suggesting he was miffed that a new discovery could not immediately be lodged in his "domain." In June 1801, von Zach had teased him by pointing out that "the new body is occupying all the astronomers of Germany."⁵² It was a veiled challenge. Laplace answered testily, chiding Piazzi for having kept his discovery private. He said, "Piazzi is not to be excused for having initially concealed it from astronomers."⁵³ Surely Laplace was irked for not have been contacted first, and to have been scooped by a young German scientist. Very quickly, however, Laplace recovered his composure and saluted the new method and its brilliant author. Gauss made his triumphant entry into theoretical astronomy, publishing a refined version of his method in 1809 as the *Theoria motus corporum coelestium* (Theory of the Motion of Heavenly Bodies). Shortly thereafter, Laplace began to respect and praise this new star of the astronomical sciences.

## Disciples

Very few of the major Enlightenment astronomers who had turned the Newtonian program into a viable mathematical science were still alive when Laplace finished the fourth volume of his *Mécanique*. Among that select group, few other than Bossut, Lagrange, and Legendre fully understood the immense labor the writing of this work had entailed. By the sheer magnitude of his labors, Laplace now had attracted a younger generation of scientists to his project, and they began to use mathematical techniques he had devised as their new starting point for mathematical astronomy. Those trained at the newly created Ecole Polytechnique were mostly interested in the details of theoretical mechanics, a subject taught there first by Prony and thereafter by Lagrange. The first to hitch their wagon to Laplace's engine were the two young proofreaders of the *Mécanique*, Biot and Poisson, whose brilliant careers were shaped by the experience.⁵⁴ After his detailed analysis of the *Mécanique*, Biot, who was named to the Collège de France in 1800, wrote an important foundation text that summarized the advances of his day and was explicitly based on Laplace's treatises.⁵⁵ Later, as Laplace shifted his research to terrestrial physics, Biot followed the leader and actively participated in the Société d'Arcueil. He remained loyal to Laplace his entire life, writing a tribute to his mentor as late as 1850.⁵⁶ Poisson also took a similar path, though he was equally intrigued by the brilliance of Lagrange, who taught him analysis. With Laplace, he developed a personal bond, probably dating from the time Poisson was studying at the école centrale in Fontainebleau, sharing the same inspirational teacher, A. L. Billy, who later tutored Laplace's son in le Mée.[57] Though eight years older, Poisson befriended Emile Laplace, with whom he kept up an association for years after the death of his father. Like Biot, Poisson was an active member of the Arcueil group, and like him he moved from mechanics to terrestrial physics, and even to probability theory, in which Laplace excelled as well.[58]

Laplace never taught at the Ecole Polytechnique, but he was closely associated with its curriculum, first as an examiner and then as a member of its council.[59] He published several short pieces in its publication, the *Journal de l'Ecole Polytechnique,* and his lectures given at the Ecole Normale were printed there in 1800.[60] A few years later, Poisson, who had become a professor at Polytechnique by 1808, also published in the *Journal,* extending and refining parts of the *Mécanique* by using more sophisticated mathematical techniques, but without any implied criticism of his mentor. Stimulated by Poisson's innovations, both Laplace and Lagrange offered further refinements.[61] The lecturer in physics, a somewhat less talented mathematician than either Biot or Poisson, Jean Henri Hassenfratz, was actually the first to lecture at the Polytechnique on astronomy, eventually putting into print a paraphrase of the *Exposition*.[62] Even though astronomy was not often specifically taught at the Ecole Polytechnique, the *Mécanique* became a model for the manner in which the cadre of engineers trained in science approached all physical problems, including practical applications.[63] During the ensuing years, this analytic turn of mind, often labeled "Cartesian," became the characteristic mark of French specialists.

Other followers began to emerge once the new Faculty of Sciences of the University was founded in 1808. There Biot was called upon to teach astronomy, and Poisson, mechanics. Among those who selected topics close to Laplace's concerns were Louis Pierre Marie Bourdon and Louis Etienne Lefébure de Fourcy.[64] De Fourcy's thesis on the attractive force of the spheroid figure of planets is filled with references to Laplace's work, giving due attention to others, like d'Alembert, Lagrange, and Legendre, who had made significant contributions. A young Genevan named Alfred Gautier, stimulated to work in astronomy by the example of Maurice, who had remained close to Laplace, also enrolled at the Paris Faculty. In his two-part thesis submitted in 1817, he reviewed Laplace's work on the moon's motion, also assessing in detail the role assumed by others. Most laborious was his treatment of the theory of planetary motions, which ends with a paean to the work of Laplace:

> The idea of bringing together in a single work the theories of all the planets . . . and to determine their motions pushing precision further than had ever been done seems at first colossal. The execution of this work, which is only a part of what Mr. Laplace has done, already seems superior to the capacities of a single individual . . . The gathering of all these labors, the generality, richness, originality, and profundity that rule the analytic methods and formulae, the importance and multiplicity of applications, their practical utility, the exactness of the tables that derive from it, insure to the illustrious author of this vast enterprise the gratitude of the learned world, the admiration of future ages, and must place the *Mécanique* in the ranks of the most imposing and durable monuments of the human mind.[65]

## Critics

Imposing as the work was, it was to be less than permanently durable. Even those who admired the monument could be critical. Most of the translators and paraphrasers, for example, found the work to be cryptic in places. And one extremist commentator challenged Laplace's entire approach to mathematics, urging a wholesale rejection of the *Mécanique*. The Polish refugee who taught mathematics in Marseilles, Józef Maria Hoëné-Wroński, launched a series of philosophical attacks against both Lagrange and Laplace, which barely received a hearing at the Institut.[66] The scientific establishment turned a deaf ear, and he became increasingly provocative and obscure, so that quickly he was completely marginalized. His writings, couched in Kantian language laced with a political message, eventually turned him into a troubled messianic philosopher.[67] More effective was the challenge raised by James Ivory in 1809 and 1812 from across the English Channel regarding Laplace's treatment of the ellipsoid shape of spheroids and their attractive powers. Gauss used a technique analogous to Ivory's to discuss a similar issue in 1812, initially without knowing the work of his English counterpart.[68] Though at first unimpressed by Ivory's contention, Laplace grew to appreciate his work, and by 1824 was praising him in a letter to the president of the Royal Society, Sir Humphry Davy, saying that "Ivory is without doubt one of our best mathematicians."[69] The papers by Ivory and Gauss led to an extended discussion that involved other Parisian scientists, chief among them Poisson, who tended to support Laplace's approach.[70] The discussion started by Ivory in particular was clearly a significant addition to Laplace's work, and was treated by the scientific community in a standard manner.

Not so with the quarrel with Plana over the moon's motion.[71] In 1818, Laplace had arranged for the Academy to offer a prize to the scientist who could improve the method of calculating the moon's position on the basis of the theory of universal gravitation—in part because he expected that one of his followers, Damoiseau, would be rewarded. It turned out that Francesco Carlini (an associate of Oriani) and Plana were already at work developing a better lunar table. Plana, director of the Turin Observatory and at one time a student of Poisson at the Ecole Polytechnique, decided to submit a joint essay for the prize with Carlini. Despite having been declared a winner, Plana was miffed because the award was also shared by Damoiseau, who had strictly followed the mathematical method earlier proposed by Laplace. Carlini and Plana had some serious reservations about Laplace's approach, and disagreements between Laplace and Plana surfaced publicly. Privately, too, resentments circulated about Laplace's alleged unfair treatment of critics who dared to deviate from the Laplacean mold. Baron von Zach, who in 1799 had spoken deferentially of Laplace as "the Newton of our era," had over the years grown more critical of the astronomer's character, in particular, his handling of foreign astronomers.[72] Laplace's colleague Legendre, whose ego had for years been bruised by Laplace's scant treatment of his work, shared a low opinion of the arrogant "Marquis" with Plana. In a letter of 1826, Legendre congratulated Plana for his courage in "proving our immortal [colleague] to be mistaken," and ironically asked if such criticism was not likely to "trouble his repose and bring about an agitation that might have dire consequences."[73] It turned out that Plana's other challenge to the *Mécanique*'s author with respect to the Jupiter/Saturn calculations, which was much publicized, was resolved shortly after Laplace's death by the discovery that Plana had inadvertently erred, reversing the sign of parts of an equation. Thus the self-assured "Marquis" had the last word, to the delight of his partisans.

A much more serious criticism was launched in the Académie in March 1828, one year after Laplace's death, by Louis Poinsot, who pointed to Laplace's limited calculations used to assert the invariable plane of the solar system. That issue was raised in the context of Poinsot and Cauchy's new ideas about linear moments in mechanics, and these quickly drew a sharp reply by Poisson, who tried to align the new findings with Laplace's work.[74] Like the discussions originated by Carlini and Plana, Poinsot's remarks and the rejoinders by Laplace's loyal allies show that research he had initiated on celestial mechanics was still the staple astronomical currency for dozens of years after publication of the *Mécanique*.

Major changes in Laplace's treatment of astronomical work came shortly thereafter, with the introduction of a host of new mathematical tools, such as Bessel functions and Hamiltonians, that greatly simplified the application of mechanical theory to positional calculations. Thus the "monument" that Gautier referred to earlier was clearly being reshaped by the generations that succeeded Laplace, as it always is in the ever-moving scene of the history of science. Successive French authors like François Félix Tisserand and Poincaré continued to hail Laplace's accomplishments, even while helping to transform the original *Mécanique* into a new "modern" version.[75] Throughout the nineteenth century, Laplace's initial commitment to determinism, the stability of the solar system, and the law of universal gravitation continued to hold sway.

## Terrestrial Physics

Confident as he had become about the fundamental principles he followed in astronomy, Laplace chose to revisit his pre-Revolutionary concerns with terrestrial physics, this time assisted by a host of young partisans. Discussions of these issues came to constitute what has aptly been called the Laplacean program in physics.[76] It was bolder than the astronomical edifice he had synthesized, but more fragmentary, and in the long run more provisional. Whereas he was able to encompass most of the central features of the solar system and build a durable structure for positional astronomy, he could not easily attain the same comprehensive model for physics. In part, this shortcoming was due to the historical situation. Newton had essentially transformed the study of the heavens with his laws of mechanics and the principle of gravitation, whereas no single figure had yet produced a comparable system of explanation for all terrestrial phenomena. It might be presumptuous for Laplace to think of himself as "the Newton of physics," but the notion must have crossed his mind. What other motive would lead him to devote so much of the last twenty years of his life to producing and initiating experimental and theoretical structures for this set of earthly phenomena? Some of these studies, after appearing elsewhere, were folded into the fifth volume of the *Mécanique,* ultimately published in 1825. Others were issued under the names of his younger colleagues, although they clearly had been stimulated and approved by the master.

The turn from astronomy came just as Laplace was completing work for his fourth volume in 1802. Laplace wanted a better handle on the laws governing the refraction of light through the earth's atmosphere, a problem involving issues that had principally been treated by meteorologists. Various chemical and physical problems had to be addressed, including the content of the gases in the air, its density under various pressures and temperatures, humidity, and of course the nature of light itself. The complexity of these issues was also very much on the mind of his friend Berthollet, who was composing a new theoretical basis for understanding chemical reactions. In the *Essai de statique chimique,* which appeared in 1803, Berthollet explicitly echoed Laplace's approach for solar astronomy. He pointed out that "the forces that produce chemical phenomena are all derived from the mutual attraction of bodies, to which the name affinity has been given in order to distinguish it from astronomical attraction. It is likely that the one and the other constitute a single property, but . . . chemical affinities are subjected to special conditions that . . . cannot be derived from a general principle."[77] He went on to explain that his treatise would deal with these conditions in order to ascertain the general laws of chemical activity. These were the very laws that Laplace was grappling with in treating atmospheric refraction. Berthollet discusses the role of heat, light, electricity, solubility, and mass, paying special attention to gases. In writing his treatise, Berthollet consulted his colleagues—in particular Laplace, who contributed two extensive notes dealing with Mariotte's law.[78]

During the next year another, more elementary, general treatise was published in which the contribution of Laplace is also acknowledged.[79] The book was a departure from traditional texts. Haüy, the academician whose work on crystallography Laplace had hailed in 1784, was commissioned by Bonaparte to prepare a textbook in physics for use in secondary schools, the new lycées. It was written assuming correctly that students could not handle the calculus. As a result, it did not incorporate the quantified techniques recently used by the cadre of scientists trained at the Ecole Polytechnique, many of whom had been influenced by Laplace. Haüy's treatise, though sprinkled with ideas about short-range molecular forces, remained mired in verbal and geometric terms that had been the staple of older texts. A program to reform education and to introduce more modern mathematical techniques was already in place at the Ecole Polytechnique, where Laplace's domineering personality at the governing council had overtaken Monge's original program featuring descriptive geometry.[80] It was time to deal with the hiatus between the old and the new physics.

This gap seemed at first filled by the Prussian professor of physics in Berlin, Ernst Gottfried Fischer, who had written *Lehrbuch der mechanischen Lehren* (Manual of Mechanistic Doctrine) for students. Under the influence of Laplace and Berthollet, Biot and his wife, Françoise Brisson, worked together to translate the work which appeared in 1806.[81] The title chosen—*Physique mécanique*—was a pointed reference to Lagrange and Laplace's famous books. In the preface, dedicated to Berthollet, Biot mentions that his mentors at Arcueil had both lamented that physics was not sufficiently cultivated in France in a proper manner, similar to astronomy and chemistry. He averred that "what has harmed physics in France is that it has been seen as a science of demonstration rather than research. We have been content to offer the public a series of brilliant experiments instead of concentrating on establishing exact laws of the phenomena and their relationships, which can be accomplished only by mathematical reasoning."[82] Laws of nature could be best expressed by stating them functionally, using the language of analysis rather than the visual language of geometric drawings in textbooks. The heart of the new physics was to be advanced on paper rather than primarily in the laboratory. It was an important turning point in the history of physics; the study of physics was henceforth to be quantitative and analytic.[83]

Laplace worked at first behind the scenes, principally through his support of Biot and Poisson, both of whom joined the group meeting at Arcueil, which was first held at the home of Berthollet, and after 1806 conducted in the gardens adjoining the properties owned by Berthollet and Laplace. The Société d'Arcueil, we have seen, was a magnet drawing the finest scientific minds of France and abroad. The nature of the unrecorded, freewheeling discussions makes it difficult to ascertain which participant in the group contributed which part of the program, but it is clear that they were all influenced by the example and personality of Laplace.

## Speed of Sound

According to Robert Fox, the starting shot for Laplacean physics began with a paper by Biot on the speed of sound written in 1801 at the direct instigation of Laplace.[84] The paper was in fact not much concerned with the mechanism of short-range molecular forces; instead it principally addressed the mathematical techniques required to close the gap between the theoretical values of the speed of sound based on Newton's analysis and the experimental data, which were off by 10 to 20 percent (depending on which figures were taken as true). Most likely this route was followed because of an earlier paper submitted to the Académie by another, somewhat obscure, mathematician named Marc Antoine Parseval des Chênes, on the integration of a series used to arrive at a formula for the speed of sound. It was a paper refereed favorably by Laplace, and later by Lacroix.[85] Laplace's innovation was to suggest that the phenomenon of adiabatic compression needed to be considered to explain the discrepancy, but it took until the 1820s for the issue to be considered fully, and even then, it was not totally settled.[86] Laplace had already turned back to physics with his preoccupation with atmospheric refraction. The program in terrestrial matters that he had largely set aside to write the bulk of the *Mécanique* was revived with new perspectives in line with the new mathematization of physics.

It is in practice difficult to pinpoint which of Laplace's interests in terrestrial physics led him to return to his previous convictions about short-range forces between molecules. Nor is it evident that he immediately assumed and later held on to a uniform physical model for each of the imponderable substances then populating descriptive physics. Electricity, magnetism, light, heat, and gases seemed similar to the ethereal medium often postulated by contemporaries to make sense of attraction and repulsion. These substances were regularly thought to pervade all matter without fundamentally changing its character—that is, they were known through their effects, but gave researchers only clues about their inner nature. For example, "caloric," the term used to refer to the matter of heat, was according to its dictionary definition, "a very subtle fluid, rare, elastic, weightless and pervading the entire universe; it penetrates all bodies with more or less ease."[87]

Laplace and Lavoisier's agnostic stance about the nature of such substances (articulated in their classic "Mémoire sur la chaleur") was philosophically sound, but difficult to respect fully when setting up equations to express the laws under which they operated. This was particularly thorny with respect to repulsive forces manifest in electricity, magnetism, and heat—forces that separated bodies without actually taking up room between matter. One could be resigned to allowing for attraction without possessing a physical model, given the countless confirmations of the laws of universal gravitation. But what about small-range forces of repulsion? Laplace might have fruitfully used the Boscovichean concept of point centers of force to skirt the issue, but he was never comfortable with this maverick Jesuit's approach. He was more inclined to imagine caloric as a material entity than relying solely on a mathematical force function, even if it could not be directly ascertained experimentally. Perhaps this is because heat, unlike gravity, dilated substances when it was added, which made it seem like a material construct. Laplace and Lavoisier had pursued this idea in the winter of 1781, without ever publishing their results of measurements of the expansion of solids under the effect of heat.[88] Instead of fully dealing with the physical character of the phenomena he was investigating, Laplace preferred pursuing mathematical techniques he had developed for celestial mechanics—techniques that could be applied in the earthly domain.

## Capillarity

Few such techniques were as fruitful as those he used to attack the problem of capillarity. When he publicized Laplace's first attempt at a "Mémoire sur les tubes capillaires" (Theory of Capillary Action) in the *Moniteur,* Biot hailed it as an extension of the work presented in the *Mécanique*.[89] After providing a clear, nonmathematical explanation for the lay public, he praised it to the skies:

> The theory that we have just expounded is a fitting appendix to the *Mécanique céleste*. The genius of inventiveness shines at every step, guided and enlightened by a deep understanding of the laws of physics and chemistry . . . Here Mr. Laplace subjects capillarity to calculation.

Let us hope for the advancement of science that he continues . . . A subject just as beautiful, and perhaps even more difficult, is electric phenomena, for which a multitude of precise experiments lend themselves to the laws of calculation. The effects of chemical affinity, so numerous and varied, are yet another topic of research that seem to belong to him . . . Men that nature has endowed with true genius are so rare, that it is almost a duty for them to persevere.

He ends his glowing account with a Latin verse, appropriately taken from Lucan's *Pharsalia; or, The Civil War:* "Nil actum reputans, si quid superesset agendum" (Thinking nothing done, while aught remained to do).[90]

Biot also added a similar account of the new theory of capillarity to the translation of Fischer's *Lehrbuch,* without providing the analytic details. Nonetheless, he stresses that Laplace arrives "at conditions of equilibrium . . . by methods worked out in the *Mécanique céleste* to calculate the attraction between a body and a spheroid."[91] While there is reference to short-range forces, the main emphasis is the fruitfulness of applying analytic techniques already worked out for astronomy. This was indeed the way Laplace portrayed his own work to others. In a letter written early in 1806, he wrote that he "saw with pleasure the spread of the domain of analysis, seizing the most subtle and important subjects of physics and chemistry."[92] One of Laplace's youngest disciples, Alexis Petit, went so far as to treat only the mathematics of this new theory in his doctoral dissertation at the Faculté des Sciences in 1811.[93]

While Laplace relied on a physical intuition about the forces acting to define the shape of the meniscus, he put his full trust on analysis derived from differential equations.[94] Even the disparity between theory and experiment, brought to light by trials executed by Haüy and Jean Louis Trémery, did not dissuade him. He went back to the analytic drawing board and produced several new theoretical derivations that were completed only after his death (by Poisson in his 1831 *Nouvelle théorie de l'action capillaire* [New Theory of Capillary Action]).

By that time, the English natural philosopher Thomas Young had already taken an alternate tack to the Laplacean approach, centering his focus on the physical model, based on water tension, and largely eschewing the complicated mathematical manipulations adopted by the Laplacean school.[95] Initially, Young was both disappointed that Laplace had taken no notice of his work, and elated that he, Young, had first arrived at conclusions similar to those Laplace had reached. In the second supplement to book 10 of the *Mécanique,* published in 1807, Laplace acknowledged Young's independent research, but criticized him for failing to link capillarity to short-range molecular forces. This exchange profoundly annoyed Young, who launched a trenchant campaign against Laplace personally, and generalized his ire to the "French school's" excessive reliance on abstruse mathematics at the expense of physical models based on experimental science. In a famous review of another Laplace production in the new *Quarterly Review,* Young gave vent to his deep resentment: "We then objected to him [Laplace] a want of address or of perseverance in the management of his calculations, presuming that the principles, on which they were founded, were capable of being applied, with much greater precision, to the phenomena in question: our suspicion has since that time been justified by an essay of an anonymous author in this country, who, without any great parade of calculation, appears to have afforded us a general and complete solution of the problem, which Mr. Laplace had examined in particular cases only."[96] He continued his accusations later in the article: "We complain also, on national grounds, of an unjustifiable want of candour, in not allotting to the observations of different authors their proper share of originality. What has a man of science to expect from the public, as a reward for his labours, but the satisfaction of having it acknowledged, that he has done something of importance toward extending the sphere of intellectual acquirements?" The tone turned sardonic when he added: "A Turk laughs at an Englishman for walking up and down a room when he could sit still; but Mr. Laplace may walk about, and even dance, as much as he pleases, in the flowery regions of algebra, without exciting our smiles, provided that he does no worse than return to the spot from which he set out."[97]

## Theories of Light

Nowhere was the tension between physical models and mathematical theory more important than with new developments in optics, pioneered by Laplace's followers at Arcueil. The conflicts it engineered have often been portrayed principally as a debate between emission partisans and those who favored the new undulatory approach.[98] It clearly pitted two theories against one another. But Young's evaluation, even if worded aggressively, was close to the mark. When approaching physics problems, the Laplacean group banked more on their mathematical prowess at the expense of developing models that closely reflected direct observation.

What began as a great triumph for two members of the Arcueil group, Biot and Etienne Malus, pushed Laplace into a new mathematical tour de force to transform Huygens's law for the wave-like propagation of light, recently revived by the British scientist William Wollaston, into one consistent with the corpuscular theory.[99] In one of his cleverest juggling acts of analysis, Laplace reduced the wave conception to the principle of least action, much as he had relied on it in the first volume of the Mécanique. In his presentation to the Académie shortly after Malus offered his discovery of polarization, Laplace praised Christiaan Huygens for establishing empirically his law, but faulted him for not providing a proper cause behind his undulatory theory. By "proper," Laplace meant a law "in accord with the general principles of mechanics," in this case the principle of least action.[100] The revision that Laplace devised to address this shortcoming made refraction dependent on short-range forces between particles in the ray of light and the crystal. Laplace considered this idea so important that he published a short version immediately in the Journal de Physique, and then the full text in the Mémoires de la Société d'Arcueil, instead of waiting for its appearance in the Mémoires of the Académie.[101] Despite the bitter denunciation of its profound significance by Young, the French scientists were especially taken with this new tack, and immediately set to work to extend the theory. In a series of papers starting in 1812, Biot applied it successfully to explain polarization, double refraction, and some color phenomena.[102] By the time he wrote his textbook on physics, Biot could rejoice that light had been incorporated into his dream of a single broad explanation for physical phenomena that stemmed from intramolecular forces.

The jubilation very quickly turned to controversy, however, when the achievements of the Laplacean program were called into question by one of their own followers, Arago. He had collaborated with Biot in experimental work on refraction in 1806, and now teamed up with an independent engineer named Fresnel, who carried out some innovative experiments on aspects of the diffraction of light rays that were left unexplained by the Laplacean theories. In a presentation to the Académie in October 1815 and some additions shortly thereafter, Fresnel argued strongly for a wavelike theory to explain the dark and light band phenomena. Unlike the approach used by Young, whose work he had not known, Fresnel used mathematical techniques he had learned as a student at the Ecole Polytechnique to work out his proposals. Arago strongly supported him to the consternation of Biot. The two young scientists had had a previous falling out that now turned into a full public conflict.[103] In 1819, Fresnel won the prize on diffraction offered by the Académie, stating that "one cannot arrive at the complete explanation of the phenomena, and that only the theory of undulation can take account of all those that the diffraction of light presents."[104] The Laplacean program was unraveling, much to the dismay of Biot.

What was Laplace's reaction? We know that eventually he admitted the superiority of the undulatory theory, but he held back his public declaration until 1822.[105] Unlike Biot, who had composed a major treatise on physics in 1816 that was unified by the assertion of short-range forces to explain all local phenomena, Laplace turned out to be less dogmatic. An attitude expressed by Fresnel in his later writings, largely influenced by Fresnel's friend André-Marie Ampère, must have coincided with Laplace's deeply held views about the difficulty scientists as mere mortals have about understanding the essence of nature.[106] Fresnel was cautious about claiming he had reached "truth," emphasizing that he was merely proposing a hypothesis that provided a maximum of explanations using a minimum of causes. This was a concept dear to Laplace's philosophy of science, which he had tried to apply to all physical phenomena. Fresnel's work, limited to optical phenomena, was in this case more secure than the edifice erected by the Laplaceans.

### Theories of Heat

A similar story of the fragmentation of the Laplacean program was simultaneously developing with respect to caloric theory. Another former student of Polytechnique, Fourier, well trained in the mathematics that Laplace admired (though not residing in Paris near enough to participate in the Arcueil group), proposed a new way of conceiving the concept of heat in a major treatise submitted in 1807. As Ivor Grattan-Guinness has carefully pointed out, the novelty of Fourier's treatment was how he had conceived of radiation in successive lamina and treated the notion mathematically.[107] In the process he invented the famous

Fourier series for integrating the diffusion formula, which eventually supplanted the procedures employed by Biot and Laplace. Fourier won the prize competition set by the Laplaceans in 1809, but his work did not appear in full until many years later.[108] It has been assumed that the opposition of the Laplacean clan was responsible for the delay, but the issue is much more tangled. True, Biot and Poisson were openly opposed to Fourier's ideas, and may well have dragged their feet about publication of this winning prize essay. Fourier himself tried to counter their objections before fully pronouncing himself. But although Laplace had been critical of some details of Fourier's approach, as is evident in their correspondence in 1808, he was on the commission that awarded Fourier the prize in 1809, and later was his supporter.[109] As is the case for Fresnel, Laplace appreciated the mathematical prowess demonstrated by Fourier, even though he was not entirely comfortable with all its intricacies. Although this work was later taken to favor the dynamical theory for the nature of heat that Laplace and Lavoisier had avoided in 1783, Fourier refused to pronounce himself categorically against the caloric theory assumed by most of the Laplaceans, as well as by a Genevan colleague of Laplace, Prévost.[110] Laplace himself tried to save the principles of caloric theory in a series of papers he wrote in the 1820s, but to no avail.[111] Like the passing of the corpuscular theory of light, the inroads into the old caloric theory signaled a new fissure in the program for establishing a comprehensive physics of short-range molecular forces. The series of attacks from Young and Arago, as well as the new direction taken by Fourier, all converged to snuff out Laplace's dream of becoming "the Newton of physics."

But even if he failed to hitch all of terrestrial physics to his astronomical program, Laplace remained the most influential force in mathematical physics for generations after his passing. His "monument," the *Mécanique*, may not have been etched in stone, but it served as a standard against which all new work would be measured.

## 10

# Probability and Determinism

Today Laplace is generally noticed by historians of philosophy for his graphic assertion about determinism, and for what has been designated as Laplace's "daemon," the super-intelligent being he postulates.[1] The locus classicus for this statement is in the introduction to his *Théorie analytique des probabilités* published in 1812 and repeated in the popular *Essai philosophique sur les probabilités* two years later. We have noted that the idea surfaced much earlier, in 1773, and that it was embedded in a rich mathematical and philosophical context and tradition that make him more its carrier than its originator. What made him return to this idea in the midst of his continuing concern with a modern version of Newton's program in celestial mechanics? On one level, the answer is simply that having established the stability of the solar system, he wanted to record the "correct" philosophical grounds on which his work was based, in the expectation that genuine scientists would adopt his approach. For Laplace to have selected this time to set down his convictions in a foundational treatise on mathematical probability theory, however, is actually a more complex and significant matter. The popular version was not meant as a text in systematic philosophy, argued from metaphysical principles. Yet it spoke to basic issues that were in the air at the end of the Enlightenment and derived their thrust in part from topics that were current following the onset of the French Revolution. In the opening pages to his treatise, Laplace pointedly refers to "the eternal principles of reason, justice and humanity," and calls on philosophers "to direct their attention to a subject worthy of their concern."[2]

The classic problem about causes of events and the inevitability of their unfolding was obvious to all witnesses and commentators of the recent political upheavals. How could an understanding of the laws of nature contribute to illuminating these troubling questions? Laplace, like other professional scientists, was convinced that knowledge of a specialized kind should assist mankind in reaching sensible conclusions, and he was prepared to formulate precepts that would be applicable to society. For him, probability theory was a valuable tool whose principles needed to be set down not only for mathematicians, but also for the benefit of the educated public and especially for decision-makers in the post-Revolutionary world. Hence it was not merely a self-serving gesture to dedicate his detailed manual on probability to the current political governor Napoleon. The highly technical treatise was envisioned as one more occasion for the emperor to reaffirm his belief that science was a valuable social activity deserving continuing support from society at large, and from his government in particular, because of its potential utility. The dedication the scientist penned may sound fawning, but it also needs to be considered seriously:

> TO NAPOLEON-THE-GREAT. The favor with which Your Majesty deigned to accept my dedication of the *Traité de mécanique céleste* prompted me to dedicate this work on the calculus of probabilities to you. This subtle mathematics bears on the most important issues of life, which are, for the most part, only problems of probability. In this regard, it will be of interest to Your Majesty, whose genius knows how to appreciate and so justly encourage all that contributes to the progress of knowledge and to public prosperity.[3]

Napoleon responded in kind when news of the gift Laplace had offered reached him. In a letter dictated from Belorussia during his Russian campaign, the emperor referred to probability as "the first of the sciences." He added that Laplace's publications "contribute to the renown of our nation. The advancement and perfection of mathematics is intimately linked to the prosperity of the State."[4]

Laplace was not the first to advance such claims in the name of this science. In 1785, Condorcet had produced a major treatise championing probability theory's general utility. In the *Essai sur l'application de l'analyse à la probabilité des décisions* (Essay on the Application of Calculus to the Probability of Decisions), he wanted to demonstrate how this mathematical technique could guide society to arrive at more rational choices. He had hoped to develop the principles for a truly scientific society, or as he termed it, to establish the "moral and political sciences."[5] The times before the Revolution were not propitious. Laplace's ambition to station this relatively new mathematical tool at the center of attention was no less sincere than Condorcet's hopes had been, and probably more justified in view of the progress the subject had made since the 1780s. Just as Laplace and his mentor d'Alembert proclaimed virtue in the study of the heavens, Laplace and his former colleague Condorcet vaunted social utility in the practice of probability. Moreover the presence at the Institut of a set of proto-social scientists originally in favor with the regime suggested that the time was now better suited for a scientifically grounded movement.

### The Path to the *Théorie Analytique des Probabilités*

Producing the *Théorie analytique* so late in his life may appear singular, but it was stimulated by a train of thought that began during his days as a lecturer at the Ecole Normale in 1795. Laplace had devoted the last of his lectures to the subject of probability, and had seen this lecture republished several times.[6] It was an idea the public associated with him just as much as celestial mechanics, which he had popularized in the successive editions of the highly acclaimed *Exposition*. To complete his standing as the grand synthesizer of scientific progress, he needed to produce an analogous work. The lectures at the Ecole Normale stood merely as the first draft of his popular presentation on probability; they needed amplification to satisfy Laplace and his larger public.

There were other factors that turned Laplace once more to probability theory. The creation of the Institut National des Sciences et des Arts in 1795 put him in close touch with French leaders of the literary and philosophical world, with whom he had had little contact prior to the Revolution. The Institut met periodically in a large gathering of all its Classes, including academicians from nonscientific sections who voted on the admission of new members proposed separately by each Class. As a conscientious elector, he needed to stay abreast of current cultural trends, particularly those related to science. A mark of his success in this endeavor was his selection as the first representative of the Institut to address the legislative body in 1796.[7] From his perspective, the Institut's most salient innovation was the new "Second" Class of Moral and Political Sciences, which held in its midst partisans of the science of *idéologie* headed by Pierre Jean Georges Cabanis and Antoine Louis Claude Destutt de Tracy.[8] For a short but significant period at the turn of the century, the *idéologues* exercised a strong cultural influence on French society. Behind their varied specific concerns, they shared a conviction that education and reform should be based on the same philosophical principles that their colleagues in the physical and mathematical sciences had successfully employed.[9]

Inspired by Condorcet and Condillac, the *idéologues* were in search of a way to create a viable rational science that would undergird human affairs. Laplace had already encountered this approach at the Ecole Normale, particularly in the spirited lectures on human understanding offered by Dominique Joseph Garat, who asserted that the methods deployed by his mathematical colleagues were of universal application. Garat touched on a variety of historic and metaphysical issues that were to stimulate discourse among members of the Institut, particularly in the Second Class of Moral and Political Sciences. Laplace befriended its members socially and followed their debates at a distance. Though confining himself publicly to the mathematical sciences in the First Class of the Institut, he privately contemplated the larger issues raised by the *idéologues*. Their most articulate scientific colleague was Cabanis, who had married the sister of Condorcet's widow and now befriended Laplace. Trained in medicine, Cabanis was centrally concerned both with the solidity of conclusions reached by physicians and the nature of laws that related the inorganic world with living matter. Most of these concerns reflected current discussions sparked by Xavier Bichat's radical assertion that, contrary to laws in the physical world, living matter obeyed variable laws.[10] Under such a scheme, Cabanis wondered how a philosophical physician could be sure of the causes of diseases he was treating, or even what laws governed normal human activity. The lectures on probability theory at the Ecole Normale stimulated his thinking and brought him into a dialogue with Laplace at the Institut, raising metaphysical issues not so easily settled by the new probabilistic techniques that Laplace had wanted to popularize.[11]

Early in February 1798, Cabanis sent Laplace three copies of *Du degré de certitude de la médecine* (On the Degree of Certainty in Medicine), one of which was intended for the Institut's library.[12] This exchange was not their only interaction. Cabanis presented a series of memoirs to the

Institut, later published as an influential book entitled *Rapport du physique et du moral de l'homme* (On the Relations between the Physical and Moral Aspects of Man), which kindled Laplace's interest. In these rambling philosophical discourses, Cabanis took up a vast array of topics about human action and their causes as was commonly understood. Like Destutt de Tracy, who had broached a similar topic in his "Mémoire sur la faculté de penser" (Memoir on the Faculty of the Mind) two years earlier, Cabanis asked how the mind gave the body commands, or to use the language of the times, how sensations, acting through the nervous system, were the real cause of activity.[13]

Distant as these considerations seem to be from Laplace's goal to write a philosophical essay on probability, they nevertheless revived ideas Laplace had entertained about causality during his student days at Caen. Laplace still rejected the facile explanation that God was the ultimate cause of natural or human action. One widely known expression of this conviction was his famous encounter with Bonaparte in 1802, when Laplace uttered the phrase "I have no need of that hypothesis" in answering the question of where God fit in Laplace's system of the world.[14] (This oft-repeated phrase may not have been spoken verbatim, but Herschel's diary records an event that reports the gist of the exchange.[15]) Napoleon and other contemporaries frequently noted Laplace's dismissal of a function for the Supreme Being in natural philosophy.[16] It was never clear from these comments whether Laplace had no use for the concept of an interventionist God in astronomy, or denied his existence altogether.

Laplace's astronomical colleague Lalande knew better and gave him an entry in his 1800 edition of Maréchal's *Dictionnaire des athées* (Dictionary of Atheists).[17] He was in a position to know Laplace's inner beliefs and is likely to have recalled his denial of God's existence that so outraged Guettard years before the Revolution. The paragraph Lalande chose to characterize those beliefs is much less threatening, suggesting that the more mature scientist had become less aggressive. In it Lalande relates that for his colleague, belief in God was not meant for the educated and rational elites. Lalande himself was less nuanced, openly preaching atheism to the common people in the streets of Paris, holding forth near the statue of Henri IV on the Pont-Neuf.[18] We do not know if Laplace objected to being included in the dictionary. After all, the listing could not be taken too seriously if, among the other atheists, Lalande listed Bonaparte and Jesus Christ himself.

Just at this time, First Consul Bonaparte was finalizing discussions with Pope Pius VII that led to the Concordat of 1801. Lalande had to be silenced, and Delambre and Laplace were delegated to convince the aged astronomer to cease his public proclamations of atheism.[19] Laplace himself never set forth his own opinions in print, keeping to his habit of announcing only his views on scientific matters. Yet everyone close to him knew he was significantly opposed to the principles of both organized religion and the Christian faith that he had abandoned early on.

Laplace's not mentioning the Supreme Being in his treatises on probability fits with his philosophical commitment solely to laws that can be inferred from natural phenomena. In a mathematical treatise he could assert to Napoleon that he did not need or want to treat religious matters. Yet in his *Essai philosophique,* he could not help but deride the monk Guido Grandi, Gottfried Leibniz, and John Craig for thinking that mathematics could be used to support the notion and role of God in nature.[20] To him, these were all delusions to be avoided by rational thinkers. Behind such occasional eruptions, Laplace privately articulated a much more extensive set of beliefs, sparked anew by the discussions surrounding *idéologue* principles. These beliefs were not publicly proclaimed, but nonetheless form an important part of his concerns.

## Private Reflections

As he was reworking his Ecole Normale lectures in preparation for the *Essai philosophique,* Laplace set down on paper more extensive remarks on all these subjects. The twenty manuscript pages that have survived through the efforts of his family and the manuscript's custodian, the Académie des Sciences, deal in succession with religious and philosophical matters: causality, volition, the power of the mind, and miracles as attested by Scriptures.[21] He had no intention to see these pages published, but the ideas he set down privately are consistent with notions advanced in a guarded manner in the *Essai philosophique.* They reveal how deeply his thinking was affected by the cultural discussions of the first fifteen years of the new century, and how concerned he remained with issues he had once faced in his youth. While occupied with formulating his views on probability for public consumption, he was absorbed with fundamental metaphysical concerns triggered by these views.

Since the manuscript is not dated, nor specific enough to refer to contemporary events, one does not know precisely when it was written, nor to what end.[22] My assumption is that the cultural debates sparked by Cabanis and his colleagues, and by Bichat, revived ideas that had engaged him long before, when he chose to forsake the clerical career on which his family had embarked him. It is likely that he opened a meaningful dialogue with his physician, the physiologist François Magendie, who subsequently wrote a spirited defense of determinism for the life sciences that reflected Laplacean stands.[23]

Several themes in the manuscript are so traditional that they might have been penned any time during the Enlightenment. The first section on the New Testament immediately challenges its literal veracity. Laplace's use of biblical exegesis and historical criticism to attack traditional Christianity was part of a late-seventeenth-century practice that is well documented, and his long quotation from a minor English commentator, Henry Dodwell, suggests that Laplace recalled some of the theological discussions and readings current during his student days at Caen.[24] To reject as hearsay the content of the Apostle's preachings as Dodwell did was not sufficient for Laplace. He wanted to account for how the stories told in the New Testament had come to hold sway in more modern times. A host of recent writings provided him with the clues he needed in order to explain, with the use of astronomical information, how myths had become doctrines of Christianity. The manuscript reveals that he was conversant with the views of Charles François Dupuis on the origin of religious cults as well as the speculations of Edward Gibbon, who is cited by name.[25] In the private reflections, Laplace always ends with the assertion of the credulity of the common man throughout the ages.

More closely related to probability is a section on "mysteries" that echoes some of David Hume's positions and his arguments about the nature of causality.[26] In the manuscript, Laplace argues like Hume that causality cannot be established simply from the constant succession of events in which the earlier one is assigned as the cause of the subsequent ones. To establish causality, it is necessary to determine the connecting link between the events. In the case of human volition assumed as the cause of bodily movement broached by the *idéologues,* one needs to understand the laws of what he calls "psychology" to be able to explore how will directs action. Like Cabanis, he admits that our knowledge of these laws is deficient, but he assumes that they exist, just as surely as those operating in the physical realm.[27] Despite the paucity of our knowledge, he expects that scientific research in these "moral" sciences will eventually be able to establish their veracity:

> What are we to do in the midst of these uncertainties? Observe the phenomena that the outer and inner senses can make known to us, determine their mutual relationships, and ascend to the level of general laws for these relationships. Physical phenomena perceived by the outer senses have laws that are peculiar to them, and that constitute the subject matter of natural philosophy. Intellectual phenomena likewise have special laws that constitute the subject-matter of psychology ... All these laws are immutable, and derive from the unknown nature of the substances they direct. The caprice of our will, even when we judge it to be most independent, is subject to these laws, and it is only by a series of illusions that we attribute to it the power of determining itself, and of moving by the sole act of its own command the parts of our body.[28]

The implication is that determinism pervades the world of the living, much as it does the inorganic realm. Though never expressed in public by Laplace, this notion surfaces in the contemporaneous writings of his physician, Magendie.

Laplace's remarkable faith in the existence of constant laws in nature despite our ignorance of their essence is the driving force behind his advocacy of probability theory. As he indicates repeatedly, probable knowledge is the most we can hope for, given our inability to arrive at ultimate understanding. Nature follows constant laws, and the scientist's task is to infer these laws from phenomena as much as possible, thereby coming close to divining these natural regularities. Bayesian induction becomes the essential tool for inferring laws in nature from ascertained data. His entire epistemological program rests on the wise employment of this powerful tool, which he was proud to have shaped in 1773. He takes it as his mission in the *Essai philosophique* to convince the public that highly probable knowledge is far better than the variety of illusions to which mankind has fallen into over the centuries.

In the first edition of the *Essai philosophique*, there is a short section devoted to the "illusions that [commonplace] estimation of probabilities" often engender.[29] There, Laplace repeats the standard argument against the misapprehensions of gamblers who do not understand the odds against them, or assume that chance favors them. He also argues passionately that all kinds of superstitions have been mistakenly adopted whenever obvious causes are not manifest. In the manuscript, but not in his publications, Laplace asserts that religion itself owes its popularity to man's willingness to accept mythical explanations when reason fails to provide convincing ones. This train of thought leads into a denunciation of miracles, which he understood all too well are among the central foundations for Christian theology. In the process, both transubstantiation and the Resurrection of Christ are thoroughly trashed and mocked. For him, they are striking examples of man's unwillingness to accept his limitations. In a stunning reversal of Pascal's position, Laplace pictures Christianity as a misleading myth rather than an answer to his epistemological predicament.[30] Beyond this, he especially takes umbrage at the views of the English mathematician John Craig, whose *Theologia Christianae principia mathematica* (The Mathematical Principles of Christian Theology) of 1699 was an attempt to establish the truth of Scriptures by a mathematical argument that was faulty.[31]

Laplace's attention to both the philosophical and technical issues about probability were fed by other minor events as well. In 1800, for example, he was asked to examine the draft of a book on probability sent to the Institut by Daniel Encontre, an instructor of mathematics from Montpellier, who later taught Auguste Comte.[32] A few years later, he was visited by a Genevan colleague, Prévost, who had published several articles on probability, some with Simon Antoine Jean Lhuilier, while both were working in Berlin. Laplace also shared an interest in heat theory with Prévost, and Laplace greeted with interest Prévost's translation of Dugald Stewart's philosophical treatise that, like Condorcet's writings, mixed scientific and philosophical topics with ease.[33] Stewart, like Condorcet and Prévost, was concerned with the theory of knowledge and recognized the need to connect testimony with belief through the use of probability. Other minor figures who wrote on probability may also have been known to Laplace in the early years of the century.[34] The subject crept into his consciousness sufficiently often to set him to work, driven alternately by technical and metaphysical concerns.

## The Role of Gauss

The detailed mathematical treatise that Laplace published in 1812 was most significantly affected by his interaction with the young German mathematician Carl Friedrich Gauss. No other scientist of the era was more influential in advancing the science of probability that Laplace had cultivated actively since the 1770s. As was pointed out earlier, their first interaction was over astronomical issues, concerning the orbit of Piazzi's discovery of Ceres. The successful method that Gauss employed for determining the position of the heavenly body was initially somewhat obscure, but involved a manipulation of error theory with which Laplace had always been concerned.[35] The two mathematicians began an exchange of letters in 1804, with Laplace obviously recognizing the merits of his young correspondent, who was elected that year as a corresponding member of the Institut. The two shared an interest not only in astronomy, but also in pure mathematics. In fact, it was when Laplace thanked Gauss for sending him a paper on number theory on 28 January 1808 that he alluded to the similar subtlety required to treat probability theory.[36]

Before he was fully aware of Gauss's use of probability theory, Laplace was busy expounding on its application to issues he had treated in the *Mécanique*. He had offered some examples in a paper published in the *Journal de l'Ecole Polytechnique* in December 1809.[37] Then, on 9 April 1810, Laplace presented a new memoir to the Institut on a subject he had struggled unsuccessfully with earlier. The numerical solution to the problem of finding the mean value taken from a series of observations involved a lengthy set of calculations that failed to reveal how close to the true value they actually were. To have an approximate value without knowing the probability that it was close to reality was unsatisfactory. As Charles Gillispie has pointed out, Laplace first tried to transform integrals into convergent series, hoping to create a better tool to cope with the problem, and applying it in particular to the case of some ninety-seven comets for which reliable information was available. Calculations of the likelihood that they behaved similarly to planets, rotating in the same direction and in a plane close to the ecliptic, indicated they were unlikely to be linked to his hypothesis about the origins of the solar system.[38] It was just at that time that Laplace was about to move his nebular hypothesis from a footnote in one edition of the *Exposition* to the main body of the text in a subsequent one.[39]

The memoir he presented in April 1810 was entitled "Mémoire sur les approximations des formules qui sont fonctions de très grand nombre, et sur leur application aux probabilités" (Memoir on Approximating

Formulas That Are a Function of Large Numbers, and on Their Application to Probability). There Laplace used the central limit theorem, but without realizing that it was in any way related to the least square method, whose priority was being contested by Gauss and Laplace's French colleague, Legendre. All this fell neatly into place once Laplace had studied and absorbed Gauss's remarkable *Theoria motus corporum coelestium in sectionibus conicis solem ambientium* (Theory of the Motion of Heavenly Bodies Moving about the Sun in Conic Sections), which appeared in 1809. Laplace thought it so important that, as president of the Bureau des Longitudes, he devoted an entire meeting on 20 September 1809—and part of the next one a week later—to discussing this book. To concentrate on one subject in this way was an exceptional measure. According to the minutes of the Bureau, he explicitly commented on the parts dealing with the probability of errors.[40]

At first, Laplace apparently did not fully grasp the work's relationship to his concern with finding mean values for a set of observations of comets. Gauss's treatise was studied again in December 1809, winning the Lalande medal awarded by the Institut. But it was only after July 1810 that Laplace became fully aware of the way it simplified his fundamental notions.[41] In his book, not only did Gauss give a mathematically elegant explanation for finding positions of new bodies like Ceres and Pallas, but he also offered a new way of looking at the error curve—a way that became known as the Gaussian distribution. Moreover, he demonstrated how this error distribution was mathematically tied to the least square method.[42] Seeing the light, Laplace penned a brief supplement to his 1810 memoir, in time for it to appear in the same volume, in which he took full advantage of the Gaussian discovery, integrating it into his understanding of the theoretical basis for probability theory.[43] As Stephen Stigler has pointed out in detail, this was a crucial step in the construction of the *Théorie analytique,* which could now be written as a complete theory based on a sounder footing.[44] By placing error theory on a Gaussian distribution, Laplace had integrated the principle of least squares into the now powerful tool of probability. He immediately recommended its use to his scientific correspondents.[45] It was only a short step to the fashioning of the treatise, which was preceded by a burst of publishing activity that he presented in 1811 to the Bureau des Longitudes and the Institut.[46]

It was in these papers that Laplace justified in full making his generating functions a key to the renovated science of probability, stating that with this central tool, "one can find with ease the limits of the probability of consequences and causes indicated by events, when these are numerous, and the laws according to which this probability approaches its limits as the number of events is multiplied."[47] In the presentation on definite integrals made to the Institut on 28 April 1811, he waxes eloquently about the "remarkable" results brought about by the use of the least square method—which he attributes both to Legendre and Gauss, without resolving their priority controversy—and announces that he intends to make the calculus of generating functions the basis for his projected new treatise.[48] He indicates in vivid language that the law of large numbers is central to his deterministic philosophy because "this research, the most delicate part of the theory of chance, warrants the attention of geometers for the analysis technique it demands, and of philosophers, for showing how order is ultimately established, even for those events that appear entirely given over to chance, and for revealing the hidden causes that are nevertheless constant, and on which this order [régularité] depends."[49]

Like the *Mécanique*, the 1812 treatise was not to stand as a canonical text in its initial version. It evolved continually as Laplace's younger colleagues, particularly the mathematicians Lacroix and Poisson, read and commented on it.[50] There were a few mathematical mistakes that were quickly corrected with an errata prepared by his Arcueil neighbor's talented son, Cauchy.[51] Laplace was quick to encourage and to promote talent when he recognized it, and to make use of capable younger minds when it suited him. In this particular case, Cauchy happened to be the son of Laplace's chief secretary at the Senate, his neighbor in Arcueil, and a contemporary acquaintance of his own son, Emile.

## The *Essai Philosophique sur les Probabilités*

For all the brilliance of the technical treatise of 1812, the *Essai philosophique* of 1814, which was originally situated as the introduction to the *Théorie analytique*, does not reach the same level of coherence of his earlier popular *Exposition*.[52] The text was worked over continuously from its first version at the Ecole Normale to the first separate edition in 1814. Each time, Laplace altered it by adding new paragraphs, rearranging their order, and expanding various philosophical messages that he wanted to impart to his audiences. After the 1814 publication, six other amended editions appeared, indicating that he was never fully satisfied with its composition. In the end, it reads more like a series of aphorisms, heavily illustrated with a shifting set of examples, than a reasoned popular exposition of an entire domain of mathematics.

Following some classical assertions about the theory of knowledge, Laplace sets down ten principles that both denominate and prescribe how the subject is to be pursued.[53] Immediately thereafter, he launches into a discussion of techniques for quantifying probability. He was obviously anxious to turn the discursive character of his lectures at the Ecole Normale into a recognizable handbook based on mathematics. The students and his adult readers were meant to absorb the fundamental notion that commonplace chance needed to be treated like other sciences, logically and in a numerative manner, so that false intuitive prejudices would be supplanted by a reasoned science.

The very definition of measuring probability as a fraction of actual (or as he calls them, "favorable") events divided by all possible events sets the tone. Following this initial assertion, Laplace indicates how combinations of probable events must be manipulated mathematically when events are either independent or cumulative. His third principle alerts the reader to another of Laplace's goals in presenting his philosophical views to a wider public. He starts his explanation of this principle by pointing out that even in the case of independent events where the simple product of probabilities yields the probability of a specific succession of events, the value diminishes as the number of events increases. If one fails to heed this fact, Laplace warns, one will be easily fooled. Only by adhering to the mathematical logic of probability theory can one avoid common illusions. By analogy, he adds that historical testimony is weakened as it passes from one generation to the next. Hence historians must consider ancient information with a strong dose of skepticism. Though he does not explicitly point it out here, testimony from the Bible is among the oldest, so is therefore among the most suspect. To accept biblical stories as probable, one should require more independent witnesses for corroboration. In his private musings about miracles, he develops this argument to reject the key theological principle of the Resurrection. It seems likely that his religious skepticism fueled much of his repeated attacks against the unenlightened common man in the *Essai philosophique*. He apparently enjoyed preaching the power of the new calculus of probabilities to destroy what he considered engrained misconceptions and prejudices.

With the next principles, Laplace returns to this main message in another fashion. Those who fail to sift their mistaken beliefs through the correct new theory are bound to derive illusory conclusions. For example, the expectation that past events alter the probability of independent future events is commonplace. To drive this point home, Laplace focuses on the meaning of the concept of "extraordinary" events, a topic that Condorcet had already examined decades earlier.[54] By offering hypothetical examples of drawing differently colored balls from an urn, he tries to persuade his readers that the errors stem from organizing events into classes, instead of treating events individually. Laplace must have realized that this approach was not sufficiently convincing; he returned to the issue numerous times in his *Essai philosophique*.

By the time Laplace reaches a sixth principle, he links these basic precepts for handling the probable to his dictum about causality: "It is the fundamental principle of the analysis of chance to move from events to causes."[55] In the very next section, entitled "Analytical Methods," Laplace repeats in prose many arguments made in the technical mathematical papers he had prepared in his youth. He explains how probability theory turns on the solution of difference equations, expressed differentially when one variable is under consideration, and with partial differentials when several variables are at issue. These were the very concerns that he had investigated in his early years in Paris. As was well known from the writings of Brook Taylor and de Moivre, the solution to these problems required a deft manipulation of sums of series, a subject greatly expanded by d'Alembert and Lagrange. Laplace attempted to explain in prose how his "generating function" would yield even more general and mathematically reliable solutions. Few lay readers could be expected to follow these explanations without first knowing the contents of the *Théorie analytique*. Indeed, in one of the editions of the *Essai philosophique*, Laplace blurts out his frustration: "more ample details on this matter would be difficult to convey without recourse to mathematics."[56] Most such lay readers took his assertions on faith. After all, Laplace had become France's leading authority in mathematics following Lagrange's demise in 1813.

Judging from the continuous reworking of his text for a dozen years, Laplace was never satisfied. Though the last edition has served as a classical statement of a brand of probability theory, the *Essai philosophique* is not a polished, definitive tract—instead it reflects both the ongoing research Laplace and a few contemporaries were conducting, and his own uncertainties about the most effective way to present the subject to the general public. Unlike the *Exposition,* which recapitulated a century of well-established ideas, the *Essai philosophique* helped introduce a relatively new topic, one transformed by Laplace from a way to develop theories of gambling into a respectable mathematical activity carrying its own broad principles.

Probability theory was a tool meaningful for political as well as for scientific ends, and hence required a judicious presentation to two audiences. Before the Revolution, Laplace had already explored in technical writings both political arithmetic and natural philosophy, and assumed that various lessons for both purposes could be drawn from a proper understanding of the new calculus. The more he was drawn into political activity as minister of the interior, senator, and later peer of the realm, the more he felt obligated to instruct his colleagues about the merits of using the new mathematics to arrive at more equitable and rational decisions. He spoke to that effect on occasion as a member of the Napoleonic Senate and as a peer. Under the intellectual sway of Condorcet, he fully believed that progress in these domains should be achieved by the intimate collaboration of scientists and of administrators who would apply scientific methods.

A telling example was Laplace's treatment of demography, a subject of intense interest to economists and government officials, both before and after 1789.[57] Laplace had followed mathematicians of the early eighteenth century like John Arbuthnot, de Montmort, and others in focusing on the relative number of male and female births in various communities as an instance of an odd, but apparently regular, phenomenon of nature whose exceptions demanded an explanation.[58] Birth statistics were intimately tied to the calculation of the purported population of France, data considered essential for a rational system of taxation.[59] It was thus an appropriate field of investigation for a probabilist who wanted to exemplify the utility of the new science for matters of state. Other topics of political arithmetic such as annuities and mortality tables were equally convenient subjects for illustrating its potential role.[60] For example, Laplace was anxious to show that even matters of public health, such as measures to eradicate smallpox by vaccination, had a calculable effect on demography, and hence on the average length of life, which in turn affected governmental and financial affairs.

Among the most innovative discussions about the utility of probability theory initiated by Laplace were various systems for recording the true wishes of an electorate. Laplace had been familiar with election procedures in the Académie, and was alerted to their theoretic import by papers presented by Borda and Condorcet in the 1780s, as well as by his interest in the reform of his institution in 1785, 1792, and 1803.[61] The issue was broached again by Pierre-Claude-François Daunou at the Institut in 1803, when new statutes were being discussed. There he cited remarks that Laplace had made in his lectures at the Ecole Normale, stimulating the probabilist to extend his views in the third edition of the *Essai philosophique*.[62] In each successive edition of the work, Laplace would add or subtract paragraphs, changing the wording of certain sections in search of the most precise and effective way of conveying his message to the public. In letters to nonscientific friends to whom he sent complimentary copies of the *Essai philosophique*, he delighted in pointing out sections of the work that would appeal to a layman. This contrasts with his comments accompanying complimentary copies of the *Exposition*, which dwelt principally with its philosophical import.

When providing examples of the importance of probability theory in the physical sciences, Laplace principally referred to his own research, some of which was being inserted as supplements to the *Théorie analytique* that appeared frequently after 1814. There is a striking use of the first-person singular that shows up in the second half of the *Essai philosophique*. Laplace was not shy about vaunting his own discoveries in lunar theory, in the Jupiter/Saturn problem, and in the satellites of Jupiter, where error theory played a capital role. In the third edition of the *Essai philosophique*, he launches a detailed discussion of tides, indicating that important data collected at Brest at the behest of the Académie starting in 1801, and numerically analyzed by Bouvard, permitted him to publish a sophisticated theory to the Institut in 1820, and to discuss it in detail at the Bureau des Longitudes.[63] Laplace was also quite proud of his probabilistic considerations related to barometric readings, many of which were based on the work of his Arcueil colleague Ramond de Carbonnières.[64] This was not the only example of expressions of self-assurance made at the end of his career. He offered himself as an exemplar of how wielding the new mathematical tools could lead to a deeper understanding of nature and man.

## Speculations

Laplace also took advantage of multiple editions of the *Essai philosophique* to express opinions extending far beyond the realm of established scientific truths. He speculated on the power of probability theory to set man straight on religious and human affairs, and asserted that future researchers in the natural sciences would benefit from paying attention to his new calculus. By analogy to Newton in his Queries of the *Opticks,* Laplace wanted to offer suggestions for those who would come after him. The aging scientist was spewing out ideas in a fitful manner, which largely explains why the work underwent so many changes and remains a rather unsystematic essay. At the same time, this last major work allowed him to leave an account of his deeply held opinions about the nature of mankind and even of other living organisms. In the often rambling paragraphs meant to illustrate the power of probability theory, Laplace revealed a lifetime of opinions and seized the chance to moralize on life's meaning. In a number of places, there is a curious mixture of precepts and hopes linked to his career and his times.

None of his comments are more revealing than those on politics and governance. The 1812 version of his philosophical principles was produced just as Napoleon was attempting to extend his empire to the East. The first separate edition of the *Essai philosophique* appeared before the emperor was sent to Elba and without the ornate dedication quoted at the beginning of this chapter. The second edition was written after the downfall of Napoleon, and contains a veiled criticism of his conduct:

> It follows from this theorem that for a series of events indefinitely prolonged, the operation of regular and constant causes will in the long run prevail over irregular causes . . . Thus favorable and numerous chances being constantly linked to the practice of the eternal principles of reason, justice, and humanity that produce and preserve societies, there is a great advantage to adhere to these principles, and a great inadvisability to deviate from them. If one considers history and his own experience, one will conclude that all evidence supports this result. Take for example the good will secured by governments that have

acted according to them, and the benefits they received from the scrupulous adherence to promises they made: what authority they obtained within [their country] and what esteem abroad! Notice by contrast the depths of misfortune into which peoples have been cast by the ambition and perfidy of their leaders. Every time a great power intoxicated by the love of conquest aspires to universal domination, the sense of liberty among the unjustly threatened nations breeds a coalition to which it always succumbs. Likewise in the midst of the multiple causes that direct and restrain various states, natural limits operating as constant causes ought in the end prevail. It is thus important for the stability as well as the prosperity of empires to remain within these limits to which they are constantly reverted by the action of these causes; just as is the case when the waters of the seas whose floor has been lifted by violent tempests sink back to their level [bassins] by the action of gravity.[65]

In the fourth edition, Laplace adds a sentence that further develops these views: "Consider the benefits that stem from [public] institutions founded upon reason and the rights of man by peoples who have established and preserved them."[66] About the political revolutions that he had witnessed, he formulated a set of principles derived from physics to favor evolutionary change:

Let us apply to the political and moral sciences the method founded upon observation and calculation, which has served us so well in the natural sciences. Let us not offer fruitless and often injurious resistance to the inevitable benefits derived from the progress of enlightenment; but let us change our institutions and the usages that we have for a long time adopted only after extreme caution. We know from past experience the drawbacks they can cause, but we are unaware of the extent of ills that change may produce. In the face of this ignorance, the theory of probability instructs us to avoid all change, especially to avoid sudden changes which in the moral as well as the physical world never occur without a considerable loss of vital force.[67]

In these few paragraphs, Laplace expressed in a cryptic manner the views he had formulated after experiencing the Revolution and the Empire. He believed that the stability of nature, as revealed through scientific findings, helped to preserve the human species. Such views were also of a piece with his steadfast character.

With respect to his other digressions into the human sciences, he appears on first reading out of character. Nevertheless, these odd comments turn out to be consistent with his strong beliefs about the laws of nature, and related to his ruminations against divinity and its operations in the natural world. Moreover, his views were shaped by personal experience as a member of the Institut, as well as by current unsolved problems in human behavior. They were further informed by readings published in Geneva, or conversations with his Genevan acquaintances Prévost and Maurice. One suspects that they were also heavily colored by the opinions of Cabanis and perhaps his friend and personal physician Magendie.

Toward the end of a section in the second edition concerned with the uses of probability theory to assist in determining probable causes, Laplace reflects on the tools one could wield to distinguish between a variety of hypotheses even in the organic world. Among these are the role of current electricity and animal magnetism, the sun and the moon, and even the presence of metals and running water for the functioning of the nervous system. After warning his readers that extreme caution must be exercised in investigating these matters, he suggests that probability theory may be able to discriminate accidental from real causes, as it did for him in celestial mechanics. The effects of these various factors, he opines, may well be quite weak, but must not be dismissed out of hand. In a striking pronouncement, he writes: "Even though they [the causes] have not manifested themselves in several instances, one ought not to reject their existence. We are far from knowing all the agents of nature and their mode of action, so that it would be unphilosophical to reject the phenomena only by virtue of our inability to explain them, given our current state of knowledge."[68] Considering the novelty of Volta's battery, which first produced current electricity, as well as the continuing discussions about galvanic action, animal magnetism, and other claims that may seem outlandish today, these ideas were perfectly appropriate coming from a leading scientist, especially since he presented them with his usual brand of skepticism. Moreover, as one of the Institut's judges who was to award a prize for a paper on the influence of galvanic action, Laplace tried to keep an open mind.[69] The most original topic he discussed in the fourth edition of 1819 was very likely stimulated by remarks in Cabanis's *Rapport du physique et du moral de l'homme* and a visit from Prévost in 1808, which led him to explore some of the literature in the area he was to name *psychologie*. He devoted some twenty-eight pages to this subject about which he had never previously written.

In his unpublished manuscript written years earlier, Laplace adopted Cartesian dualism and used the critical distinction between inner and outer senses. He knew how to extract laws in the material realm by studying phenomena reached by the five "outer" senses. How should one deal with the world of ideas and feelings, which he believed also operated according to constant natural laws? By focusing on the inner senses, he assumed, man can begin the analysis of this other realm that existed as clearly as the tangible one. Common sense reveals that mankind, and even some animals, have instincts, habits, memory, intelligence, and passions. Their study could be dubbed *physiologie intellectuelle* (mental physiology) in contrast to the *physiologie visible* (manifest physiology), which was receiving renewed attention at the Académie des Sciences from his physician Magendie and through a new prize offered by the institution.[70]

Laplace adopted the term *psychologie* to name this portion of understanding that had barely begun to be explored. The word was coined by the Genevan natural historian Charles Bonnet, whose work Laplace cited on several occasions. It also corresponded to the science of the *res cogitans* (intellectual realm) that Descartes had tried to promote in the seventeenth century. At numerous places in the *Essai philosophique*, Laplace had referred to ways that mankind had been led astray by prejudices and passions—ways that demanded correction with the new mathematical tool he was championing. His excursus on psychology was inserted at the end of a section on "the illusions that [commonplace] estimation of probabilities" engenders.[71]

Taking his cue from Bonnet, Laplace discusses successively in this section phenomena such as sympathy (whose operation he likens to harmonic resonance in string instruments), which leads to union among individuals of like persuasion; associationism, which links events by repetition or through verbal signs that help recall connections; memory, which remains active even when impressed in one's youth, and which can bring vivid images to the consciousness; habits, which become ingrained and are followed unconsciously; and attention, which gives priority to some experiences above others and leads to odd visual illusions. Inevitably, Laplace's comments lead him into another expression of his doubt about how philosophers like Pascal demonstrate the existence of

God.[72] Here Laplace returns to his favorite comments about how the common man is easily misled. The entire passage about psychology is offered as an explanation for how fears and prejudices may be reinforced by inner experiences that act powerfully on the individual. He concludes by admitting that his remarks are imperfect, but "ought to throw attention of philosophical observers to the laws of the sensorium or the intellectual realm, which it is as important to grasp as those in the physical world."[73]

## Reception of the *Essai Philosophique*

Laplace's *Essai philosophique* did not go unchallenged. Among the first to object to its deterministic character was its original proofreader, Cauchy. During most of the young mathematician's life, he kept his disapproval of Laplace's work in check, possibly because of his initial loyalty to his father's friend and because he could find no mathematical argument to attack. Gradually, however, as he found himself in conflict with Laplace as a professor of analysis at the Ecole Polytechnique, and as his Catholic principles firmed up, he became one of Laplace's important detractors. What gave him heart was the open disapproval by the accomplished Italian mathematician Paolo Ruffini, who published a long essay countering Laplace's use of probability theory for the "moral sciences."[74] The explicit motivation for Ruffini's objections was founded on a fundamental disagreement about the applicability of the mathematical theory to the realm of human activity. Like Cauchy, Ruffini was a deeply religious Catholic who opposed the materialistic scientists, whose arguments threatened man's free will. In 1806, Ruffini had written a treatise on the incorporeity of the soul that won him the plaudits of the pope and election to Pontifical Academy in Rome.[75] Ruffini was in touch with Oriani, Plana, and above all Cauchy, who esteemed him for his original work in algebra, and all of these scholars had reason to applaud an open attack on the antireligious views of Laplace.[76] The mounting criticism of Laplace's assumptions gained its full force in the 1840s, especially when Antoine Augustin Cournot published in 1843 his subtle and epoch-making views in the *Exposition de la théorie des chances et des probabilités* (Explanation of the Theory of Chance and of Probability).[77] Despite these criticisms, Laplace's two treatises on probability performed as inspirational landmarks that put social statistics on the map and helped to raise the enterprise to the status of a major mathematical discipline.[78] His concluding words in the *Essai philosophique* are an accurate and perceptive assessment of his hopes for the *Essai*'s place in the modern world:

> The theory of probability ultimately is only common sense reduced to calculus. It makes us appreciate with exactness what sound minds sense by instinct, often without realizing it. If one considers the analytic methods this theory has occasioned, the truth of the principles on which they are based, the refined and delicate logic demanded for solving problems, the establishments of public utility based upon it, . . . [and that] it supplements most happily the ignorance and weakness of the human mind . . . we will see that there is no science more worthy of our reflection and none more useful to be included in our system of public education.[79]

## 11

# The Waning Years

Laplace lived for more than a dozen years past the publication of his last major treatise, the *Théorie analytique des probabilités*. He was close to seventy-eight years old when he passed away on 5 March 1827. Until almost the very end, he was in full command of his mental powers, exercising his administrative functions with his customary diligence. He attended meetings of the Académie des Sciences and the Bureau des Longitudes, forcefully offering his scientific opinions, and from time to time penning a new article that immediately appeared in print. He continued to exchange letters filled with detailed information with foreign scientists in Italy, Germany, Switzerland, and Great Britain, showing that he was fully abreast of current issues. He had abandoned none of his habits, even as his health began to restrict his physical activity. His vision was attenuated, and what might be diagnosed today as hypertrophy of the prostate prevented him from attending protracted formal ceremonies.[1] He ate less as his energy level ebbed. There were already signs in 1813 that Laplace was getting old.

## The Passing of the Empire

As Laplace aged, he drifted toward more conservative positions on all fronts. This shift coincided with the advent of the political restoration of the monarchy and with the accumulation of honors bestowed on him by Louis XVIII despite Laplace's well-known support of the emperor. Laplace was taken as a visible fixture of France's scientific renown, honored as much for his symbolic value in the midst of military defeat as for his personal accomplishments. For those critical of the new regime,

Laplace was immediately portrayed as a rank opportunist, a turncoat who would willingly abandon his loyalties to suit the changing political fashions. His behavior was called that of a "weathercock," in French, a *girouette*.[2] It is true that he rallied quickly to the new regime, just as the Paris region was being invaded by coalition soldiers. Shortly after Napoleon was sent to Elba, Laplace was deputized as the leading member of the Senate to welcome Louis XVIII to Paris. But this transfer of allegiance was not pictured by Laplace as treasonable or opportunistic, and indeed, the events demand a fuller explanation of his actions.

Like many in France, Laplace had followed the debacle of Napoleon's Russian campaign with serious misgivings. The Laplaces had lost their only daughter in September 1813, and now feared for the safety of their remaining offspring, Emile, who was posted to the eastern front with the emperor. Fortunately, a minor wound suffered near Dresden kept their son off the battlefield. Although Emile refrained from discussing the military situation in letters to his parents, it was clear that Napoleon had overextended himself, putting the nation at peril. It was at this juncture that the scientist's loyalty began to weaken. Although he still had easy access to Napoleon, Laplace's personal relations with the emperor cooled significantly. An exchange related by Chaptal particularly cut him to the quick: "On his return from the rout in Leipzig, he [Napoleon] accosted Mr. Laplace: 'Oh!, I see you have grown thin—Sire, I have lost my daughter—Oh!, that's not a reason for losing weight. You are a mathematician; put this event in an equation, and you will find that it adds up to zero.'"[3]

As the French retreated and the coalition forces advanced, the homeland was invaded. Unwelcome Polish troops were billeted at their Arcueil home, to the special annoyance of Madame Laplace. By early April 1814, Paris was surrounded, and Senator Laplace took refuge with his colleagues at Saint Germain-en-Laye, awaiting orders from Napoleon.[4] Before any arrived, Laplace moved once more westward to Evreux, and thence near Lisieux to the Château de Mailloc, which belonged to his son-in-law, Adolphe de Portes.[5] Meanwhile, Talleyrand was negotiating a truce with the allied powers involving the dismissal of Napoleon as head of state and the return of the monarchy. A rump Senate, without Laplace or his friend Berthollet, voted on 2 April to dethrone Napoleon in the hope of salvaging what they could for the defeated country.[6] Napoleon was taken prisoner and sent off to Elba.

After these sudden and dramatic events, Laplace returned to his home in Paris on 12 April and met with the Senate.[7] He and Berthollet affixed their names to the 2 April register after the fact, legitimizing the abdication of the emperor. Thus the official record shows that Laplace approved of Napoleon's deposition, even though he had been absent when the vote was taken. From his perspective, after 12 April, he was merely ratifying a fait accompli. As chancellor of the Senate, Laplace was naturally delegated to lead a welcoming committee for the returning Louis XVIII. The new monarch immediately rewarded the senators by naming many of them—including most prominently Laplace—to the newly created Chamber of Peers.

Laplace had visibly crossed party lines, and when Napoleon returned from Elba for the Hundred Days, Laplace was shunned by the emperor and refrained from supporting him. Emile Laplace had also made a switch, transferring from being aide-de-camp of the emperor to a post as the general secretary of the company of dragoons of the Orléans family.[8] They were not alone in accepting Napoleon's downfall, which was dictated by the victorious enemy. Moreover, Laplace was eager to recover peace and stability for the nation, which he had always favored, including when he had approved of Bonaparte's coup d'état in 1799. If it meant abandoning his recent benefactor in favor of a new ruler, he was prepared to change his allegiance. Above all, he wanted a reversion to a semblance of stability so that scientific pursuits might continue unabated. In his correspondence with foreigners in Britain, Piedmont, and Geneva, he lamented the disruption that political change always seemed to cause. Like many pragmatic politicians, he favored a return to normalcy.

Laplace felt comfortable with the conservative elites who took power after Waterloo. They in turn embraced him as the leading scientist of their era, who legitimized France's continuing cultural role in Europe. Laplace took advantage of his reputation, for example, by asking Metternich to support longitude measurements from Milan to Fiume in the Hapsburg lands.[9]

## Serving the Restoration

Shortly after the new regime took power, Laplace was named president of a commission to reform the curriculum of the Ecole Polytechnique. The commission's principal goal was to demilitarize the school associated with Napoleon, and to place it under the stewardship of the minister of the interior. Laplace carried out this goal with diligence, and used the opportunity to advance his preference for the teaching of analysis over geometry, thereby completing the transition from the school's original curriculum (which Monge had backed strongly in 1794).[10] It was at this time that Cauchy, the son of Laplace's neighbor at Arcueil and the proofreader of Laplace's work on probability, was named professor at the Ecole Royale Polytechnique and delivered his series of brilliant lectures on analysis and mechanics. They capped the reform of the teaching of analysis that Laplace had always sought, and went far beyond his expectations. Laplace initially urged that they be published.[11] Later, Laplace and others in the administration objected to its excessive length and inappropriateness as a text for students.[12]

There was another issue being discussed in 1816. An anonymous author, thought to be Félicité Robert de Lamennais, wrote a scathing pamphlet denouncing François Guillaumé Jean Stanislas Andrieux's course in belles lettres taught at the Ecole, in which his curriculum purportedly echoed his anti-Royalist and anti-religious sentiments.[13] Andrieux was a personal friend of Laplace. Yet in the proposed reform of the institution, there is no advice given about individuals, only about the curriculum. Laplace's commission steered clear of suggesting personnel changes, but the directors of the school nevertheless dismissed Andrieux, most likely because of his political and religious views. In a letter to the widow of Lavoisier, Laplace pointedly indicated that "he was not consulted about the choice of personnel," and even claimed not to be aware of the manner in which decisions were made.[14] This disavowal is difficult to accept, but indicates quite clearly that whatever personal views he may have harbored, Laplace abstained from offering them in his official capacity. He understood and respected the difference between private preferences and public responsibilities.

In June 1816, the administration appointed him as president of another commission whose task was to establish a new official map of France, one that would capitalize on the surveying advances made since the beginning of the Revolution. Laplace responded with alacrity, and initiated a series of exchanges with ministers eager to take advantage of the latest scientific techniques. Laplace had long been an advocate of a program to map the nation accurately, not merely for the sake of improved cartography and geodesy, but because it would offer an equitable basis for taxation according to the cadastre. Throughout the revolutionary period, there had been plans to create a corps of well-trained surveyors to serve the nation-state, each of which had run afoul of administrative difficulties. Some groups were associated with the Ecole Polytechnique, others with the military.[15] For the rest of his career, Laplace became a champion of the science of geography, urging that more resources be devoted to map-making and enlisting the aid of other European scientists to map the world and to connect their findings to French meridian and longitude measurements. In 1820, he provided assistance to Gauss, who was directing a meridian measurement between Skagen in Denmark and Lüneburg in what was then Hannover.[16] He was disappointed that little progress was made linking British and French geodetic data, despite his efforts to promote a junction across the Channel.[17] Locally, he was among the influential supporters of the new Société de Géographie de France.[18] To his thinking, this activity was one of the important byproducts derived from a more accurate understanding of the shape of the earth, which he had closely analyzed in the *Mécanique*. He strongly supported its most prominent French practitioner, the scientist Louis Puissant who, it turns out, was elected to Laplace's seat in the Académie des Sciences upon his death.

In addition to these appointments, which cemented his loyalty to the French state, Laplace was rewarded with several public responsibilities that he also took seriously. As a noted member of the Chamber of Peers, he delivered several opinions on administrative issues he felt competent to discuss, but whose long-range significance has turned out to be minor. In 1817, he urged that the land tax code be linked to accurate geodetic data. He relied on his command of probability theory to urge changes in the jury system and to suggest the abolition of government lotteries. He spoke intelligently and with mathematical figures on the budget, and on financing the public debt.[19] Each time, he made use his renown as a scientist to gain authority on political issues, but without taking an identifiable party stand; though far from innocent about politics, he stayed clear of partisan affiliations. Laplace understood full well how complicated and important were decisions in the public realm. In one letter to the political economist Jean-Baptiste Say, he reflected on the heavy responsibility of his office and on its intrinsic problems: "[I admire] your writings that leave nothing to be desired for the literary glory of France in one of the most useful branches of human knowledge, whose importance I understand, for each year I have to deliberate on weighty political issues. I appreciate the dangers of old errors and our ignorance of true doctrines."[20]

## The Académie Française

Only in one of Laplace's many activities has overt partisanship been suggested, although the evidence for such a bias is not decisive. Laplace was elected to the Académie Française on 11 April 1816, replacing Michel Louis Etienne Regnault de Saint Jean d'Angély, who had been removed by royal fiat along with others for their political transgressions during the Revolution.[21] Even though Laplace was nominated and elected by academic colleagues at the literary society and approved by the new powers, it is likely that ministers were consulted before his name was put forward. The minister of the interior who engineered the composition of the new Royal Academy and ratified the election was none other than Vincent Marie Viénot Vaublanc, who had been a student at the Ecole Militaire when Laplace taught there.[22] Critics of the regime took this appointment as evidence that the *girouette* was once more being rewarded for rallying behind the political agenda of the monarchy. As if to avoid personal embarrassment for the new academician, Laplace was not asked to pronounce the customary eulogy of the former member to whose eighth chair he had been assigned. It would have proven quite awkward to prepare the eulogy, given the personal circumstances. Laplace had hosted the now-expelled Comte Regnault at his Arcueil home only four years earlier.[23]

Laplace's nomination to the literary Académie was easily justifiable, given the success of his *Exposition* and *Essai philosophique,* both of which were genuinely appreciated for their prose. Moreover, Laplace joined the impressive list of scientists who, the century before, had been elected to the Académie Française, including Fontenelle, Maupertuis, Dortous de Mairan, d'Alembert, Buffon, Bailly, Condorcet, Vicq d'Azyr, and Louis-Antoine de Bougainville. When he enrolled in the renewed Académie in 1816, he was welcomed by colleagues, many of whom he knew from Napoleon's Senate and at the Institut. Some remembered that Laplace had been sounded out to become a founding member by planners who had hoped to reestablish this Académie as a literary society in 1800, 1803, or something akin to it in 1807.[24] A substantial number were also part of his social entourage, joining his family for dinner parties. They included most prominently men of letters like Andrieux, Nicolas François de Neufchâteau and Pierre Antoine Noël Daru, each of whom was, at one time or another, a servant of the government. He felt at ease joining in the clubby atmosphere prevalent among the famous forty members.

Laplace was elected chancellor or director of the Académie several times during his ten-year tenure. In that capacity, he was able to assist the executor of the will of the wealthy Baron Antoine Jean-Baptiste Robert Auget de Montyon who, with the backing of his old friend Laplace, bequeathed a large donation to establish an annual award known as the Prix de Vertu. Laplace was to read the first speech honoring the recipients in 1821.[25] Six years later, on 16 January 1827, he presided over the contentious discussion to protest the government's Peyronnet project to restrict freedom of the press, which was initiated by Jean Charles Dominique de Lacretelle, Joseph-François Michaud, François-Juste-Marie Raynouard, and Louis Nicolas Lemercier, all avowed academicians of a liberal persuasion.[26] According to one account, Laplace and two other peers abstained from voting on whether to present the "respectful" petition to Charles X.[27] A contemporary royalist newspaper reports the names of eight members who spoke against the motion, including Laplace.[28] This account does not reveal the nature of their opposition, which could have been procedural as well as substantive. The motion to protest the government's restrictions against the press nevertheless carried by a vote of eighteen to six, and the company directed its presiding officer to present the remonstrance to the monarch. But Laplace never discharged the obligation in person. Rumor had it that because Laplace was opposed to the petition, he refused to carry out the will of the Académie. The most recent biographer repeats and embellishes the rumor, claiming that he became "the bête noire of the liberals."[29] It is quite likely that Laplace did not approve of the petition, but was prevented from fulfilling his duties as presiding officer for other reasons. For instance, the minister informed the Académie that the king would not deign to receive the petition, and Laplace could hardly have overridden the will of the monarch, even if he had been inclined to try. Coincidentally, Laplace became deathly ill and was unable to attend meetings of the Académie for the rest of January 1827. A month and a half later, he passed away.

There is a posthumous event that illuminates this episode. Laplace's successor at the Académie Française was to be Pierre Paul Royer-Collard, an outspoken liberal defender of freedom of the press. In his acceptance speech before the Académie, Royer-Collard made no allusion to Laplace's alleged approval of the Peyronnet law, nor his failure to follow through with the body's explicit desire to present the petition to the monarch. He did, however, comment on Laplace's deep feelings in a way that helped to make sense of his nonpartisan public behavior: "The sciences were his whole life, and the only passion he acted on. He saw in their progress that of general enlightenment, and in them a guaranty of the public good . . . The deep preoccupation M. de Laplace had with higher studies is the only excuse—if one were needed—for having silently traversed the good and bad days, without fervor or anger, [sailing] above our hopes and fears."[30] Whatever views Laplace may have held privately on this issue or others, they remained removed from the brisk political polemics that agitated the Restoration era. He could remain on good terms with men of different persuasions, taking refuge in his only lasting passion, science.

## The Venerable Scientist

While at the Académie des Sciences, Laplace felt not only at home, but as one of its oldest veterans, very much in command. Their proceedings reveal that he was called on repeatedly to perform important and delicate tasks, many of which involved creating and awarding prizes for research scientists. The secretaries of the scientific assembly were men he had personally launched on their careers; they knew his abilities and trusted his fairness in assessing the work of others. Delambre held his secretarial post until 1822 and was followed by Fourier, who would eventually read Laplace's eulogy. An anonymous donor, befriended by Laplace—and who turned out to be the wealthy Baron de Montyon—offered large sums to create prizes for statistical research and for experimental physiology.[31] Another prize was set up to reward the inventor of the newest machine that would minimize pollution.[32] Laplace served on all the commissions to draft the wording defining the special topics for each prize, and served on several committees to adjudicate the submissions. He was also repeatedly appointed to examine essays eligible for the physics prize, which he channeled into topics on caloric, diffraction, and mechanics—all subjects of current deliberation with which he stayed abreast. Laplace was especially pleased in 1822 to announce the awarding of a prize in mathematics he had initiated: Hans Christian Ørsted was to be recognized for his discovery of the action of current electricity on magnetism, which launched a new era in classical physics. No wonder the scientific community admired and feared Laplace as the most powerful arbiter of the physical sciences for some twenty years following the founding of the Société d'Arcueil. He held the reins of power and judiciously exercised his views on contemporary physics at the Académie, just as he had in astronomy at the Bureau des Longitudes.

Laplace's opinion on appointments for a variety of scientific posts in the educational system and the learned societies was sought out and often strongly influential. In retrospect one can confirm that he favored men of talent, even when their scientific views did not concur with his own. The men he supported included Biot, Fourier, Cauchy, Dulong, Damoiseau, and his personal physician, Magendie. One of his fellow mathematicians who knew him well, Louis Benjamin Francœur, wrote: "He honored all who cultivated the sciences and granted them encouragement and advice. He used the influence of his immense renown and public appointments only to support men who had an aptitude for serious study; his whole life was devoted to protecting the learned and to extending the domain of science."[33] Among foreigners, he recognized and applauded the accomplishments of Volta, Gauss, Ørsted, Humboldt, Ivory, John Herschel, Davy, and Charles Babbage, all of whom made substantial contributions to their chosen disciplines.[34] He was also pleased to exchange views with popularizers of his doctrines, above all the talented Scottish feminist Mary Somerville, whom he and his wife graciously entertained at Arcueil in 1817.[35] So impressed was he with creative work undertaken in Britain that he seriously entertained taking a trip to England with his granddaughter Angélique de Portes when he was well into his seventies. In letters to his foreign correspondents, he praised the land that had produced Newton, and that had since developed a social system of governance much more stable than the ones he had experienced in France.[36]

## Personal Character

As he became a popular scientific personality and a member of the government establishment, Laplace was sought out by many foreigners.

Some of them remarked on his character and demeanor, thereby offering us a glimpse at his personality. He was by nature a cautious individual, little prone to initiate human contacts unless they offered some obvious advantage. With those who could further his scientific program, however, like Oriani and Gauss, he made singular efforts to assist them in their careers. He found ways to protect Oriani's Milan observatory even as General Bonaparte's troops advanced in their Italian campaign. When Gauss faced tax problems with the new French authorities, he smoothed things over.[37] No doubt he also provided assistance to the Maurice family and to the physicist de Saussure while Geneva was temporarily annexed to France. These moves were generally sincere and genuine, but seemed always gauged on the scale of scientific merit. He would willingly help serious competent scientists when he could, but he had little patience with solicitors who had no credentials and was pitiless with would-be scientists, whom he summarily dismissed from consideration. The examples of Jean-Paul Marat before the Revolution and Hoëné-Wroński during the Empire are in this regard characteristic. He paid no heed to visionaries like Pierre-Hyacinthe Azaïs who had unwarranted scientific pretensions.

The prominent British chemist Davy commented that Laplace appeared proud and self-assured in 1813 when they first met in Paris.[38] Many who attended meetings at the Institut and watched him from afar spoke of his distant attitude, which they took as haughtiness or self-satisfaction. This was the initial view of the astronomer Johann Friedrich Benzenberg, who was in Paris in 1804.[39] Baron von Zach also privately chafed at Laplace's imperious attitude. In 1814, he reported the low esteem others held of him for his "arrogance, conceit and insolence."[40] Generally, Laplace was outwardly formal and courteous, but he rarely warmed to strangers. A Prussian composer visiting Paris in 1802 found him reserved and polite, reminiscent of an old regime courtier. He seemed to be "a bit stiff."[41] But some who witnessed this austere bearing also noted his stoic mien: "Laplace had a very venerable appearance—a slender figure, small sharp features, a high prominent wide brow, white locks hanging straight down his temples, and a benevolent droop of the lower lip. I heard him speak only once, and but a few words."[42]

Laplace did grow strongly attached to a few select foreigners—notably the secretary of the Royal Society, Blagden, who was also a close friend of Berthollet and visited Arcueil often, and the younger Genevan astronomer Maurice, with whom he shared long exchanges. Blagden, who died during a visit to France, was also quite fond of Laplace, leaving instructions for the purchase of an expensive ring in his will.[43] By contrast, Laplace's earlier cordial feelings toward Deluc cooled considerably once the Deluc began to oppose the new chemistry of Lavoisier and tried to link modern geological explanations with a strict adherence to the Bible.

When the scientific issue was one close to his heart, too, Laplace could be very forthcoming. The same Benzenberg who was at first put off by Laplace's bearing changed his impression after discussing his experiments on the earth's rotation with the scientist, whom he praised as focused and able to encompass the "whole triangulation of human knowledge."[44] According to Davy, who dined with Laplace in 1820, Laplace considerably softened his tone in the Restoration period.[45] It is telling that this change in demeanor came once Davy had attained a prominent rank, as president of the Royal Society of London. Clearly Laplace prized titles and honors. He was always mindful of status, whatever the regime in power—royal, revolutionary, or imperial. The story related by Davy that Laplace pinned his medal of the Order of the Reunion on his dressing gown may be apocryphal, but it is symbolically appropriate.

Stability, steadfastness, and perseverance were hallmarks of Laplace's approach to his work and became ingrained in his person as well.[46] There was a steady rhythm to his life. When his wife was at his side, they enjoyed a full social life with the families of men he knew from his official capacities at the Chamber of Peers, from the two academies to which he belonged, and from Emile's military friends. Madame Laplace was active in supporting her husband's career, and was a great asset to his social success in high society. They entertained individuals of all persuasions, even the more devout individuals who knew Laplace's aversion to revealed religion. His wife was a conventional practicing Catholic whose living quarters were adorned with religious artifacts.[47] Often during the summer months when he returned to Arcueil, Madame Laplace would leave him to take their granddaughter Angélique to visit with her close relatives in Poissons, near Joinville. The correspondence between the spouses is filled with household matters, family news, and concern for each other's health and well-being.[48] When posted near Paris, Emile would often join his father and work with him on applied problems of probability, which were likely destined for future editions of the *Théorie analytique* or the *Essai philosophique*.[49] In this way, Laplace led a regular, tranquil, and happy existence.

April 1823 was the fiftieth anniversary of Laplace's election to the Académie des Sciences, a singular achievement for a country boy who had taken a bold step by opting for a career outside of the Church, one that had propelled him from initial uncertainties to the full honors and rewards of old age. As dean of the mathematical academicians, he had outlasted most of those who were initially his colleagues in 1773. All of his close contemporaries he worked with had passed away: Bailly, Bochart de Saron, Condorcet, Lavoisier, and Vicq d'Azyr as victims of the Revolution; the mathematicians Borda, Cousin, Lagrange, Bossut, and Monge, who died, respectively, in 1799, 1800, 1813, 1814, and 1818; as well as his neighbor and close friend Berthollet, who had just expired in 1822. Of the original academicians who had greeted him in 1773, only four were still alive, but none was particularly close to him: the anatomist Antoine Portal, the chemist Balthazar Georges Sage, the botanist Antoine Laurent Jussieu, and the last of the scion of the Cassini family of astronomers, Cassini IV, with whom Laplace had an especially antagonistic relationship.[50] For the celebration that took place on 24 April, sixteen of his academic colleagues were invited, and reportedly stayed up until one o'clock in the morning singing, reading rhymed tributes, and making countless toasts. It was a special event that Emile, stationed in Spain, sorely regretted missing.[51]

## Newton, God, and the Laws of Nature

It was during this same period that Laplace began to inquire about the private life of Isaac Newton, no doubt comparing the Englishman's long career with his own. One issue that preoccupied him was why Newton made his unexpected pronouncements on religious matters in the second edition of the *Principia*, views that Laplace did not in the least share. He was anxious to understand what had provoked Newton's assertive comments about God's role in the universe, and he assumed that his illness following the fashioning of the *Principia* in the mid-1680s was the cause of this "derangement." Laplace wanted to assure himself that this was an aberration of old age, or a form of mental disorder, and to that end he sent letters and various emissaries to England to ascertain the events that had led to this wayward turn of mind.[52] He could not imagine that someone as brilliant as Newton could possibly have been genuinely committed to such religious beliefs for his entire life. What must have surprised him most was Newton's willingness to link these deeply personal views to his scientific work. Laplace had always striven to separate his private sentiments from his public utterances. It is worth noting that Laplace had not always been able to keep his own religious opinions from creeping into his writings as he grew old. Despite all his confident assertions, Laplace apparently had not completely freed himself from the trauma associated with his own conversion away from Christianity.

Publicly, Laplace maintained his agnostic beliefs, and even in his old age continued to be skeptical about any function God might play in a deterministic universe. Nonetheless, he was curious about contemporaries' notions on this subject, and was stimulated by the engaging conversations he had with the Genevan astronomer Maurice, who belonged to an evangelic movement attempting to bring wayward Christians back to the fold. They talked at length, and Maurice lent Laplace essays by the Scottish Presbyterian Thomas Chalmers and possibly by the Belgian philosopher Louis de Potter.[53] Earlier, he had Gilbert Burnet's *A Rational Method for Proving the Truth of the Christian Religion* read to Laplace to show him how Christians argued for miracles.[54] Nevertheless they failed to convert him. Maurice reported Laplace's comment that "from the very first lines [of Burnet's work], it was clear that the author had firm convictions," hence his conclusions were not convincing in spite of his commendable logic. Maurice adds that nevertheless "little by little the basis for his beliefs were modified, and I believe what held him back most was his profound conviction about the invariability of the laws of nature, which would not permit of supernatural events."[55] Apparently Laplace would not yield on this point. Yet he was willing to acknowledge the social value of Christianity for indigenous populations. In particular, he approved of the civilizing influence of the faith for the native Tahitians. He once remarked to Maurice: "You know Christianity is quite a beautiful thing . . . the only religion that has always accompanied true civilization and led these men to adopt a proper behavior, and turn them to learning and to law-abiding liberty."[56] He had made similar remarks after reading Baron von Humboldt's account of his voyages to the Americas.

Maurice was convinced that Laplace was on the road to redemption. In his account of Laplace's last illness, he chose to find additional signs that the great man was edging back toward Christianity. The narrative he penned shortly after Laplace's demise is gripping, and entirely devoid of any pretense:

Thus he was well disposed [toward Christianity] which seemed to announce a genuine desire to believe, when in February 1827 he was stricken with an inflammation of the lungs to which he succumbed on Monday 5 March at 9 in the morning. During the five weeks of his illness, I saw him regularly; he wanted me close by, taking my hand and repeating "as soon as I am better, you will often come to dinner with me and we will *talk*." I cannot attribute this special friendship which he bestowed on me to anything except his *secret desire to share the happy convictions* that God had granted me. There was nothing else in me that would explain the particular preference he continually showed me, but in spite of my sincere desire, I was never able to have an extended conversation with him; his excellent wife who never left his side would not allow it because of his discomfort and general weakness.[57]

In a more detailed account entitled "Illness and Last moments of M. de Laplace," written several years later, Maurice is even more specific about his encounters with the dying scientist, trying his best to portray him as a serious candidate for conversion:

I went there at first on Sunday evening, February 4. As soon as he knew me there, he wanted to see me and I came to his bedside. Even though quite ill, he held out his hand, clasped mine, and assured me several times of the "tender friendship he had always had for me." Thereafter, when I was present, he always received me by his bed, always spoke to me, clasping me with a remarkable vigor and repeated all the pleasure he had in seeing me, and "the pleasure he expected when speaking with me as soon as we could dine together." I interpreted this, as well as the singular pleasure that my presence caused him, and by the tender way he spoke to me, as recalling the frequent conversations we had had on Christian religion for over thirty years—but especially during the last seven or eight [years of] conversations, in which he had been able to assure himself of the happiness I experienced myself in my profound conviction about these august truths. It was natural for him in his condition to be led to renew his meditations because of their import.[58]

Maurice chronicled the last moments of Laplace's life as well, and was somewhat annoyed about the discrepancy between what the mathematician Poisson said to him personally on the day of Laplace's death and what was reported in the official eulogy by Fourier delivered two years later. For dramatic purposes, Fourier publicly indicated his last utterance to have been "what we know is insignificant, what we do not know is immense," no doubt echoing Newton's alleged saying in the last years of his life, "I seem to have been only like a child, playing on the edge of the sea . . . whilst the great ocean of truth lay unexplored before me."[59] Maurice's version seems more authentic, but equally enigmatic.

> On Friday 2 March our illustrious patient began to lose consciousness of his surroundings as well as the use of his speech. I did not go there on Saturday or Sunday, but made constant inquiries about his condition. Monday at 9 A.M. I learned that he was expiring and rushed over, adding my tears to those shed by his wife, his son and the late Poisson, his most distinguished disciple. I left with him [Poisson] about an hour later and here is what he told me: "You know that I do not share your [religious] opinions, but my conscience forces me to recount something that will surely please you. The day before yesterday, I was in the living room with Laplace's son and the astronomer Bouvard. Magendie, his physician, came around noon and we tried to guide him back to his bed where he seemed to recover some of his senses. His son took his hand, the sufferer looked at him and said in a feeble voice, 'Hello, my good friend.' Bouvard came next, but was not recognized. I was next and said, 'It is I, Poisson, your student, and this person is Bouvard, whose zealous calculations brought your brilliant discoveries to light.'" "Then," Poisson added, "he fixed me with a pensive look and uttered with some discomfort the phrase: 'Ah! we chase after phantoms [*chimères*].' After this saying he turned over and never spoke again."[60]

Maurice could not help editorializing on this ultimate phrase, suggesting that "Here was a very profound saying, for him quite characteristic. It seems to point to his realization of the vanity of that which had occupied him too exclusively, for too long."[61] We do not have other sources to know the exact course of events, but clearly there was no bedside conversion, no effective last rights administered by a Church official, even though the devout Madame Laplace called in the deacon from Arcueil and the nearby Missions Etrangères on rue du Bac. The Catholic newspaper *La Quotidienne* announced that Laplace had died in the arms of two *curés* (priests), implying that he had a proper Catholic end, but this is not credible.[62] To the end, he remained a skeptic, wedded to his deterministic creed and to an uncompromised ethos derived from his vast scientific experience.

# Conclusion

Who was Laplace, this enigmatic figure that played so prominent a part in his era, but has largely been neglected by chroniclers of science? Despite his public renown, Laplace remained a private man, guarded and somewhat detached from all but his closest friends and disciples. From his very early years, deprived of his mother's love, he repressed strong displays of emotion and rarely provided outward signs of deep attachments. We know he had no taste for the romantic age, preferring authors like Jean Racine, and favoring classic operas of the Old Regime over the new generation of novelists.[1] His personal life was untainted by adventures or heroic exploits of the sort experienced by scientists who measured the meridian or accompanied Napoleon across the Mediterranean to Egypt. Even the expansive raconteur Arago, who knew Laplace well at the Société d'Arcueil and the Académie des Sciences, could not find tales to make over his placid personality into that of an eccentric innovator.[2] Nor could his close follower Biot do more than offer a generous anecdote to add color to his patron's plain existence.[3]

Pierre Simon traveled little, never having ventured south of the Loire nor to any foreign land where he would have been acclaimed as a cultural icon. Nothing in his life seems to have been decided on the spur of the moment. He was a consistently rational and deliberate man, almost as steady in his behavior as in his vision of the stable solar system. In the face of a political world churning around him, he remained serene and resolute, continuing to pursue his professional agenda, as well as protecting his family and attending to its welfare. As a result of his accumulated titles and appointments, he became a wealthy man, and as was customary left all his possessions to his wife, who, being younger than Laplace, outlived him in comfort for some thirty-five years.[4] To his satisfaction he knew his bachelor son, Emile, would succeed him as a peer, and that his beloved daughter Sophie's child married into a branch of the distinguished Colbert family. Their descendants have retained the title of marquis that he proudly wore.

To all appearances, there was only one dramatic episode in his life that determined his career, related with scrupulous care by his eulogist Fourier.[5] It was when the twenty-year-old Norman would-be abbé dared to present his critical ideas to the most prestigious French savant of the day, d'Alembert. The encounter, which he recalled in detail and shared with Fourier, radically changed his life, setting him on his highly successful career. But the decision he then took to abandon religion for science left scars that never completely healed. Even on his deathbed, Maurice tried to rescue him for Christianity, and Laplace was open to hearing about arguments concerning divinity. Almost without exception, he remained silent in print about his religious sentiments, while in private, he set down the unmistakable convictions that reinforced his youthful decision to abandon the Church.

Once settled in Paris, Laplace carved out an independent existence, mindful of the examples set by d'Alembert, Lagrange, and Condorcet, but always putting his own stamp on the work he presented to the Académie des Sciences, where he was welcomed at an early age. He successively forged a research agenda in celestial mechanics and probability theory, making serious progress before the Revolution in elucidating puzzles of the physical universe. Demonstrating the near stability of the solar system, and developing the probabilistic calculus to discern how close was the astronomical data to the Newtonian laws of nature, he was convinced that the physical universe followed invariant rules. He worked relentlessly to close the gap between observational facts and astronomical theory, enlisting trustworthy colleagues to assist him. Throughout his life, he never wavered from this pattern, confident that empirical evidence would vindicate the universal law of gravitation.

It was this Newtonian "truth" that led Laplace to seek an explanation for gravitation by investigating the role of various imponderable fluids, eventually teaming up with Lavoisier in a series of epoch-making papers on calorimetry. It opened his eyes to the likelihood that terrestrial physics could one day be incorporated into the set of broader principles he found in the heavens. On his own and with younger physicists at the Société d'Arcueil, he investigated the tides, capillary phenomena, and double refraction, but without ever bringing about the synthesis with celestial mechanics he had anticipated. His efforts nevertheless opened the way for the younger generation to manipulate mathematical analysis in order to grapple with these elusive phenomena. A master of the calculus, he wielded it as a precise tool to grasp the nature of the physical world.

Marriage and the coming of the Revolution ushered in a reorientation in his behavior and personality. At age forty, he became more sociable, coming out of his shell as a brilliant technician to begin a more visible life as a public figure. Unwilling to engage in the volatile politics of the era, he nonetheless found a significant role as a technocrat, helping to promote the decimal metric system and to rebuild the scientific establishment following its collapse during the Terror. After his brilliant public lectures at the Ecole Normale, the creation of the Bureau des Longitudes, and the formation of the Institut National des Sciences et des Arts, Laplace was recognized as the embodiment of French science, an honor he began to exploit for the benefit of its continued advancement. Named for a brief time minister of the interior by Bonaparte, he entered the legislative body as a senator, and later as a peer. His public renown was cemented by the publication of several editions of the popular *Exposition du système du monde*, and later by the *Essai philosophique sur les probabilités*. As an international celebrity, he was paraded by successive governments as the symbol of French mastery in the scientific realm.

Flattered and proud as he surely was, Laplace persisted with his research, authoring dozens of pioneering articles until well into his seventieth year. The respect he had earned for his dazzling prowess did not mask his recognition that progress demanded continuing hard work. He could still appreciate and profit from the work of younger contemporaries, whom he recognized and praised, whether or not they were countrymen. His faithful disciples Biot and Poisson carried on his research programs, and the English mathematician Ivory and the brilliant German Gauss received his accolade. During the Restoration, some of Laplace's assertions were challenged, notably his beliefs about the nature of heat and light, and he resisted capitulating to the new views until credible mathematical theories were elaborated. He also resisted challenges to his astronomical and mathematical work from figures like Plana and Ruffini, whose ideas were not always treated with the respect they deserved. But even they realized how central and significant were his overall scientific accomplishments.

Laplace was a man of his time, schooled in Enlightenment philosophy and following its common practices. His career was launched with customary patronage and, as his prowess grew, he offered his protection to those he deemed worthy. Long before the Revolution toppled the Church, he had opted for a totally secular version of life, engaging with military rather than religious institutions. When the Revolution arrived, he was prepared for turning education into a vehicle for rewarding talent and hard work, especially if these were manifested in mathematical aptitude. He seized the opportunity provided by new governments to advance his convictions about the proper role that science ought to play in society. The classical century of science he ushered in not only recognized his specific scientific accomplishments, but also held him up as a symbol for secular progress.

Laplace's legacy in science was unparalleled until the Einsteinian revolution. In astronomy, his nebular hypothesis and his version of celestial mechanics became the starting point for theoretical research, though both were superseded by the time Henri Poincaré revised mechanics. In physics, new theories and tools of observation rendered many of Laplace's suppositions obsolete, but not his mathematical tools, which were taught and used by generations of young scientists throughout the nineteenth century. A similar fate was in store for his work on probability, which laid the foundation for a whole century of new perspectives.

Laplace's style in science and his personal behavior had in common a disdain for what Auguste Comte called the mythological and metaphysical stages of history.[6] He turned his back on the remnants of Aristotelian discourse about essences and the philosophical disputes that had engaged the generation of Maupertuis, Euler, and d'Alembert. In his published pronouncements, at least, he eschewed metaphysics, and was in a profound way a positivist *avant la lettre*. While he may have privately entertained questions of why the world is as it is, he thought the only certain knowledge worth publishing by a professional was to be obtained by formulating mathematical statements for observed regularities in nature. For Laplace, the immediate goal of scientists was the setting down of mathematically expressed empirical laws that might eventually lead to a deeper understanding of the universe. If he needed a philosophical creed, it could have been "From phenomena to laws, and from laws to causes." It was his abiding hope that the study of nature would lead ever closer to the discovery of these causes through the formulation of invariant laws.

# Abbreviations

| | |
|---|---|
| AAdS | Archives de l'Académie des Sciences, Paris |
| AHES | *Archive for History of Exact Sciences* |
| AIHS | *Archives Internationales d'Histoire des Sciences* |
| AN | Archives Nationales, Paris |
| Bancroft | Bancroft Library, Laplace papers, University of California, Berkeley |
| BI | Bibliothèque de l'Institut, Paris |
| BN | Bibliothèque Nationale, Paris |
| CIPCN | *Procès-verbaux du comité d'instruction publique de la convention nationale*, ed. James Guillaume (Paris, 1891–1959), 7 vols. |
| DSB | *Dictionary of Scientific Biography,* ed. Charles Gillispie (New York, 1970–1980), 18 vols. |
| *Essai philosophique* | Laplace, *Essai philosophique sur les probabilités* (Paris, 1814) |
| *Exposition* | Laplace, *Exposition du système du monde* (Paris, an 4 [1796]), 2 vols. |
| HSPS | *Historical Studies in the Physical Sciences* |
| Gillispie | Charles Coulston Gillispie et al., *Pierre-Simon Laplace, 1749–1827: A Life in Exact Science* (Princeton, 1997). |
| Hahn | Roger Hahn, *The Anatomy of a Scientific Institution* (Berkeley, 1971). |
| *Mécanique* | Laplace, *Traité de mécanique céleste* (Paris, an 7 [1799]–1825), 5 vols. |
| O. | *Œuvres* |
| O.C. | *Œuvres Complètes* |
| PV, Institut | *Procès-verbaux des séances de l'Académie tenues depuis la fondation de l'Institut, jusqu'au mois d'août, 1835* (Hendaye, 1910–1922), 10 vols. |
| RHS | *Revue d'Histoire des Sciences et de leurs Applications* |
| *Théorie analytique* | Laplace, *Théorie analytique des probabilités* (Paris, 1812) |

# Draft of a Letter to Laplace by Jean-Etienne Guettard

It is disgraceful to think that the Academy of Sciences is an "anti-Christian body established to overthrow the Christian Church," as the late Mr. Jars said, or as you asserted in his name, or as you said for yourself, echoing the words of others. When you delivered this profanity the other day at the home of Mr. de Villers, in the presence of several academicians and Mrs. Lavoisier, and insulted the Academy, and went so far as to assert that you personally had abjured this religion, there were voices raised in remonstrance. For my part, joining with them, I shuddered and feared that God, armed with his bolts of lightning, would crush us all: you for your impudence and blasphemy, and us for not having made you realize as much as we ought to have the enormity of your proclamation, and for not having deserted you as one would someone afflicted by the plague.

One has to have been an abbé like you to be so brazen and sacrilegious as to deny the existence of God the way you denied it at a dinner given by Mr. De Séjour, at which you said with unbelievable effrontery that "if there is a God, he should strike you dead the very moment you denied his existence." You were in effect imitating another unbeliever who said, "If there is one [God], let him strike me before I place my finger in this hole. Since he has not exterminated me, he does not exist." He did not punish this godless person any more than you. Yet God is eternal and the godless are but a speck of dust, a plaything of the winds. They raise their heads arrogantly, pass on, and disappear. He does not exist, that is all you believe in. His matter, so say you, turns back into the totality [of things]. According to you and your kind, what animates us is a fire, a subtle matter, a vital and material principle, which losing its activity or finding obstacles in its path, ceases to put in motion the springs of your body, just as a watch loses its spring, runs down, and breaks down. What one calls life vanishes and the material after its breakdown is used to reassemble a fellow creature.

Of what delirium are you possessed when you speak this way? How can you resist the inner feeling that raises you above matter, that lifts your longing beyond all times? Which, as affluent as you may be on this earth, will make you thirst for greater rewards, which, as miserable as you may be, will sustain you in adversity through the hope of happiness after death? You will say these are the result of emotions, the consequence of the principles of religion we have been steeped in since our most tender age, and which it is almost impossible to shed at maturity. These principles are very potent; they are manifest since they entail, despite ourselves, a conviction that cannot be lost but by the greatest of efforts that are equivalent to those we need to persuade someone that two and two are not four, and that straight lines do not exist.

*Source:* AAdS, dossier Guettard, Chabrol donation, box 2. The letter was written before Guettard's death in 1786.

# Four Nonscientific Manuscripts by Laplace

## On the Books of the New Testament

I have endeavored to learn about the authenticity of these books, and my research on the subject has led me to this conclusion by Dodwell, one of the most erudite men in these matters:[1]

> Canonical writings lay hidden in the book-boxes of individual churches, or even of private men, until a later epoch—the era of Trajan, or perhaps Hadrian,[2]—therefore not reaching the notice of the Catholic Church. Or, if by chance they had appeared in public, they would still at that point have been obscured by such a multitude[3] of Apocrypha and pseudepigrapha, that they could not have been distinguished from them, without new investigation and new evidence. And from that new evidence, through which genuine Scripture was distinguished from Apocryphal books and writings falsely ascribed to the Apostles, derives all the authority which the true Apostolic writings thereafter possessed and today possess in the Catholic Church. But even so, that later attestation of the Canon was subject to the same problems as our own traditions of the elders whom Irenæus saw and heard.[4] For it was removed from its origin by such a great span of time, that its credibility was limited by the number of people who had been in contact with those still remoter periods.
> 
> But certainly before that age, which I have said was that of Trajan, the specific number of books[5] had not yet been established which thereafter would have to be employed in deciding sacred matters of faith; nor had the spurious writings of the heretics been rejected, nor the faithful warned to be on guard thenceforth against using them.

Moreover, the true Apostolic writings were customarily bound together with Apocryphal books in the same volume, with the result that there existed no popularly recognized mark or ecclesiastical judgment as to which should be preferred to which. We have extant even now exceptionally brilliant authors of these times,[6] Clement of Rome, Barnabus, Hermas, Ignatius, Polycarp (who doubtlessly have written in the order in which I have named them):[7] all more recent than the extant New Testament writings,[8] with the exception of Jude and John.[9] But you will not discover in Hermas even a single passage of the New Testament, and in the rest, not even one writer of Gospel mentioned by name. And if they happen to display passages which resemble things in our Gospels, you will for the most part, however, find them to be so altered and interpolated, that it cannot be known whether they took them from our own, or from other, Apocryphal Gospels. But, in several instances, these same writers also make use of Apocryphal writings, which certainly do not exist in the currently known Gospels. Thus it is certain that until then, no distinction between the Apocryphal and the Canonical books of the New Testament had been established by the Church, particularly if we may also admit this further observation, that they affix no censure nor any other mark to the Apocryphal books, whence the reader can gather that they set less store by the Apocrypha than by the true Gospels. One is therefore prone to suspect that any agreements with our own texts did not come about because of any particular plan, in accordance with which it had been decided that questionable things were to be confirmed by reference to canonical writings. Further, one suspects that even these similar readings could have been taken from Gospels other than the ones which we possess. But why should I confine myself to speaking of the Canonical books?[10] It is not even certain that the Gospels had become known to the Church and were in widespread use among the clergy, to judge by later Canonical writings themselves. The authors of that age were not in the habit of adorning their works, mosaic-like, with citations from the New Testament. This was a custom of later writers, and it was also their custom in the case of those Scriptures which they themselves recognized, for they very frequently do cite the books of the Old Testament. Doubtless, they would have cited the writings of the New Testament also, had those, too, been received into their Canon. St. Paul, *Acts* XX: 35, cites a maxim of our Lord. If he adduced it from a written source, it was certainly from no Gospel which we possess. Thus the Gospels lay hidden in the corners of the earth where they had been written, so that not even the later writers of Gospel became aware of what their predecessors had written on the same subjects. Otherwise there would not have been so many apparent contradictions,[11] which almost from the first establishment of the Canon have exercised the minds of learned men. Had St. Luke seen the genealogy of the Lord in Matthew, he certainly would not have produced another having almost nothing in common with it, and at that, advancing not the slightest reason for such a different scheme. And when he declares in his preface that the reason for a new composition is that he himself approached the work supported by the stories of eyewitnesses,[12] this clearly implies that the authors of the Gospels which he had seen were bereft of such support. They were far from being "eyewitnesses" themselves, and no wonder, as they did not even trouble, in their turn, to question eyewitnesses[13] with any care. Accordingly, it is implied that they were untrustworthy, and that their credibility was rightly called into doubt. Thus it is plain that the authors of Evangelical History seen by Luke must have been other than the evangelists whom we have. (*Dissert. 1ª in Irenæum* § 38.39).

The apologists for the Christian religion are forced to concur with this conclusion by Dodwell, which, for me, seems to prove that the first Christians had several beliefs in common, such as the mission of Jesus Christ announced by St. John the Baptist, his preaching, death, and resurrection, and that on all details each was free to believe as he wished. But as the number of Christian congregations increased they organized according to a common plan, and among the large number of religious writings in general use, they selected the most widely accepted ones to make up the Scriptural Canon. I think that at first they were not very surprised by the contradictions found in these works. But as the respect they were later to hold for them increased to the point of regarding them as divinely inspired, they tried to reconcile them by means that will never satisfy an unprejudiced man of common sense. St. Matthew and St. Luke report the birth of Jesus Christ in a wholly dissimilar manner.[14] Not a word is said by St. Luke of the coming of the Magis, the flight of the Holy Family into Egypt, and the massacre of the innocents ordered by Herod.[15] Can we reasonably suppose that he would have omitted events of such importance if he had given credence to them? On the contrary, his narrative seems to exclude them by these words: "When the days of the purification of Mary were fulfilled according to the law of Moses (this time period occurred 40 days after the birth), Joseph and Mary took Jesus up to Jerusalem to present him to the Lord."[16]

In an attempt to reconcile the two genealogies of Jesus by St. Matthew and St. Luke, it has been asserted that the first one was that of Joseph and the second one, of Mary.[17] But if St. Luke had claimed to have written the genealogy of Mary, would he not have said so? Would he have said that Joseph was the son of Heli, if he had only been his son-in-law, and if Heli had been the father of Mary?[18]

St. John makes no mention of the institution of the Eucharist in the Last Supper before the Passion of Jesus Christ, an institution that the three other Evangelists report in a almost identical terms;[19] but he only states that during this supper Jesus left the table and washed the Apostles' feet.[20] There is no overt contradiction between the Evangelists on this point: a negative testimony does not contradict a positive testimony, but it weakens it when it deals with events of such importance that a historian would very likely have reported them if he had judged them worthy of credence. In this instance his silence becomes a strong proof against their having occurred.

St. John makes no mention of the institution of the Eucharist in the Last Supper before the Passion of Jesus Christ, an institution that the three other Evangelists almost always report in similar terms.[21] He tells us that during this supper Jesus rose from the table to wash the Apostles' feet. There is no overt contradiction between the Evangelists on this point, since a negative testimony does not contradict a positive testimony. But it invalidates it when the event is of such importance that those who have omitted it would very likely have reported it, if they had given credence to it. Their silence then becomes a strong proof against the occurrence of this event.

Several incidents of the Passion and of the Resurrection of Jesus Christ reported by the Evangelists are very difficult to reconcile. According to St. Matthew, the two thieves crucified with Jesus Christ cursed him.[22] Only one cursed him, according to St. Luke, and was sharply rebuked by the other man.[23] According to St. Matthew, Jesus appeared to Mary and to Mary Magdalene, who on the morning of the day of his Resurrection had gone to see his tomb. He ordered them to tell the Apostles that they would see him in Galilee.[24] According to St. Luke, these women caught sight near the sepulcher of only two men clothed in dazzling raiment, and they did not at all see Jesus, who appeared, for the first time, only to two of his disciples going to Emmaus.[25] These disparities and the many similar variations found in the life and death of Jesus

Christ prove that the early Christians attached little importance to such details, and even to several fundamental particulars of our faith, such as the Ascension of Jesus to Heaven after his Resurrection, which St. Mark and St. Luke report,[26] and about which St. Matthew and St. John do not speak at all.

Several critics have been led by a close inspection of the Gospels to think that there existed among the early Christians a primitive Gospel that was copied almost literally in several texts by our Evangelists, and which probably contained only the most commonly accepted matters. In the midst of the darkness covering the early era of Christianity and the contradictions found in the Gospels, it is difficult to form a clear notion of its origin and its founder. It only seems very probable that a Jew, filled with enthusiasm, preaching a gentle and pure morality and the equality of all men, promising eternal rewards—especially to the poor—and directing his disciples to preach his doctrine to all peoples without distinction, became the leader of a sect sufficiently widespread to have aroused the jealousy of other sects, and to have led the priests, whose enemy he had declared himself to be, to have him condemned to death. The entire remainder of his life seems to me either uncertain, or invented, or filled with absurd tales such as that of the temptation of Jesus by the devil, which is reported by St. Matthew;[27] as for the healing of diseases, possessions by the demon, and the great gathering of people to be healed, we see in this only what has often occurred in our own day. Popular credulity accepts without examination the most exaggerated accounts, sometimes strengthened by extraordinary healings that the influence of a very excitable imagination on the animal constitution can effect. As for genuine miracles, such as resurrections, our Gospels are far from offering the necessary certainty to establish their actual occurrence.

I shall take as an example the most important of all resurrections that of Jesus himself. Three things are needed to establish it, namely the Crucifixion of Jesus Christ, his death, and his reappearance after his death. His existence and his Crucifixion seem rather likely to me; they are supported by the testimony of Tacitus,[28] though it can be said that this great historian only reported a popular rumor on a subject that he thought to be of too little importance to clear up. It does not seem entirely unlikely to me that Jesus Christ did not die during the Passion. The four Evangelists agree in stating that Joseph of Arimathea asked for and obtained from Pilate the body of Jesus Christ and that having taken it down from the cross he wrapped it with a cloth and placed it subsequently in the sepulcher.[29] But if the disciples of Jesus removed it from the tomb during the night, as the rumor in Jerusalem had it, according to St. Matthew,[30] could it not be possible that his death had only been apparent, and have we not seen a great many examples of executed men who were brought back to life? I know that according to the Gospel of St. John a soldier, to make sure of Jesus's death, pierced his side with a blow from his lance;[31] but this circumstance, omitted by the other Evangelists, still does not suffice to establish his death. Despite the difficulties that this explanation of the reappearance of Jesus after his Crucifixion may present [presents], one should no doubt accept it sooner than believe in the Resurrection, if it were truly patent that he reappeared to his disciples; but this, in my opinion, is insufficiently proven. If there is anything at all certain in the sum of human knowledge, it is the invariability of natural laws. Never has this foundation of any kind of certainty been belied by well-attested facts. And every time that we have been able to subject to a rigorous examination facts that seem to contradict it, they have been shown to be false, or else dependent on hidden circumstances that we have been able to uncover, and sometimes to natural but still unknown causes, whose investigation they have stimulated. Nature often presents us with extraordinary phenomena whose causes are unknown to us. Thus an excited imagination produces effects that appear supernatural, but which are basically only the consequences of physiological laws worthy in all respects of the attention of physicians who are philosophically inclined. These effects have aroused considerable awe in all periods of history. The natural apprehensiveness of the human mind inclines it to seek the source of any matter that, in its opinion, seems to deviate from everyday laws, and to collect explanations tending to assuage the mind. In the centuries of superstition and ignorance nothing satisfied this purpose better than the belief in supernatural agents, to which all minds were predisposed by the prejudices of childhood, education, and prevailing opinion.

The first and most infallible of the principles of criticism, therefore, is to reject miraculous facts as untrue. By adhering to it, one will always be in agreement with the sensible part of mankind. It would be absurd to apply the common rules of criticism to these facts and to discuss them as if they were ordinary facts. A history tainted with miracles should, by that fact alone, arouse a great deal of suspicion, which it would nevertheless be unfair to carry to an extreme. Thus the miracle of the healing of Pascal's niece, recounted by Racine in his *History of Port-Royal*, should definitely not diminish credence in the other facts that this great man, a poet as veracious as he is admirable, has recorded in this history.[32] We would, therefore, be able to follow the progress of Christianity in the Acts of the Apostles, if we were quite certain of the authenticity of these Acts—the only writing we have on the history of the first years succeeding the death of Jesus Christ. But it is clear from the Dodwell passages cited how much the Christians in the beginning fabricated spurious works, and altered those in existence; and Lucian, in his essay on the death of Peregrinus, portrays this impostor arousing the liveliest enthusiasm among the Christians of Asia, explaining their books and writing some himself.[33] When he was imprisoned for having adhered to their cause, they lavished him with assistance and their . . . [illegible] with a dispatch, states Lucian, that is impossible to describe. The nature of the sectarian spirit is to stamp on the sensorium a state of mind that allows itself to be easily seduced and to exaggerate the probabilities favorable to it, by attenuating contrary probabilities. The Apostles and their historians are depicted as simple men, full of candor, incapable of deceit by their very ignorance, and proclaiming matters so striking that they would have been denied on the spot, had they not been true. In addition, one gives complete credence to all that they have reported, and even to miracles that one considers must have entered into the designs of God in order to establish the religion that he wanted to give to mankind. The means of verifying miracles are lacking, and no doubt if this verification had been made at the time when it was possible to do so, one would have discovered the illusions and frauds by which their faith had been established.

Chance has given us the means of verifying the following fact of nearly the same antiquity as the miracles in the Gospels, but which is much better authenticated, and whose improbability, though incomparably smaller than that of a miracle, should have sufficed at least to counterbalance the testimonies supporting it. "If we are to believe St. Justin," states the abbé de Condillac (*Cours d'étude*, Book XV), "Simon the Magician was received in Rome like a god; a statue of him was erected with the inscription *Simoni deo sancto*.[34] The saint himself saw this statue, which was still in existence around the year 150. St. Clement of Alexandria, St. Irenæus, St. Cyril of Jerusalem, Tertullian, Eusebius, and Theodoret vouch for the same fact, and St. Augustine adds that this statue had been erected by the public authorities.[35] Here is a well-verified fact, and which seemingly has never been contested by the pagans. On the isle of the Tiber, however, in the same spot where St. Justin asserts having seen this statue, a statue was unearthed in 1574 with the inscription, which still exists, *Semoni deo sancto*. That was the name of the divinity who presided over oaths. This discovery has led to the conjecture that St. Justin, preoccupied with Simon the Magician, was reading too quickly and fell under a misapprehension." The true inscription reads *Semoni sanco deo fidios sacrum*. In his *Decline and Fall of the Roman Empire*, Gibbon has worked out in a very philosophical manner the causes of the progress of Christianity.[36]

## On Mysteries

I have on several occasions made the following objections against mysteries to very able theologians: they have only answered it in a manner that is vague and as insignificant as the very things that they seek to defend. We can believe in a proposition only insofar as we have a conception of each of its terms. When I am told that the columns of a building are white, I have the idea of column and of whiteness; I can thus give my assent to this assertion, which a man blind from birth could not. To him the statement thus expounded would have no sense. When a mystery is asserted, we are like this blind man. For example, if I am told that God exists in three persons, I can neither believe in this proposition, nor reject it, since I am no more able to attach a concept to the word "person" than the man blind from birth can to the word "whiteness." The objection has been raised that there are mysteries in nature, such as attraction, magnetism, etc. Here, however, the word "mystery" has a different meaning: it signifies an unknown cause of known effects. Observation and calculation have shown that all parts of matter attract one another, proportionately to their masses and inversely to the squares of the distance between them. We have a distinct concept of physical molecules; and we conceive of their mutual attraction by the initial movement brought about. Hence we are able to believe in this attraction. There is no mystery here; only in relation to its cause is there a mystery.

The most eminent theologians have truly sensed the need to attach some concept to the word "person" in the mystery of the Trinity. St. Augustine in his book *The City of God*, and following him, Bossuet in his *Universal History*, tell us that it is sufficient to descend into one's own self in order to have an image (as a matter of fact, a very imperfect one) of this mystery.[37] We recognize there, in effect, our own being, our thought or our intellectual faculties, and our self-love which stems from this. These three things can, in their opinion, represent to us the three persons of the Trinity.[38] But this explanation, if it were true, would wipe out this mystery, which no longer would be anything but the existence of a supreme intelligence combining to the utmost degree all the properties of the human soul—a fact that has never been viewed as a mystery. According to this explanation drawn from Plato's reveries, the mystery of incarnation consists of the hypostatic union of the divine intelligence with [mankind] human nature; so that there are in Jesus Christ two natures and two wills. All this elicits a smile in any reasonable man free from prejudice.

But of all mysteries, the most inconceivable and absurd is that of transubstantiation. It offends at the same time reason, experience, the testimony of all our senses, the eternal laws of nature, and the sublime ideas that we ought to form of the Supreme Being, ideas that this mystery perverts in the strangest way. According to this mystery, the actual substance of bread and wine is not destroyed, but is turned into the body and blood of Jesus Christ. We define substance as the basic properties of a being and of its modifications.[39] Thus matter can be in motion or at rest; it can change its shape, but at the same time we know that the basic traits of all these accidents cannot become a different substance. Otherwise it would itself be a modification. Moreover, since this mystery is taking place at different places at once, the same body would be in different points of space at the very same moment—a fact that runs counter to all our experience. All our senses—touch, sight, hearing, taste, and smell—continually show us bread and wine during the Eucharist; all their physical and dynamic properties, their weight and their chemical affinities, do not appear changed, while we witness none of those that the body of Jesus Christ should show. The consecration of bread and wine would thus bring about a succession of illusions and miracles renewed at every moment, and forever. Obeying the voice of a frail mortal, the sovereign lawgiver of the universe would suspend the laws that he has established, and which he seems to have maintained invariably, and all this, in order to transfer instantaneously the body of his own son to the spot designated by the will of the priest. I know nothing in ancient and modern religion more incredible than the sum of all these things. The faith that it is given by entire nations and by otherwise very enlightened men is, in my opinion, the most striking example of the influence of the prejudices of education, of prevailing opinion, and of habit on our judgments. When we see men such as Bossuet and Arnauld expending their nightly labors to maintain such strange paradoxes, and mingling them with matters of genius, we ought to pity the human mind, and forgive its weakness for the errors that most of those offended by them would have shared had they been placed in the circumstances that disseminate these errors.[40] I have often wondered what is the greatest known truth and the greatest known absurdity. Upon reflection, it has seemed to me that the greatest truth is universal gravitation. Transubstantiation seems to me the greatest of all absurdities.

## On the Notion of Power

We derive the notion of power from the consciousness of what takes place within us when we act. Let us see what consciousness tells us about this matter. A man driven by hunger sees food within his reach, and he extends his hand to seize it. It is obvious that the image of the food, depicted on his retina, and transmitted by the optic nerves to the sensorium, arouses in it motions that revive motions previously connected to the perception of the ability of food to appease hunger, and that awaken this perception. These motions combine with those that are accompanied by the sense of need, and from this combination result motions accompanied by the desire for the food and the will to seize it; through the intermediaries of the nerves, these motions transmit to the muscles the force necessary to move the hand toward the food. The existence of all these motions of the sensorium is incontestable; yet we do not sense them; we do not know the substances to which they belong. We are only conscious of the feeling of hunger, the perception of food, the desire and the will to seize it—all of which accompany these diverse motions. But how do these motions of the sensorium give rise to these perceptions? What is their influence over the motions that follow them? According to the Cartesians, who view animals as automatons, these motions are carried out in their sensorium by virtue of the general laws of nature, without being accompanied by perception, feeling, or volition. It is likewise by virtue of the laws of the physical world that, according to Leibniz, these motions are brought about. At the same time, however, perceptions, feelings, and the volition concurring with them develop in our soul according to the special laws of the spiritual world. Their correspondence with the motions of the sensorium is what he calls *preestablished harmony*.[41]

A powerful analogy challenges the Cartesian hypothesis. An almost overwhelming inclination leads us to believe that in animals, motions so diverse and in all respects so similar to ours in the same circumstances, are accompanied by the sensations and feelings that we experience.

In the Leibnizian hypothesis, the will does not bring about any motion. All it does is to accompany those that correspond to it, and whose cause it deems itself to be. The power that it claims for itself is but an illusion similar to the one that leads it to view itself as being determined by itself and by its own energy whenever the frequently fugitive reasons for its determination have disappeared. Every time we observe a group of events that is always followed by a group of different or similar events, we are inclined to consider the first as causes of the second. It is in fact the only thing capable of giving us a notion of what we call *cause*. But the uniform succession of these events can only be fortuitous. Thus the lunar syzygies constantly precede high tides in our harbors, while the quadratures constantly precede low tides. Yet we know today that lunar phases are not the true cause of these phenomena and that they are only the circumstances that always accompany them but that do not lead to their formation. It is therefore an illusion to attribute these phenomena to the phases of the moon; and we can fear even more a similar illusion in our tendency to view the will as the cause of the movements of [the limbs of] our body, because the will does not modify them in any way. It adds no new force to those of the sensorium, which acts like a spring, without displacing the common center of gravity of our body and of surrounding bodies.[42]

We undoubtedly experience a large number of illusions that, most likely, have not all been recognized. Is there a stronger illusion than the one that leads us to attribute to bodies the colors they are seemingly endowed with? Yet various means have convinced us that this is not so, and that colors are pure sensations. If we think it impossible, however, not to admit that the painter's will guides the brush that draws a face on the canvas, we must then consider the effort we make in moving bodies and overcoming the resistance they offer us as a real force subject to the laws of dynamics, but acting within us without giving information on either its direction, or the substances on which its action is exerted, or the motion it imparts to them.

Leibniz conceived his preestablished harmony because of the impossibility of explaining the action of the mind over matter. But we know matter only very imperfectly; we are even less informed about what we call *mind*. Our outer senses show us matter as an extended substance, impenetrable and capable of motion, whose various parts are continually changing with respect to one another according to laws, the most general of which are now well-known. But the inner nature of the forces that bring about these changes is completely unknown. There is certainly a special mode for a body in motion that distinguishes it from a body at rest, and by which it would indefinitely continue to move in a straight line in a given direction, if there were nothing opposing it, a mode that it transmits to the bodies that it meets, according to the simplest law imaginable. But this mode, to which we have given the name *force*, is known only through its effects. The universal force causing physical molecules to gravitate to each other at a distance, proportionately to their masses and inversely to the square of the distance between them, is likewise known only through its effects. The same is true for all intra-molecular forces that are manifest in physical, chemical, and physiological phenomena. Matter, which several philosophers view as inert and passive, develops a prodigious activity and astonishing properties in such imponderable fluids as light, heat, magnetism, and electricity. Does not all this seem to indicate common qualities and special properties in all these entities? Are there not between them essential differences based on their nature, and on which the great differences in their behavior depend? Could it not be that the intellectual qualities perceived by the inner sense belong to them, in such a way that the substances termed spiritual are among those whose effects are disclosed by the our outer senses, and whose intimate properties are revealed to us by the inner sense when these properties are placed in circumstances favorable to the development of these properties?

What are we to do in the midst of these uncertainties? Observe the phenomena that the outer senses and inner senses can make known to us, determine their mutual relationships, and ascend to the level of general laws for these relationships. Physical phenomena perceived by the outer senses have laws that are peculiar to them, and that constitute the subject-matter of natural philosophy. Intellectual phenomena likewise have special laws that constitute the subject-matter of psychology; finally, whatever the principle of the connection between these two kinds of phenomena may be, they have mutual relationships that can be observed through the connection of the senses that correspond to them. All these laws are immutable, and derive from the unknown nature of the substances they direct. The caprices of our will, even when we judge it to be most independent, are subject to these laws, and it is only by a series of illusions that we attribute to it the power of determining itself, and of moving by the sole act of its own command the parts of our body.

These illusions have thrown men into the strangest errors. Accustomed to look on the will, when guided by the mind or aroused by the passions, as the real cause of the phenomena that it commands, they have attributed all phenomena to similar causes, which they have endowed with all the qualities of human nature, but whose power they have exalted in proportion to the size of the effects brought about. That was sufficient for the tranquility of the imagination, which is always curious to learn the cause of everything that strikes our senses. This tranquility is easily achieved by most men, who rarely venture beyond first questions. Thus, in the present instance, no inquiry was made at first into the origin of these invisible causes, and the oldest author to have written on theogony—Hesiod in his poem[43]—asserted that the gods and men are both created by unknown natural forces; we can therefore consider this theogony, and paganism in general, as if it were a kind of superstitious atheism. Extraordinary phenomena or great disasters made them invoke imaginary phantoms, which were believed to be their cause. It was thought they would be appeased by the most bizarre measures and often by the cruelest sacrifices. To them were erected temples, whose ministers—among the oldest civilized nations—sought to bestow a regular pattern on religion. As observers of celestial bodies, and noting the influence of the sun on the seasons and on all terrestrial products, they turned it into a major divinity, to whom they subordinated the planets and the stars, making them secondary divinities. Subsequently concealing under the veil of allegory celestial phenomena and the operations they attributed to these celestial bodies, they made up fabulous stories of gods and heroes, which they offered to the credulous ignorance of the people. These fables, having been passed down from century to century, and the tradition of their allegories having been lost, were, because of their remote antiquity, generally accepted as truths by the very ministers of these religions. Such is the origin of all theogonies, whose bases are astronomical and whose traces Christianity has preserved. For example, it is very likely that the birth of Jesus Christ was placed at the winter solstice to indicate the return of the sun toward the equator. Certain philosophers, who rose to the level of general speculations, have increasingly curtailed the number of these divinities, and they have finally reduced them to a single one to whom some have attributed the general order of the universe; others, the very creation of the beings that compose it. In the process of attributing to divine will the power that they had ascribed to the rule of the human will, they extended this power to encompass nothingness, from which all of nature was made to emerge. Some have even held that unless this creation were continually renewed, the universe would revert to nothingness. That is how far false ideas on power and freedom have misled the human mind.

## On Causality

We are inclined to regard as an event's cause that which it always follows. Thus when the collision of a body with a billiard ball is always followed by the billiard ball's motion when nothing opposes it, we consider this collision as the cause of this motion. Again, in the case of the movement of the arm following the volition of moving it, we consider the action of the will to be its cause. But this connection between two events, in which one always follows the other, is not sufficient to prove that the first is the cause of the second; we must further show how it brings it about, since observation shows us phenomena that are always followed by other phenomena, of which nevertheless they are not the causes. As an example, I will mention the tides: the higher ones always follow the lunar syzigies in our harbors, while the lower ones follow the quadratures. It would appear, therefore, that the lunar phases are the causes of tides. Yet we are quite certain today that this is not so. We know that tides are brought about by the attraction of sea waters toward the centers of the sun and moon. The theory of universal gravitation clearly displays the connection of this attraction with the phenomena: it shows that the tides are highest toward the syzygies, and lowest toward the quadratures, and always happen concurrently with the variations in tides without being their cause, and that their principal variations stem from the respective position of the two celestial bodies, a position on which the variations of lunar phases also depend, which, in turn, happen concurrently with the variations in tides without being their cause. Thus the concurrence of phenomena establishes that earlier ones are the cause of subsequent ones only insofar as we see how they bring them about, and this is not what happens with movements that follow acts of the will.

To make sure of this, it suffices to reflect on what takes place within us as we formulate these acts. A man driven by hunger sees food within his reach, and he extends his hand to seize it. It is obvious that the image of the food depicted on his retina, and transmitted to the brain by the optic nerve, arouses in it motions that, by reviving the motions previously connected to the perception of the nutritive quality of the food, awaken this perception. These motions combine with those that the sense of need gives rise to, and from this combination result motions accompanied by the desire for the food and the will to seize it. Through the intermediaries of the nerves, these motions transmit to the muscles the forces necessary for the hand to seize the food. The existence of all these motions of the sensorium is incontestable; yet we do not sense them. We are only conscious of the feeling of hunger, the perception of food and its nutritive quality, the desire and will to seize it—all of which are connected in an unknown manner to these diverse motions.

According to the Cartesians, who view animals as automatons, these motions are carried out by virtue of the general laws of nature, without being accompanied by perception, feeling, or volition. It is likewise by virtue of the laws of the material world that, according to Leibniz, these motions are brought about in man. At the same time, however, perceptions, feelings, and volition develop in our soul according to the special laws of the nonmaterial world: their correspondence with the motions of the sensorium is what he calls *preestablished harmony*. The Cartesian hypothesis runs counter to analogy. An overwhelming inclination leads us to believe that in animals, all these motions, so diverse and in all respects so similar to our own, are accompanied by sensations and feelings that are nearly similar to those that we experience. In the Leibnizian hypothesis, the will does not bring about any motion: all it does is to precede those that follow it, and whose cause we deem it to be. The power that it claims for itself is but an illusion similar to the one that leads us to view the will as being determined by itself and out of itself whenever the frequently fugitive reasons for its determination have disappeared. The hand of the artist, even in his most marvelous creations, is not guided by his intelligence; it is moved by virtue of the general laws of motion combined with the dispositions that the artist's sensorium has received from nature and from all the causes that have modified it. The operations of his mind take place at the same time as his work, but they follow different laws, and are related to them only by a rapport of harmony. The power that we attribute to them is but an illusion similar to the one that leads us to view the will as being determined by itself and by its own energy whenever the reasons that determine it have disappeared.

Leibniz conceived his preestablished harmony to handle the impossibility of explaining the action of the mind over matter.[44] But we know matter only very imperfectly. We are even less informed about what we call *mind*. Our outer senses show us matter as an extended substance, impenetrable and capable of motion, whose various parts are continually changing with respect to one another according to laws, the most general of which are now well-known. But the inner nature of the forces that bring about these changes is completely unknown. There is certainly a special mode for a body in motion that distinguishes it from a body at rest, and by which it would indefinitely continue to move in a straight line in a given direction, if there were nothing opposing it, a mode that it transmits in a collision according to the simplest law imaginable. But this mode, to which we have given the name *force,* is known only through its effects. This holds true for the universal force that causes physical molecules to gravitate to each other at a distance inverse to the square of the distance between them, and in general, for all the forces that appear in physical, chemical, and physiological phenomena. Matter, which several philosophers view as inert and passive, develops a prodigious activity and astonishing properties in such imponderable fluids as light, heat, magnetism, and electricity. Does not all this seem to indicate common qualities and special properties in all these entities? Are there not between them essential differences based on their nature and on which the great differences in their behavior depend? Is it impossible that the qualities that our inner sense makes us discover in ourselves belong to some of those whose existence is revealed to us by the outer senses?

What are we to do in the midst of these uncertainties? Observe the phenomena that our outer and inner senses reveal to us, determine their mutual relationships, and ascend to the level of general laws for these relationships; finally, discern among the observed appearances those that stem from illusions. Their number is very large, and we probably shall never succeed in recognizing them all. One of the strongest is the illusion that transfers to the bodies the colors whose sensations they make us experience: it is easy to recognize it by various means and to assure oneself that these colors are within us. We have just observed that in the case of the hypothesis of preestablished harmony, an illusion leads us to attribute the control of his brush to the will of the painter. This illusion is no stronger than the preceding one. If, however, we deem it an impossibility, we can then view the act of will of the artist whose hand sketches a face as identical to the effort we make in moving bodies and in overcoming the resistance they offer us. We are conscious of it, without being conscious of its direction, of the molecules on which it immediately acts, and of the motion that it imparts to them. It is subject to the dynamic principle of the equality of action and reaction, and it cannot move the center of gravity of a system of bodies of which we are a part.[45]

It acts like a spring that would be released by the causes that determine the will.

This is what the inner senses and experience teach us about the power of volition. One is far from recognizing in it this power that the vulgar attributes to the will on our limbs, which seem to obey its single command alone. This illusion has had strange consequences that, modified from century to century since the beginnings of society, still rule mankind. In the beginning and with an apprehensive curiosity, men inquired into the causes of phenomena, especially those that concerned their preservation and their happiness. As they found within themselves the causes of a great number of effects, they assumed that observed events were due to similar causes, and they peopled the Earth and the heavens with visible or invisible intelligent beings, who, through their wills, produced these events. This hypothesis, which seemed so natural and simple, and was sufficiently soothing to the imagination, was accepted by all peoples, with various modifications, and became the basis for different religions. Everywhere at the time of great disasters they invoked these imaginary beings and to appease them, they had recourse to the most bizarre means, and often to the cruelest sacrifices. To them were erected temples, whose ministers bestowed a regular pattern on religion. As observers of celestial bodies, they noted the correspondence of their appearances with the seasons they assumed that they were the causes of these variations, and they deified them, making a major divinity out of the sun, on account of its great influence on terrestrial products. Subsequently concealing under the veil of allegory celestial phenomena, and the operations they attributed to them, they made up fabulous stories of gods that they offered to credulous, ignorant men. These fables, having been passed down from century to century, and the tradition of their allegories having been lost, were, because of their remote antiquity, generally accepted as actual truths by the very ministers of these religions. Such is the origin of all theogonies, whose bases are astrological and whose traces Christianity has preserved.[46] For example, it is very likely that the birth of Christ was placed at the winter solstice, by analogy with the return of the sun toward our climes.

At first little concern was expressed about the origin of divinities. Hesiod, the oldest author to have written on theogony,[47] asserted that the gods and men are created by unknown natural forces. We can therefore consider paganism as a superstitious form of atheism. Certain philosophers who subsequently raised it to the level of general speculations have adopted a first, unique eternal cause—which, by the sole command of its will, has fashioned the universe, and in the opinions of some men, has made it emerge from nothingness. Some have gone so far as to think that, were it not for an incessantly renewed creation, the universe would constantly revert to that state.

*Source:* All four of these manuscripts in Laplace's hand are in the AAdS, dossier Laplace. The translations from the French and the Latin are by Roger Hahn.

## Illness and Last Moments of M. de Laplace

I went there for the first time on Sunday evening, February 4. As soon as he knew me to be there, he wanted to see me, and I came to his bedside. Even though quite ill, he held out his hand, clasped mine, and assured me several times of the "tender friendship he had always had for me." Thereafter, when I was present, he always received me by his bed, always spoke to me, clasping me with a remarkable vigor and repeated all the pleasure he had in seeing me, and "the pleasure he expected when speaking with me as soon as we could dine together." I interpreted this, as well as the singular pleasure that my presence caused him, and by the tender way he spoke to me, as recalling the frequent conversations we had had on Christian religion for over thirty years—but especially during the last seven or eight [years of] conversations, in which he had been able to assure himself of the happiness I experienced myself in my profound conviction about these august truths. It was natural for him in his condition to be led to renew his meditations because of their import.

Never was he more cordial than on Saturday, February 24. He was about; I was alone with him for a long time; he often held out his hand and spoke some kind words. I presented him to his son-in-law De Portes, and to his second wife. A third party would have taken me for his son-in-law or his son. Since after these visits he was tired, his head dropped and I thought him asleep. Mrs. Laplace being there, I took the occasion to leave discreetly, but he called me back with an almost unintelligible voice to hold out his hand, squeeze it, and to say goodbye with such an expressive look on his face that I was as troubled as touched by it. He wanted me to dine with him, but I answered that I was already engaged with Legendre, who would surely learn with pleasure of his expected recovery. This response seemed to please him and he nodded approvingly very markedly with his head.

On Tuesday the 27th at 3 P.M. he was up and I spent three hours with him, but he was in fact less well. I presented him to Cuvier, whom I had been asked to call, and he was able to speak with him with all his wits. I stayed along with him for a good time after Cuvier's departure. He then wanted to go back to bed, but first took three turns around his room supported by his wife and me. He was still rather strong, and this was to be his last outing. Mrs. Laplace undressed him while I kept him up, but when we laid him on his bed I had a lot of trouble because of my arm. He was complaining a lot, and showed some absence of mind. We quickly asked to fetch his doctor, but since only a lady servant was sent, I wanted to run shortly thereafter, and I was already on the stairway when he recalled me "wanting to have me by his side always" until the arrival of his doctor. Thereafter I did not speak to him.

His condition worsened the next day, the 28th. On March 2 there was no more hope. However on Saturday the 3d, he was a bit better. He received his son, Poisson and Bouvard. Poisson told me that he had first said hello to his son in a broken and unnatural voice; and that Poisson said to him: "Mr. Laplace, here is your good friend Bouvard, whose calculations put your beautiful discoveries on Jupiter and Saturn in such good light, discoveries whose fame will never die" (and this to try to discover if he grasped his state of affairs . . .). "In his healthy and clear voice" after a moment of silence [he said,] "We chase after chimera" . . . and then turned over never to speak again. Here was a deep thought for him, and rather characteristic. It seemed to indicate that he glimpsed from then on the real vanity [of activity] with which he had for too long been singly preoccupied. Poisson judged that he knew he was near the end.

From then on, he lost his wits. In the night of Sunday 4th to the 5th, he uttered a death rattle at three in the morning. The vicar from Arcueil was called, who in concert with the vicar from the Missions [Etrangères] administered the sacraments at 7 A.M. in the presence of Poisson and his wife, though he was not conscious of what was happening around him; then, without pain, he passed away at 9:05 A.M. on Monday March 5 1827. Newton had died on Monday March 20, 1727.

*Source:* Text written by Jean Frédéric Théodore Maurice, private archives of the Maurice family, Geneva.

# Notes

## 1. Norman Beginnings

1. The family, which maintained some records despite fires and wars, has been generous with its hospitality and information. Most of the records, once located at the Château de Mailloc near Orbec, are presently in at the Bancroft Library of the University of California, Berkeley. Mailloc entered the family estate when Laplace's son-in-law, Adolphe de Portes, purchased it in 1812. Laplace's remains, originally interred at Père Lachaise in Paris, were moved to Saint Julien de Mailloc in 1888.
2. Georges Abel Simon, "Les origines de Laplace: Sa généalogie, ses études," *Biometrika* 21 (1929): 217–230; and "Laplace: Ses origines familiales et ses premiers débuts," *Normannia* 10 (1937): 477–496, are the indispensable studies underpinning the first two chapters of this biography. Nevertheless, consultation of the sources used by the abbé Simon has in a few cases yielded a different reading.
3. Simon's description of the Carrey family differs in his two articles; see ibid.
4. Henry Le Court, *Généalogie de la famille Le Cordier* (Laon, 1909), pp. 45–46; Romuald Szramkiewicz, *Les régents et censeurs de la Banque de France nommés sous le Consulat et l'Empire* (Geneva, 1974), pp. 54–59; and Serge Sochon, *Pierre Simon Laplace, 1749–1827: Un savant issu des lumières* (Paris, 2004).
5. Population figures are inferred from the 1713 census reproduced in Pierre Gouhier et al., *Atlas Historique de Normandie* (Caen, 1967–1972), vol. 2; and the *Statistique générale de la France: Territoire, population* (Paris, 1837), p. 268, which gives the population of larger towns in 1789. In the former, Beaumont is listed as having 122 *feux taillables*, 2 *feux privilégiés*, and 1 *communauté* [religious].
6. Alain Huet, "Annebault et Bourgeauville aux XVIIe et XVIIIe siècles: Contribution à l'étude démographique du pays d'Auge," *Annales de Normandie* 22 (1972): 277–300.
7. Bernard Garnier, "La mise en herbe dans le pays d'Auge aux XVIIe et XVIIIe siècles," *Annales de Normandie* 25 (1975): 157–180; and Alun

Davies, "The New Agriculture in Lower Normandy, 1750–1789," *Transactions of the Royal Historical Society*, ser. 5, vol. 8 (1958): 144–145.
8. For a less optimistic view of life in Normandy, see Robert M. Schwartz, *Policing the Poor in Eighteenth-Century France* (Chapel Hill, N.C., 1988).
9. Nels W. Mogensen, "Structures et changements démographiques dans vingt paroisses normandes sous l'ancien régime: Une analyse sociale," *Annales de Démographie Historique* (1975): 352 and 357.
10. Marriage contract notarized by Maître Louvet, 16 June 1744, Archives du Calvados, 8E 28474.
11. Although the abbé Simon lists her name as Marie Viel, I read the signature as Marie Piel. In the body of the contract, it is given as Pielle, a phonetic spelling of the same name. Piel is a common surname in this region of Normandy.
12. Some details about the manor are given in Henri Le Court, "Tourgéville-sur-Mer et ses fiefs: Glatigny et ses seigneurs," *Annuaire des Cinq Départements de la Normandie* 52 (1887): 245–246. For descriptions, drawings, and photographs of Glatigny, see Arcisse de Caumont, *Statistique monumentale du Calvados* (Caen, 1859), vol. 4, p. 234; Philippe Déterville, *Grands et petits manoirs du pays d'Auge* (Condé sur Noireau, 1982), pp. 148–150; and Jean Bureau, *Sites et manoirs du pays d'Auge* (Caen, 1968), pp. 14–15.
13. Rental contract notarized by Maître Louvet, 24 December 1744, Archives du Calvados, 8E 28474.
14. Archives du Calvados, C 12069, lists lands bought through notaries at Pont-l'Evêque, Rouen, and Touques.
15. Simon, "Laplace: Ses origines familiales et ses premiers débuts," pp. 483–486. My reading of the sources cited by Simon shows that Bonneval was the largest landowner, with a revenue of 3,400 livres, followed by Leperchey with 2,200 livres, and Louis Bretocq and Nicolas Bretocq, each with 1,200 livres.
16. *Biographie moderne ou galerie historique, civile, militaire, politique et judiciaire* (Paris, 1815), vol. 2, p. 200; unchanged in the second edition (Paris, 1816), vol. 2, p. 229.
17. Philippe Devillard, "Notes pour servir à l'histoire de la vie, de la carrière et de l'œuvre de Pierre Simon de Laplace, 1749–1827," in *Le lycée technique d'état Pierre Simon de Laplace* (Caen, 1972), pp. 8–9 provides good evidence that he was born at the town inn, and offers a contemporary description of the premises on which he was born, which have since been torn down.
18. Archives du Calvados, 4E 123.
19. Marriage contract notarized by Maître Feral, 10 October 1769, Archives du

Calvados, 8E 28507; and Devillard, "Notes." Laplace corresponded with one François Jacques Louize who wrote about family affairs; see F. J. Louize to Laplace, 28 February 1824, in Bancroft, box 10, folder 12.

20. Marriage contract notarized by Amelot of Annebault, 9 November 1788, cited in Archives du Calvados, C 11982, fol. 70r. and C 11971, fol. 142r.
21. Archives du Calvados, 4E 124, 30 June 1788; and contract between Marie-Anne and Pierre Simon Laplace, 13 July 1788, in Bancroft, box 10, folder 12. Pierre de Laplace died on 29 June 1788.
22. Archives du Calvados, C 12088, fol. 44v. His date of death was 22 December 1759.
23. Archives du Calvados, 4E 123. Her last signature is for the baptism of Jean Pierre Le Herissey in 1757. The marriage contract is cited in Mogensen, "Structures et changements démographiques." According to the local historian Philippe Devillard in *Le pays d'Auge* (December 1987): 14, Marie Anne Sochon died on 21 November 1768 and was buried the next day in Beaumont-en-Auge. But his statement that she died at age twenty-five does not allow her to have had children starting in 1745.
24. The death of a Marie Anne Sochon on 3 March 1785, with a Guillaume Mauduit as heir, is recorded in Archives du Calvados, C 12088, fol. 52v. I do not believe this was Pierre Simon's mother, but probably a cousin. In the 1769 marriage contract, his mother is referred to as *feue* (late).
25. Archives du Calvados, 4E 123. My reading of the baptismal certificate is in small ways at variance with that printed in Baldassare Boncompagni, "Intorno agli atti di nascita e di morte di Pietro Simone Marchese di Laplace," *Bullettino di Bibliografia e di Storia delle Scienze Matematiche e Fisiche* 15 (1882): 464; and by Simon, "Laplace: Ses origines familiales et ses premiers débuts," p. 487.
26. Simon, "Laplace: Ses origines familiales et ses premiers débuts," p. 487.
27. Ibid., p. 486. In the 1769 marriage contract, Halley is described as a cousin of the Laplace family, though the link is not explained further.
28. Nels W. Mogensen, "La stratification sociale dans le pays d'Auge au XVIIIe siècle," *Annales de Normandie* 23 (1973): 211–251; and Huet, "Annebault et Bourgeauville aux XVIIe et XVIIIe siècles," p. 281.
29. Henri Le Court, "Nobles ou vivant noblement à Pont-l'Evêque dans l'espace de quinze à vingt ans, depuis 1742," *Annuaire des Cinq Départements de la Normandie* 53 (1888): 228–230.
30. Arcisse de Caumont, "Notices biographiques: Sur M. Bretocq," *Annuaire des Cinq Départements de l'Ancienne Normandie* 23 (1857): 484–487; and Simon, "Laplace: Ses origines familiales et ses premiers débuts," p. 486.
31. On his death certificate, Pierre de Laplace is still listed as mayor.
32. The administrative and geographical links of Beaumont with the rest of

Normandy are listed in Célestin Hippeau, *Dictionnaire topographique du département du Calvados* (Paris, 1883), pp. xxxii–xxxvi.
33. Yves Nédelec, *Le diocèse de Lisieux au XVIIIe siècle* (Paris, 1954).
34. Beatrice F. Hyslop, *L'apanage de Philippe-Egalité, duc d'Orléans, 1785–1791* (Paris, 1965), pp. 43–44.
35. Simon, "Laplace: Ses origines familiales et ses premiers débuts," p. 488.
36. Caumont, *Statistique monumentale du Calvados,* vol. 4, pp. 222–223; and Georges Abel Simon, "Les prieurs de Beaumont-en-Auge," *Bulletin de la Société Historique de Lisieux* 27 (1926–30), 147–175.
37. For all details about this school, I have relied on Archives du Calvados, 2D 1175; Armand Bénet, "Documents relatifs à l'histoire des lycées et collèges: Collège de Beaumont-en-Auge (Calvados)," *Revue de l'Enseignement Secondaire et de l'Enseignement Supérieur* 5, no. 8 (1886): 344–356; Pierre Le Verdier, "Exercices du Collège de Beaumont-en-Auge (1770–1773)," *Bulletin de la Société de l'Histoire de Normandie* 7 (1893–1895): 362–373; and Victor Le Fort, "Le collège et l'école militaire de Beaumont," *La Revue Illustrée du Calvados* 7 (1913): 55–58 and 75–78. See also the current synthesis in Marie-Madeleine Compère and Dominique Julia, *Les collèges français, 16e–18e siècles* (Paris, 1988), vol. 2, pp. 95–98.
38. Ibid.
39. Léon Puiseux, *Notices sur Malherbe, La Place, Varignon, Rouelle, Vauquelin, Descotils, Fresnel et Dumont D'Urville* (Caen, 1847), p. 34. The source for this piece of information was probably Pierre Aimé Lair (1769–1853).
40. Robert Lemoine, "L'enseignement scientifique dans les collèges bénédictins," *Enseignement et diffusion des sciences en France au XVIIIe siècle,* ed. René Taton (Paris, 1964), pp. 101–123.
41. For details, consult the articles by François de Dainville and Pierre Costabel, in *Enseignement et diffusion des sciences,* pp. 29–100.
42. Le Verdier, "Exercices du Collège de Beaumont-en-Auge (1770–1773)," pp. 374–382.
43. Ibid., p. 378.
44. Jean Itard, "Les opinions de l'abbé de la Chapelle sur l'enseignement des mathématiques," *RHS* 5 (1952): 171–175.
45. Jacques Fabre de Massaguel, *L'école de Sorèze de 1758 au 19 fructidor an IV* ([Toulouse], 1958); and "L'enseignement à l'école de Sorèze sous Louis XVI," in *Le règne de Louis XVI et la guerre d'indépendance Américaine: Actes du Colloque International de Sorèze, 1976* (n.p. [1977]), pp. 322–325.
46. Simon, "Laplace: Ses origines familiales et ses premiers débuts," p. 491.
47. Uncle Louis's aptitude is described in François Boisard, *Notices biographiques, littéraires et critiques sur les hommes du Calvados* (Caen, 1848), p. 177.

48. Le Fort, "Le collège et l'école militaire de Beaumont," p. 76.
49. Information on the careers of other scientists is obtained from *DSB*. For a discussion of educational and career patterns, see Charles B. Paul, *Science and Immortality: The Eloges of the Paris Academy of Sciences, 1699–1791* (Berkeley, 1980).
50. Jacques Louis Moreau de la Sarthe, "Discours sur la vie et les ouvrages de Vicq d'Azyr," in *Œuvres de Vicq d'Azyr* (Paris, an 13 [1805]), vol. 1, p. 4.

## 2. From Scholasticism to Higher Mathematics

1. On Caen, see Jean-Claude Perrot, *Genèse d'une ville moderne: Caen au XVIIIe siècle* (Paris, 1975).
2. Jean Charles Victor Pouthas, "La constitution intérieure de l'Université de Caen au XVIIIe siècle," *Mémoires de l'Académie Nationale des Sciences, Arts et Belles-Lettres de Caen* (1908): 289–295.
3. Quoted in Léon Puiseux, *Notices sur Malherbe, La Place, Varignon, Rouelle, Vauquelin, Descotils, Fresnel et Dumont D'Urville* (Caen, 1847), p. 62.
4. Archives du Calvados, D 1047, fol. 43r. and D 1016, fol. 2r.
5. Puiseux, *Notices*, pp. 36–37.
6. Victor Le Fort, "Le collège et l'école militaire de Beaumont," *La Revue Illustrée du Calvados* 7 (1913): 76.
7. Jean Baptiste Thomas Gabriel Hébert, *Notice historique sur M. [François] Moysant* (Caen, [1814]).
8. Jean Charles Victor Pouthas, "Les collèges de Caen au XVIIIe siècle," *Mémoires de l'Académie Nationale des Sciences, Arts et Belles-Lettres de Caen* (1910): 153–154.
9. Puiseux, *Notices*, p. 35.
10. Archives du Calvados, D 1047, fol. 43r.
11. Jean Charles Victor Pouthas, "La Faculté des Arts de l'Université de Caen au XVIIIe siècle," *Mémoires de l'Académie Nationale des Sciences, Arts et Belles-Lettres de Caen* (1909): 364–365.
12. Adrien Pasquier, "Biographie normande" in Bibliothèque Municipale de Rouen, MS Y 43, vol. 2; Edouard Frère, *Manuel du Bibliographe Normand* (Rouen, 1858), pp. 1, 5; and Léon Tolmer, "La collégiale du Saint-Sépulcre de Caen (1777–1791)," *Bulletin de la Société des Antiquaires de Normandie* 49 (1942–1945): 139–343; and Théodore Lebreton, *Biographie Normande* (Rouen, 1857), vol. 1, p. 4.
13. Tolmer, "La collégiale du Saint-Sépulcre de Caen," p. 211.
14. V. Hunger, *Histoire de Verson* (Caen, 1908), pp. 13–14 and 29–30.
15. Pouthas, "Les collèges de Caen au XVIIIe siècle," pp. 150–152.
16. The author of the play, purportedly published at Leiden in 1779, by

M.E.M.B.C.D.S.M., was generally thought to be Monsieur Etienne [Joseph] Mauger Bénédictin de la Congrégation de Saint Maur. Mauger was also in university circles.

17. Christophe Gadbled, *Exposé de quelques-unes des vérités rigoureusement démontrées par les géometres, et rejettées par l'auteur du Compendium de Physique (Imprimé à Caen en 1775)* (Amsterdam, 1779), pp. 6–7. The exact words are "des leçons que ses Disciples seroient trop heureux de pouvoir oublier un jour, pour me servir de l'expression d'un de ses Elèves, Membre de l'Académie des Sciences de Paris."

18. Pouthas, "Les collèges de Caen au XVIIIe siècle," p. 147.

19. Alfred Hamy, *Les Jésuites à Caen* (Paris, 1899).

20. The physics instruments, which included spheres, Torricellian tubes, Boylean vacuum pumps, and a Rohault chamber, had been used by Father Bonaceau, the last Jesuit to teach mathematics at Caen. He received the proceeds from the sale of these instruments with the help of Adam, who negotiated a low price. See Gabriel Vanel, ed., *Recueil de journaux Caennais, 1661–1777* (Rouen, 1904), p. 288. For an extended discussion of the departure of the Jesuits, see pp. 230–294.

21. Tolmer, "La collégiale du Saint-Sépulcre de Caen," p. 279.

22. *Annonces, Affiches et avis divers de la Haute et Basse Normandie,* 17 and 24 July 1767, pp. 118 and 121; and *Réponse à la lettre de M. Adam professeur de philosophie au Collège du Bois à Caen, concernant l'alignement de la rivière à la-ditte ville, publié dans les Affiches de Normandie* (Caen, 1766).

23. *Réflexions d'un logicien adressée à son professeur sur un ouvrage anonyme intitulé Mélanges de littérature, d'histoire et de philosophie* (Caen, 1766). The pamphlet was deposited in Caen's municipal library before the D-day bombings in 1944, but is no longer available.

24. Frédéric Pluquet, "Christophe Gadbled," *Annuaire du Département de la Manche* (1829): 307. His actual date of birth was 29 November 1732, and he died in Caen during the Revolution. See CIPCN, vol. 6, pp. 148, 630. A good deal of biographical information may be culled from the documents in AN, M 196, which details another controversy in which Gadbled was embroiled.

25. Le Guay died on 24 August 1759, and was replaced by Gadbled in 1760. See Archives du Calvados, D 1047, fol. 38v.

26. Archives du Calvados, D 1117, item 6, p. 9.

27. Tolmer, "La collégiale du Saint-Sépulcre de Caen," p. 283.

28. Records kept about one of Gadbled's disputes indicate he had the backing of Nollet, Le Monnier, Bézout, and d'Alembert (AN, M 196). He was also in correspondence with d'Alembert.

29. *Annonces, affiches et avis divers de la Haute et Basse Normandie*, 7 October 1763, pp. 80–81.
30. Cauvin, "Souvenirs d'un octogénaire de la ville de Caen," *Annuaire des Cinq Départements de l'Ancienne Normandie* 11 (1845): 509–522. Cauvin says that six to eight students usually attended the evening meetings.
31. L. W. B. Brockliss, *French Higher Education in the Seventeenth and Eighteenth Centuries: A Cultural History* (Oxford, 1987), pp. 384–385.
32. For details, see François Russo, "L'hydrographie en France aux XVIIe et XVIIIe siècles: Ecoles et ouvrages d'enseignement," and Roger Hahn, "L'enseignement scientifique des Gardes de la Marine au XVIIIe siècle," in René Taton, ed., *Enseignement et diffusion des sciences en France au XVIIIe siècle* (Paris, 1964), pp. 419–440 and 547–558.
33. René Jacques Le Gaigneur, *Le pilote instruit; ou, Nouvelles leçons de navigation sans maître à l'usage des navigateurs du commerce* (Nantes, 1781), p. iv. More than thirty-five ship captains and merchants of Caen signed a letter on 30 July 1764 railing against "worthless . . . speculative demonstrations" taught by ex-Jesuits, which had led pilots to turn elsewhere for instruction (AN, M 196). Another similar petition was prepared on 15 September 1764.
34. One can see the quality of Gadbled's teaching by examining the *Exercice sur la théorie de la navigation* (Caen, 1779), a ten-page printed exercise defended by student Jean François Le Cerf, which refers to the latest writings of Euler, Lexell, Le Monnier, d'Alembert, and Bézout.
35. Countless examples may be seen in Jean [Joannes] Adam, *Philosophia ad usum scholarum accommodata* (Caen, 1771–1775), 3 vols., and subsequent editions in 1780, 1784, and 1787.
36. *Gazette des Arts et Métiers* (28 December 1775), pp. 3–4; in the *Prospectus d'un traité d'hydrodynamique expérimentale à la portée des artistes* (available at the Bibliothèque Municipale de Caen, Br.C52), Adam claims he has been teaching the subject for eight years. For Bossut, see Roger Hahn, *L'hydrodynamique au XVIIIe siècle: Aspects scientifiques et sociologiques* (Paris, 1965) and "The Chair of Hydrodynamics in Paris, 1775–1791: A Creation of Turgot," *Actes du Xe Congrès International d'Histoire des Sciences* (Ithaca, N.Y., 1962), pp. 751–754.
37. I have used this term in my "Laplacean View of Calculation," in Tøre Frängsmyr et al., eds., *The Quantifying Spirit in the Eighteenth Century* (Berkeley, 1990), pp. 365–367.
38. Details on Le Canu's career are hard to obtain. After serving as professor of medicine at Caen and a corresponding member of the Société Royale de Médecine in Paris, he was employed by the navy as a public health in-

spector, especially concerned with epidemics. See Archives de la Marine, C⁷174; and Jean Charles Victor Pouthas, "L'instruction publique à Caen pendant la Révolution," *Mémoires de l'Académie Nationale des Sciences, Arts et Belles-Lettres de Caen* (1911), pp. 191–223. Marguerie's life story is better served by Prosper Jean Levot's piece in *Essais de biographie maritime* (Brest, 1847), pp. 355–370; and in Joseph and Louis Michaud, eds., *Biographie Universelle,* new ed. (Paris, 1854), vol. 26, pp. 543–547.

39. *Conclusions du général de l'Université de Caen, imprimés par son ordre, avec des observations, et des réponses aux observations, sur les cahiers de morale dictés au Collège du Mont en MDCCLXIII* (Caen, 1763), quotation on p. 1. (This pamphlet is in Bibliothèque Municipale de Caen, Br.D70.)
40. Archives du Calvados, D 1016, fol. 13v.
41. Franco Venturi, *Le origini dell'Enciclopedia,* 2d ed. (Turin, 1970); John Lough, *The Encyclopédie* (New York, 1971); and John S. Spink, "Un abbé philosophe: L'affaire de J.-M. de Prades," *Dix-huitième Siècle* 3 (1971): 145–180.
42. John Pappas, *The abbé Berthier's Journal de Trévoux and the philosophes* (Geneva, 1957) [*Studies on Voltaire and the Eighteenth Century,* vol. 3].
43. Jean Stengers, "Buffon et la Sorbonne," *Etudes sur le XVIIIe Siècle* (Brussels) 1 (1974): 97–127.
44. *Mélanges de littérature, d'histoire, et de philosophie,* new ed. (Amsterdam, 1759), vol. 2, pp. 165 and 187.
45. *Mélanges,* vol. 4, pp. 11–12.
46. Ibid. pp. 19–21.
47. Ibid.
48. *Philosophia ad usum scholarum accommodata* (Caen, 1771–1775), 3 vols.
49. *Mélanges,* vol. 4, p. 224.
50. See the important study Brockliss, *French Higher Education,* especially pp. 360–390.
51. Gurdon Wattles, "Buffon, d'Alembert and Materialist Atheism," *Studies on Voltaire and the Eighteenth Century* 266 (1989): 317–341.
52. See the critical edition by Jacques Roger of Buffon, *Les Epoques de la Nature* in *Mémoires du Muséum National d'Histoire Naturelle,* n.s., ser. C, vol. 10 (1962): 28–29.
53. "Sur le système du monde," in *Mémoires de l'Académie Royale de Marine,* vol. 1 (Brest, 1773), pp. 45–84.
54. Reported in a letter from Baron d'Holbach to Paolo Frisi dated 5 October 1767 and quoted in Pierre Naville, *D'Holbach et la philosophie scientifique au XVIIIe siècle* (Paris, 1967) p. 477.
55. In a manuscript note by Condorcet, d'Alembert states, "M. Marguerie did

not push his research any further." Paris, Bibliothéque de l'Observatoire, MS Z21.
56. Puiseux, *Notices*, p. 37. Twenty years later, the same Protestant family hired the young Cuvier to tutor another d'Héricy. See Dorinda Outram, *Georges Cuvier: Vocation, Science, and Authority in Post-Revolutionary France* (Manchester, 1984), pp. 31–37.
57. Archives du Calvados, D 1129 (1769), part 9. On its last lines, the poster reads: "Répondra M. Jacques Armand d'Hericy, de Caen, Disciple de M. Pierre Le Canu, Docteur-Médecin, Professeur de Philosophie au Collège du Mont de la très-célèbre Université de Caen, le Mercredi 19e jour d'Avril 1769, trois heures d'après midi. Le mesme fera un exercice de physique expérimentale pour le VIIIe acte public."
58. Ibid.
59. Archives du Calvados, D 1129 (1769), part 8. The thesis defense by Jacques Heleine was scheduled for 9 June 1769.
60. Archives du Calvados, D 1016, fol. 2 recto.
61. André Siegfried, "Psychologie du Normand," *Etudes Normandes* 15 (1955): 235–236.
62. Louis Van Haecke, "De la locution Normande 'p't-êt' ben qu'oui, 'p't-êt' ben qu'non!' et de ses implications diverses," *Etudes Normandes* 64–65 (1967).
63. Jean Bouisset, *Notice historique sur François-Joseph Quesnot* (Caen, 1805), p. 6.

## 3. The Gifted Mathematician

1. Joseph Fourier, "Eloge historique de M. le Marquis de Laplace, prononcé dans la séance publique de l'Académie Royale des Sciences, le 15 juin 1829," *Mémoires de l'Académie Royale des Sciences de l'Institut de France* 10 (1827): lxxxii–lxxxiii. A more elaborate story is given in Léon Puiseux, *Notices sur Malherbe, La Place, Varignon, Rouelle, Vauquelin, Descotils, Fresnel et Dumont D'Urville* (Caen, 1847), pp. 37–40, in which Laplace was asked to solve a complex problem before being received by d'Alembert.
2. "It was then that he composed a remarkable letter to the person whose favor he was soliciting on the general principles of mechanics, from which M. Laplace had often recounted portions to me." Fourier, "Eloge historique de M. le Marquis de Laplace," p. lxxxiii.
3. Jay M. Smith, *The Culture of Merit: Nobility, Royal Service, and the Making of Absolute Monarchy in France, 1600–1789* (Ann Arbor, Mich., 1996).
4. Unsigned text beginning with the words "Les mesmes choses etant supposées que dans vostre demonstration" and with the heading "Remarques"

in AAdS, dossier Laplace. The manuscript was acquired by purchase in 1978. It is being published by Jean Dhombres and Roger Hahn in "La première rencontre entre Laplace et d'Alembert en 1769: Le principe d'inertie et l'ordre métaphysique," *Revue de Métaphysique et de Philosophie* (in press).
5. *Opuscules mathématiques* (Paris, 1768), vol. 4, pp. 349–357. See also "Mémoire sur les principes de la mécanique," *Mémoires de l'Académie des Sciences* (1769): 278–286, especially pp. 280–283.
6. See J. Morton Briggs Jr., "D'Alembert: Philosophy and Mechanics in the Eighteenth Century," *University of Colorado Studies,* series in History, no. 3 (Boulder, 1964), pp. 38–56; Thomas L. Hankins, *Jean d'Alembert: Science and the Enlightenment* (Oxford, 1970), chapters 7 and 8; Pierre Costabel, "De quelques embarras dans le *Traité de dynamique,*" *Dix-Huitième Siècle* 16 (1984): 39–46; Michel Paty, "D'Alembert et la théorie physique," in M. Emery and P. Monzani, eds., *Jean d'Alembert, savant et philosophe: Portrait à plusieurs voix* (Paris, 1989), pp. 233–260; and Johan Christiaan Boudri, *What Was Mechanical about Mechanics: The Concept of Force between Metaphysics and Mechanics from Newton to Lagrange* (Dordrecht, 2002).
7. "Cela prouve que si le corps se meut uniformément au sortir de la main qui l'a lancé dans l'espace, il continuera de se mouvoir de même, si rien ne s'y oppose; mais il ne m'est point démontré que ce mouvement initial doive être uniforme. Si l'on n'a point dans l'idée que nos sens nous donnent de la matière rien qui puisse altérer sa vitesse, on n'y voit rien aussi qui doive la conserver. Nous ignorons entièrement la nature des corps, et celle du mouvement, nous ne pouvons le connaître que par ses effets, ainsi la seule expérience peut nous instruire à cet égard. Je regarde donc la loi d'inertie comme une supposition dont les résultats s'accordent très sensiblement avec les phénomènes de la nature" (spelling and punctuation modernized).
8. Puiseux, *Notices,* p. 36, indicated that "the first victories this great mathematician recorded were in another field of battle: theology; he could argue with extraordinary acumen the most complex aspects of scholastic controversy."
9. The professors were Beauzée and Douchet. See Paris de Meyzieu, "Ecole militaire," *Encyclopédie,* vol. 5, pp. 307–313 and Paris de Meyzieu and Louis Félix Guinement de Kéralio, "Ecole militaire," in *Encyclopédie méthodique, art militaire* (Paris, 1785), vol. 2, pp. 226–236. A more modern source of information is Robert Laulan, *L'Ecole Militaire de Paris: Le Monument, 1751–1788* (Paris, 1950).
10. The new curriculum was designed by the engineer Bizot. See AN, MM 669, p. 21r; and Roger Hahn, "L'enseignement scientifique aux écoles militaires

et d'artillerie," in René Taton, ed., *Enseignement et diffusion des sciences en France au XVIIIe siècle* (Paris, 1964), pp. 523–525. See also David Bien, "The Army in the French Enlightenment: Reform, Reaction and Revolution," *Past and Present* 85 (1979): 82–86.
11. AN, MM 668, p. 72r; and for the list of mathematics instructors, Hahn, "L'enseignement scientifique," p. 538.
12. AN, MM 668, p. 93v. Cousin was by far the most accomplished of the mathematics instructors when appointed. When elected to the Académie des Sciences in March 1772, he was awarded two medals by the board of directors of the Ecole Militaire (AN, MM 669, p. 32v).
13. D'Alembert to Le Canu, 25 August 1769, Lyon, Institut National de Recherche Pédagogique, collection historique, A30755.
14. AN, MM 680, fol. 9v.
15. *Exposition du plan d'études pour les élèves de l'Ecole Royale Militaire* (Paris, 1769), pp. 17–30.
16. Jean Baptiste Champagny, *Souvenirs* (Paris, 1846), pp. 43–44; and Vincent Marie Viénot, Comte de Vaublanc, *Souvenirs* (Paris, 1838), vol. 1, 47–52.
17. BN, MSS, n.a.f. 1023. See also AN, $O^1$1606, 231.
18. AN, MM 668, fol. 71r.
19. AN, MM 669, fol. 59v.
20. For more details, see the appropriate chapters of *Enseignement et diffusion des sciences en France*.
21. See Hahn; as well as Roger Hahn, "Scientific Research as an Occupation in Eighteenth-Century Paris," *Minerva* 13 (1975): 501–513.
22. *Affiches, Annonces et Avis Divers,* 4 January 1769, p. 1.
23. Delambre, "Eloge de Jeaurat", BI, MS 2041, no. 77; and AN, MM 659, p. 38.
24. Guillaume Bigourdan, "Histoire des observatoires de l'Ecole Royale Militaire," *Bulletin Astronomique* 4 (1887): 497–504 and 5 (1888): 30–40.
25. Hahn, "L'enseignement scientifique," p. 534.
26. Louis Pierre Eugène Amélie Sédillot, *Les professeurs de mathématiques et de physique générale au Collège de France* (Rome, 1869), pp. 186–187.
27. Stephen M. Stigler, "Laplace's Early Work: Chronology and Citations," *Isis* 69 (1978): 239–248. Among Laplace's papers at the Bancroft Library, box 7, folder 35, there is a sheaf of notes entitled "Liste des mémoires de mathématique renfermés dans les volumes de l'Académie de Petersbourg," which stops with the year 1772. Drawn up most likely when Laplace was at the Ecole Militaire, it catalogs many of Euler's articles.
28. John L. Greenberg, "Alexis Fontaine's 'Fluxio-Differential Method' and the Origins of the Calculus of Several Variables," *Annals of Science* 38 (1981): 251–290; and "Alexis Fontaine's Integration of Ordinary Differential Equa-

tions and the Origins of the Calculus of Several Variables," *Annals of Science* 39 (1982): 1–36.

29. AN, MM 668, fol. 8 r and v.
30. Keith M. Baker, *Condorcet: From Natural Philosophy to Social Mathematics* (Chicago, 1975); and Roshdi Rashed, ed., *Sciences à l'epoque de la Révolution française: Recherches historiques* (Paris, 1988).
31. *Affiches, Annonces et Avis Divers,* 1 August 1770, p. 124; and Paris, AAdS, *Procès-verbaux,* 21 July and 1 August 1770. A copy of the printed thesis announcement for 25 July is in the New York Public Library.
32. For the Académie's publications, see Robert Halleux et al., eds., *Les publications de l'Académie Royale des Sciences de Paris (1666–1793)* (Turnhout, 2001).
33. The results of elections are as follows, with the winner in parentheses. For associate in mechanics section: 20 March 1773 (Desmarest); for assistant in mechanics section: 16 January 1771 (Desmarest); 17 April 1771 (Rochon); 31 March 1773 (Laplace); for assistant in geometry section: 17 May 1771 (Vandermonde); 14 March 1772 (Cousin). It is also possible that Laplace was a candidate for election to associate in geometry on 26 February 1772 (Jeaurat).
34. Stigler's attempt to find a meaning behind the order of the early papers takes no account of the academic elections. See Stigler, "Laplace's Early Work," pp. 237 and 248; the list of Laplace's presentations is given in Gillispie, pp. 286–296.
35. *Mélanges de philosophie et de mathématique de la Société Royale de Turin pour les années 1760–1761* (Turin, 1762), vol. 2, mathematical section, pp. 173–298.
36. Guillaume Bigourdan, "La jeunesse de P.-S. Laplace," *La Science Moderne* 9 (1931): 381–382; and Gillispie, pp. 4–5.
37. *Mémoires de mathématique et de physique présentés à l'Académie Royale des Sciences, par divers savans, et lûs dans ses assemblées* (Paris, 1774), vol. 6, p. xix. The full text of the remark, with judicious commentaries, has been usefully reprinted by Bernard Bru and Pierre Crépel in Condorcet, *Arithmétique politique: Textes rares ou inédits (1767–1789)* (Paris, 1994).
38. Danielle Fauque, "Alexis-Marie Rochon (1741–1817), savant, astronome et opticien," *RHS* 38 (1985): 3–36; and Kenneth Taylor, "Nicolas Desmarest," *DSB,* vol. 4, pp. 70–73.
39. The astronomer Lalande, one of Laplace's early supporters, was so confident of the young mathematician's chances that he wrote to Euler's son as early as 6 December 1771 that Laplace would take the slot vacated by Fontaine. See St. Petersburg, Archives of the Academy of Sciences, 1-3-59, fol. 118.
40. Lalande to Euler's son, Johann Albrecht, 12 June 1772, in St. Petersburg,

Archives of the Academy of Sciences, 1-3-58, fol. 177r and v; and Laplace to Euler, 30 May 1772, in St. Petersburg, Archives of the Academy of Sciences, 1-3-58, fol. 176r.
41. Letter dated 1 January 1773 in Lagrange, *O.*, vol. 13, p. 255.
42. D'Alembert made an unsuccessful effort early in 1770 to have a chair at the Collège Royal, possibly the one for Arabic language, transmuted into a mathematical one that Laplace could occupy. (See the letter to d'Alembert, dated 9 January 1770, in BI, MS 2466, fols. 216–217).
43. AN, MM 669, 85v.
44. Attestation signed by Laplace on 20 May 1779, in AN, $O^1 679$, no. 457.
45. Payment of the five hundred livres award, taken from the Académie's "petites pensions," started in January 1775. (See Duc de la Vrillière to Buffon, 2 January 1775, in AAdS, dossier 1775).
46. Gillispie, pp. 297–306 provides a nearly complete list of his production.
47. Contemporaries were so interested in this feature of Laplace's personality that they studied the convolutions in his brain after his death. See Karl Pearson, "Laplace, Being Extracts from Lectures Delivered by Karl Pearson," *Biometrika* 21 (1929): 212–215.
48. Condorcet, *Essais d'analyse* (Paris, 1768), p. i; and Cousin, *Traité de calcul différentiel et de calcul intégral* (Paris, an 4 [1796]), p. xiii.
49. D'Alembert was continuing to publish his *Opuscules mathématiques* throughout the early years of Laplace's career (vols. 4 and 5 in 1768, vol. 6 in 1773, and vols. 7 and 8 in 1780). These volumes were miscellanies intended in part to continue his earlier *Recherches sur différens points importants du système du monde* (1754–1756).
50. Christian Gilain, "Condorcet et le calcul intégral," in Rashed, *Sciences à l'époque de la Révolution française*, pp. 87–147.
51. Keith Baker, "Les débuts de Condorcet au secrétariat de l'Académie Royale des Sciences (1773–1776)," *RHS* 20 (1967): 229–280.
52. Bru and Crépel are equally perplexed by the exact nature of their interaction. See Condorcet, *Arithmétique politique*, p. 692.

## 4. Setting Fundamental Principles

1. Robin E. Rider, "Mathematics in the Enlightenment: A Study of Algebra, 1685–1800," Ph.D. diss., University of California, Berkeley, 1980, vol. 1, pp. 82, 86–87, and 101–102; and *Mémoires de l'Académie Royale de Marine* 1 (1773): 1–44.
2. Gillispie, p. 288, item 18. See also Letters of Lagrange to Laplace, 30 December 1776, in Lagrange, *O.*, pp. 14 and 17; Laplace to Lagrange, 3 February 1778, in Lagrange, *O.*, p. 74; and Laplace to Gauss, 28 January 1808, in Karin Reich, *Im Umfeld der "Theoria Motus": Gauss' Briefwechsel mit*

Perthes, Laplace, Delambre und Legendre (Göttingen, 2001), p. 89. He discussed the subject in his Ecole Normale lectures.
3. Originally in the preface to his *Recherches sur différens points importans du système du monde* (Paris, 1754), vol. 1, p. iii, the views were repeated with minor modifications in his *Mélanges de littérature, d'histoire, et de philosophie* (Paris, 1759), vol. 4, pp. 224–225.
4. See the important conclusions of J. Christiaan Boudri in his *What Was Mechanical about Mechanics: The Concept of Force between Metaphysics and Mechanics from Newton to Lagrange* (Dordrecht, 2002).
5. I have treated this question in greater detail in "Laplace and the Mechanistic Universe," in David C. Lindberg and Ronald L. Numbers, eds., *God and Nature: Historical Essays on the Encounter between Christianity and Science* (Berkeley, 1986), pp. 266–270. A different presentation is by Giorgio Tonelli, "La nécessité des lois de la nature au XVIIIe siècle et chez Kant en 1762," *RHS* 12 (1959): 225–241. The most recent general discussion by Cornelia Buschmann, "Die philosophischen Preisfragen und Preisschriften der Berliner Akademie der Wissenschaften im 18. Jahrhundert," in Wolfgang Förster, ed., *Aufklärung in Berlin* (Berlin, 1989), pp. 165–228, does not mention the contest because there were no declared winners.
6. *Traité de dynamique* (Paris, 1758), pp. xxiv–xxv; see also p. xxix for his views on the "Supreme Being" and "Creator." The same text is repeated a year later in *Mélanges de littérature,* vol. 4, pp. 211–212.
7. Herbert H. Odom, "The Estrangement of Celestial Mechanics and Religion," *Journal of the History of Ideas* 27 (1966): 533–548.
8. Repelled by d'Alembert's strategy, traditional religious thinkers continued to assert the intimate linkages between Christianity and science. An extreme example is Pierre Le Clerc, *L'astronomie, mise à la portée de tout le monde* (Amsterdam, 1780).
9. Condorcet, *Le Marquis de Condorcet a Mr d'Alembert sur le système du monde et sur le calcul intégral* (Paris, 1768), p. 4.
10. Ibid.
11. Ibid.
12. Ibid., p. 5.
13. Ernst Cassirer, *Determinismus und Indeterminismus in der modernen Physik: Historische und Systematische Studien zum Kausalproblem* (Göteborg, 1937). For a discussion of the similarities and contrasts between the Condorcet and Laplace versions, see Orietta Pesenti Cambursano, "Ipotesi della intelligenza assoluta in Condorcet ed in Laplace," *Miscellanea Storica Ligure* 3 (1963): 237–256; and Roger Hahn, "Laplace's First Formulation of Scientific Determinism in 1773," in *Actes du XIe Congrès International d'Histoire des Sciences, 1965* (Warsaw, 1968), vol. 2, pp. 167–171.

14. *Mémoires de mathématique et de physique presentés à l'Académie Royale des Sciences par divers savans, et lûs dans ses assemblées* (Paris, 1773), vol. 7, pp. 113–114 [*O.C.*, vol. 8, pp. 144–145].
15. Charles Batteux, *Histoire des causes premières; ou, Exposition sommaire des pensées des philosophes sur les principes des êtres* (Paris, 1769).
16. See in particular Charles Hercule de Keranflech, *L'hypothèse des petits tourbillons justifiée* (Rennes, 1761); *Essai sur la raison* (Paris, 1765); *Suite de l'essai sur la raison avec un nouvel examen de la question de l'âme des bêtes* (Rennes, 1767–1768); *Dissertation sur les miracles pour servir d'éclaircissement au sistème de l'impuissance des causes secondes* (Rennes, 1772); and *Observations sur le Cartésianisme moderne* (Rennes, 1774). Parts of these treatises are analyzed by L. Robert, "De Keranflech, philosophe breton du XVIIIe siècle," *Annales de Bretagne* (1886–1890); and Geneviève Rodis-Lewis, "Un malebranchiste méconnu: Keranflech," *Revue Philosophique de la France et de l'Etranger* 154 (1964), 21–28; and Yves Marie André, *Oeuvres philosophiques du Père André de la Compagnie de Jésus*, ed. Victor Cousin (Paris, 1843); Antoine Charma et al., eds., *Le Père André, Jésuite: Documents inédits pour servir à l'histoire philosophique, religieuse et littéraire du XVIIIè siècle* (Caen, 1844–1856).
17. Thomas L. Hankins, "The Influence of Malebranche on the Science of Mechanics during the Eighteenth Century," *Journal of the History of Ideas* 28 (1967): 193–210.
18. See David Renaud Boullier, *Apologie de la métaphysique, à l'occasion du discours préliminaire de l'Encyclopédie* (Amsterdam, 1753); *Pièces philosophiques et littéraires* (n.p., 1759); and *Discours philosophiques: Le premier sur les causes finales; le second sur l'inertie de la matière; et le troisième sur la liberté des actions humaines* (Amsterdam, 1759). Boullier has been discussed by Anna Radier, *Un défenseur de Pascal au XVIIIe siècle: David Renaud Boullier* (Paris, 1948); Luca Obertello, *Le Idee e la Realtà: La Teoria della Ragione nei Secoli XVIII e XIX* (Florence, 1971), pp. 97–134; and John W. Yolton, *Locke and French Materialism* (Oxford, 1991), chapter 5.
19. I here borrow the term used by Henri Gouhier in his *La jeunesse d'Auguste Comte et la formation du positivisme* (Paris, 1933–1941). See also Leszek Kolakowski, *The Alienation of Reason: A History of Positivist Thought* (Garden City, N.Y., 1968).
20. Roger Hahn, "Determinism and Probability in Laplace's Philosophy," *Actes du XIIIe Congrès International d'Histoire des Sciences, 1971* (Moscow, 1974), vol. 1, pp. 170–175.
21. Léon Brunschvicg, *L'expérience humaine et la causalité physique* (Paris, 1922), chapters 25 and 26.
22. Laplace's view echoed that of d'Alembert, who had asserted: "L'Univers

pour qui saurait l'embrasser d'un seul point de vue ne serait . . . qu'un fait unique et une grande vérité." *Encyclopédie,* vol. 1, p. ix.
23. Boudri, *What Was Mechanical about Mechanics,* chapter 5.
24. See *Mémoires de mathématique et de physique* (Paris, 1773), vol. 7, pp. 113–114 [*O.C.*, vol. 8, pp. 144–145].
25. When Laplace returned to this issue forty years later, he used terminology that implies he meant determinism to apply only to material phenomena. See *Théorie analytique des probabilités* (Paris, 1812), p. 177; and *Essai philosophique* (Paris, 1814), p. 2.
26. *Mémoires de mathématique et de physique* (1773), vol. 7, pp. 113–114 [*O.C.*, vol. 8, pp. 144–145]. Laplace repeats these sentiments in his *Essai philosophique,* pp. 2–3.
27. *Mémoires de mathématique et de physique* (1773), vol. 7, pp. 113–114 [*O.C.*, vol. 8, pp. 144–145].
28. Ibid.
29. Ian Hacking, *The Emergence of Probability* (Cambridge, Eng., 1975); Stephen M. Stigler, *The History of Statistics: The Measurement of Uncertainty before 1900* (Cambridge, Mass., 1986); Lorraine Daston, *Classical Probability in the Enlightenment* (Princeton, 1988); and Bernard Bru, preface of the translation of Thomas Bayes, *Essai en vue de résoudre un problème de la doctrine des chances* (Paris, 1988) [*Cahiers d'Histoire et de Philosophie des Sciences,* n.s., no. 18], pp. 7–13.
30. This issue has recently been examined "in reverse" by Lorraine Daston in "Perché la teoria della probabilità aveva bisogno del determinismo: Le origine," *Intersezioni* 10 (1990): 541–562.
31. O. B. Sheynin, "Newton and the Classical Theory of Probability," *AHES* 7 (1971): 217–243; Karl Stiegler, "On the Origin of the So-Called Laplacean Determinism," *Actes du XIIIe Congrès International d'Histoire des Sciences, 1971* (Moscow, 1974), vol. 6, pp. 307–312.
32. Thierry Martin, *Probabilité et critique philosophique selon Cournot* (Paris, 1996).
33. "Laplace's 1774 Memoir on Inverse Probability," *Statistical Science* 1 (1986): 359–378 is a careful translation with an introduction and explanatory notes by Stephen M. Stigler. Its importance had previously been recognized by Charles C. Gillispie in "Probability and Politics: Laplace, Condorcet, and Turgot," *Proceedings of the American Philosophical Society* 116 (1972): 1–20; and since reiterated by Bernard Bru in "Postface" of the critical edition of the *Essai philosophique sur les probabilités* (Paris, 1986), pp. 251–255.
34. Thomas Bayes, "An Essay Towards Solving a Problem in the Doctrine of Chances," *Philosophical Transactions of the Royal Society* 53 (1764): 370–418. See also Andrew I. Dale, "Bayes or Laplace? An Examination of the

Origin and Early Applications of Bayes's Theorem," *AHES* 27 (1982): 23–47.
35. Anders Hald, *A History of Probability and Statistics and Their Applications before 1750* (New York, 1990), pp. 441–447.
36. Lagrange to Laplace, 30 December 1776, in Lagrange, *O.*, vol. 14, p. 66. Neither Lagrange nor Laplace ever completed this translation.
37. *Mémoires de mathématique et de physique* (1773), vol. 6, pp. 363–365 [*O.C.*, vol. 8, pp. 16–18].
38. The comment is from the middle of "Mémoire sur la probabilité des causes par les évènemens," in *Mémoires de mathématique et de physique* (1773), vol. 6, p. 634 [*O.C.*, vol. 8, pp. 41–42].
39. O. B. Sheynin, "J. H. Lambert's Work on Probability," *AHES* 7 (1971): 244–256; and "R. G. Boscovich's Work on Probability, *AHES* 9 (1973): 306–324.
40. O. B. Sheynin has given a much more elaborate and detailed account in "Laplace's Theory of Errors," *AHES* 17 (1977): 1–61.
41. Roger Hahn, "Laplace and Boscovich," *Proceedings of the Bicentennial Commemoration of R. J. Boscovich, Milan 1987* (Milan, 1988), pp. 71–82.
42. "Laplace," in *The Gallery of Portraits with Memoirs* (London, 1833–1837) vol. 2, pp. 34–39 and repeated in *The Penny Cyclopædia* 13 (London, 1833–1858), pp. 325–328.
43. Vincenzo Ferrone, "Il dibatto sulla probabilità e scienza sociali nel secolo XVIII," *Physis* 22 (1980): 24–71.
44. Buffon to Laplace, 21 April 1774, eventually published in *Comptes-rendus de l'Académie des Sciences,* vol. 88, p. 1019; and partially in the *Journal Officiel*, 24 May 1879, p. 4262.
45. For an explanation of Buffon's views, see Charles Lenay, "Le hasard chez Buffon: Une probabilité 'anthropologique,'" *Buffon* 88 (Paris, 1992): 613–627.
46. *Mélanges de littérature,* vol. 5, pp. 275–276.
47. Ezio Yamazaki, "Laplace et d'Alembert: Leurs idées de la probabilité," *Fourteenth International Congress of the History of Science, 1974* (Tokyo, 1975), vol. 2, pp. 176–179.
48. Ezio Yamazaki, "D'Alembert et Condorcet—Quelques aspects de l'histoire du calcul des probabilités," *Japanese Studies in the History of Science* 10 (1971): 59–93.
49. See Keith M. Baker, *Condorcet: From Natural Philosophy to Social Mathematics* (Chicago, 1975), chapter 3; and the comments by Bru and Crépel in Condorcet, *Arithmétique politique: Textes rares ou inédits (1767–1789)* (Paris, 1994).
50. Baker, *Condorcet: From Natural Philosophy to Social Mathematics,* p. 177 and note 203.
51. See Gillispie, "Probability and Politics," for details.

52. Condorcet, *Arithmétique politique,* p. 94.
53. *Mémoires de mathématique et de physique* (1773), vol. 6, preface, p. 18; and Condorcet, *Arithmétique politique,* p. 84.

## 5. Finding the Stability of the Solar System

1. *Mémoires de mathématique et de physique* (Paris, 1773), vol. 7, pp. 113–114 [*O.C.*, vol. 8, pp. 144–145].
2. In a letter to Jean André Deluc dated 12 May 1785 in my possession, Lalande refers to Laplace as "l'homme le plus insolent et le plus méchant que je connais." See also Arthur Birembaut, "L'Académie Royale des Sciences en 1780 vue par l'astronome suédois Lexell (1740–1784)," *RHS* 10 (1957): 157; and Jacques-Pierre Brissot de Warville, *De la vérité; ou, Méditations sur les moyens de parvenir à la vérité de toutes les connoissances humaines* (Neuchâtel, 1782), p. 335. Laplace is identified in Brissot, *Mémoires, 1754–1793,* ed. Claude-Marie Perroud (Paris, [1911]) vol. 1, pp. 198–199.
3. Records of these committee assignments are in the chronologically organized manuscript proceedings of the Académie des Sciences, kept at the institution's archives. For Péronard's "Addition à un clavecin" (1779), see Albert Cohen, *Music in the French Royal Academy of Sciences* (Princeton, 1981), pp. 56–57.
4. Dez, "Mémoire sur la théorie du jaugeage," in *Mémoires de mathématique et de physique* (1773), vol. 7, pp. 383–389; Le Turc, "Métier propre à faire différents réseaux de dentelles" (1776). According to army records, Bonaventure Le Turc, professor of fortifications at the Ecole Militaire, "borrowed" the ideas for his invention from Charmoy (see Vincennes, Archives Historiques de l'Armée, Ya163). The inventor tells a vastly different story in his *Recueil de pièces justificatives relatives à l'histoire de la persécution de trente ans qu'éprouve le citoyen Leturc* (Paris, 1797), pp. xx–xxxvi.
5. The report, dated 15 May 1773, survives in the Berlin Staatsbibliothek, Sgl. Darmstaedter, J 1796.
6. No official account of the papers submitted by Fiquière, Jacquemier, Château, Du Tertre, and Viallon was given. See, however, Jean Honoré Maure, *Solution de deux problèmes de quadrature du cercle et de la trisection de l'angle* (London, 1773), in which a negative report by Laplace appears.
7. AAdS, dossier 15 May 1773.
8. Keith M. Baker, *Condorcet: From Natural Philosophy to Social Mathematics* (Chicago, 1975), pp. 202–225.
9. Birembaut, "L'Académie Royale des Sciences en 1780," p. 157.
10. Undated draft of a letter of Guettard to Laplace originally in the Fonds

Chabrol, Château de Jozerand, which was kindly supplied to me by the late professor Michelle Goupil. The draft is now in AAdS, Guettard, Chabrol donation, carton 2.
11. The full text of this letter is given at the end of this book.
12. Marquis Ducrest to Laplace, 30 December 1786, in Paris, BN, MS n.a.f. 22738, fol. 51. In this letter, Laplace is awarded an annual stipend of 800 livres. The grant may have been connected to the new observatory built at the Palais Royal. See *Le Palais Royal: [Catalogue de l'Exposition au] Musée Carnavalet, 9 May–4 September 1988*, p. 145, available at the library of the Musée Carnavalet, Paris.
13. He was appointed with Bossut to a committee that never officially reported its findings on the matter. See AAdS, *Procès-verbaux*, 11 August 1773.
14. Roger Hahn, "Laplace and Boscovich," *Proceedings of the Bicentennial Commemoration of R. J. Boscovich, Milan 1987* (Milan, 1988), pp. 71–82.
15. A report prepared by Dionis du Séjour that reviewed the dispute with a view to defusing the personal animosity engendered by it is published by John Pappas in "Documents inédits sur les relations de Boscovich avec la France," *Physis* 28 (1991): 170–174; and "R. J. Boscovich et l'Académie des Sciences de Paris," *RHS* 49 (1996): 401–414.
16. In his long and seminal paper published in the *Mémoires de mathématique et de physique* (1773), vol. 7, p. 163 [in *O.C.*, vol. 8, pp. 201–202], Laplace explicitly acknowledges his debt to d'Alembert's *Recherches sur la précession des équinoxes et sur la nutation de l'axe de la terre, dans le système Newtonien* (Paris, 1749) and its elaboration in subsequent memoirs. He reiterates the importance of d'Alembert's pioneering attempts to find a general solution for the motion of bodies operating under Newtonian forces in book fourteen of the *Traité de mécanique céleste* (Paris, 1825), vol. 5, pp. 251–255.
17. *Essai sur les comètes en général, et particulièrement sur celles qui peuvent approcher de l'orbite de la terre* (Paris, 1775). Laplace's report in the name of the Académie appears on pp. 343–352.
18. Joseph Jérôme Le François de Lalande, *Réflexions sur les comètes qui peuvent approcher de la terre* (Paris, 1773).
19. Jean Baptiste Joseph Delambre, *Histoire de l'astronomie au dix-huitième siècle* (Paris, 1827), pp. 558–559, 586–587, and 679.
20. In 1776, Jean Bernoulli could provide names for seventy astronomers in Paris alone. See his *Liste des astronomes connus et actuellement vivans* (Berlin, 1776).
21. Roger Hahn, "Les observatoires en France au XVIIIe siècle," in René Taton, ed., *Enseignement et diffusion des sciences* (Paris, 1964), pp. 653–658.

22. Seymour L. Chapin, "The Academy of Sciences during the Eighteenth Century: An Astronomical Appraisal," *French Historical Studies* 5 (1968): 371–404.
23. The quotation is from *Traité analytique des mouvemens apparens des corps célestes* (Paris, 1786–1789), vol. 1, p. i. Laplace's review appears in vol. 1, pp. xxxvii–xxxx [sic] and vol. 2, pp. liii–lv.
24. The emphasis is mine, not Laplace's. See *Mémoires de mathématique et de physique* (1773), vol. 7, p. 173 [*O.C.*, vol. 8, p. 212].
25. Adolf Yushkevic and R. Taton, "Introduction" of Leonhard Euler, *Opera omnia: Commercium epistolicum,* ser. 4A, vol. 5, pp. 7–8; and Pierre Spéziali, "Une correspondance inédite entre Clairaut et Cramer," *RHS* 8 (1955): 226–227.
26. Leonhard Euler, "Mémoire dans lequel on examine si les planètes se meuvent dans un milieu dont la résistance produise quelque effet sensible sur leur mouvement?"; and Charles Bossut, "Recherches sur les altérations que la résistance de l'éther peut produire dans le mouvement moyen des planètes," both in *Recueil des pièces qui ont remporté les prix de l'Académie Royale des Sciences depuis leur fondation en 1720,* vol. 8 (Paris, 1771). The Bossut piece is conveniently reproduced in his *Mémoires de mathématiques* (Paris, 1812).
27. The "religious" manuscripts in the AAdS, dossier Laplace, refer to these matters in the context of a discussion of the relationship between mind and matter. The full text is given at the end of this book.
28. *Mémoires de mathématique et de physique* (1773), vol. 7, pp. 182–194 [*O.C.*, vol. 8, pp. 221–234]; Jacques Antoine Cousin, *Introduction à l'étude de l'astronomie physique* (Paris, 1787), p. 132 provides a simpler, abbreviated mathematical explanation. Lalande, in his edition of Jean Etienne Montucla, *Histoire des mathématiques,* new ed. (Paris, 1799–1802), vol. 4, pp. 120–121, offers another version of this issue. For other comments, see Henri Andoyer, *L'œuvre scientifique de Laplace* (Paris, 1922), p. 50; and Gillispie, pp. 29–30.
29. Gillispie. See also the comments by Condorcet in *Mémoires de mathématique et de physique* (1773), vol. 7, pp. ii–iv, pointedly referring to Lagrange's prize-winning essay for 1774 on the motion of the moon (published in the same volume).
30. *Mémoires de mathématique et de physique* (1773), vol. 7, p. 181 [*O.C.*, vol. 8, pp. 219–220].
31. Burghard Weiss, "Newton à la Cartes—Das 18. Jahrhundert auf der Suche nach der 'wahren' Ursache der Gravitation," in *Philosophie, Physik, Wissenschaftsgeschichte: Ein gemeinsames Kolloquium der TU Berlin und des Wissenschaftskollegs zu Berlin* (Berlin, 1989), pp. 98–122; and James Ev-

ans, "Gravity in the Century of Light," in Matthew R. Edwards, ed., *Pushing Gravity: New Perspectives on Le Sage's Theory of Gravitation* (Montreal, 2002), pp. 9–40.

32. This work has been attempted by Andoyer in his *L'œuvre scientifique de Laplace*; Gillispie; and Bruno Morando in *Planetary Astronomy from the Renaissance to the Rise of Astrophysics*, Part B: *The Eighteenth and Nineteenth Centuries*, ed. René Taton and Curtis Wilson (Cambridge, Eng., 1995), pp. 131–150.

33. Augustus De Morgan, "Laplace, Pierre Simon," *The Penny Cyclopaedia of the Society for the Diffusion of Useful Knowledge* (London, 1833–1846), vol. 13, p. 326, col. 1.

34. Stephen M. Stigler, "Laplace's Early Work," *Isis* 69 (1978): 234–254; and S. L. Zabell, "Buffon, Price, and Laplace: Scientific Attribution in the Eighteenth Century," *AHES* 39 (1988): 173–181.

35. Laplace to d'Alembert, 15 November 1777, published in *Bullettino di bibliografia e di storia delle scienze matematiche e fisiche*, vol. 19, pp. 13–15 [*O.C.*, vol. 14, pp. 346–348]. Laplace writes, "It is true that without your work and the beautiful research you published in your essay on the resistance of fluids . . . I would never have dared to treat of this matter."

36. "Recherches sur plusieurs points du système du monde," in *Mémoires de mathématique et de physique* (1775), p. 91 [*O.C.*, vol. 9, p. 90].

37. In a letter to Lagrange dated 25 February 1778, Laplace congratulated himself on satisfying his patron with these words: "I think he [d'Alembert] does not have a reason to be unhappy at the way I spoke about his work." Lagrange, *O.*, vol. 14, p. 79. The phrasing of Laplace's comment, quoted earlier—"It is true that without your work and the beautiful research you published in your essay on the resistance of fluids . . . I would never have dared to treat of this matter"—is quite similar to one that Lagrange sent Laplace on 1 September 1777 in Lagrange, *O.*, vol. 14, p. 71. This suggests it may not be a genuine sign of modesty, but rather a conventional trope.

38. In an undated letter written in 1771, Laplace privately promised Condorcet, "As a matter of fact [in the paper] I will do full justice to you, who are the first to have proposed a general method of integration." It is published in *Bullettino di bibliografia e di storia delle scienze matematiche e fisiche*, vol. 19, p. 13 [*O.C.*, vol. 14, p. 345].

39. The manuscripts of this unfinished work are listed in Christian Gilain, "Condorcet et le calcul intégral," in Roshdi Rashed, ed., *Sciences à l'époque de la Révolution française: Recherches historiques* (Paris, 1988), p. 101.

40. George Sarton, "Lagrange's Personality (1736–1813)," in *Proceedings of the American Philosophical Society* 88 (1944): 457–496; and Andoyer, *L'œuvre scientifique de Laplace*, pp. 19–32.

41. Filippo Burzio, *Lagrange* (Turin, 1993); and Maria Teresa Borgato and Luigi Pepe, *Lagrange: Appunti per una biografia scientifica* (Turin, 1990).
42. Lagrange to Condorcet, 18 July 1774. Lagrange writes: "It is rather . . . a weakness of young people to be puffed up by their early successes; but their presumption subsides as their science grows." Lagrange, *O.*, vol. 14, p. 27.
43. Letter of 3 February 1778, in Lagrange, *O.*, vol. 14, pp. 72–75.
44. There is a classic comparison made by Poisson in Laplace's funeral oration: "In the issues he treated, Lagrange seemed more often to notice only the mathematics that were involved so that he prized the elegance of formulae and the generality of his methods; for Laplace, on the contrary, analysis was an instrument he wielded for the most varied of applications, always subordinating the specific method to the goal he pursued." Siméon Denis Poisson, *Discours [prononcé aux funérailles de M. le Marquis de Laplace, le 7 mars 1827]* reproduced in *Moniteur Universel*, 20 March 1827, 2d. supp., p. 6 and in *Connaissance des temps pour l'an 1830* (Paris, 1827), pp. 19–22. A similar conclusion worded less generally is in Morando, *Planetary Astronomy*, pp. 134–135. For another view by a contemporary, see [Augustus De Morgan], "Laplace," in *Gallery of Portraits with Memoirs* (London, 1833), vol. 2, p. 38.
45. Laplace's wife to their son Emile, 14 April 1813, in Bancroft, box 17, folder 11.
46. The footnote reads: "I had found this method at the end of 1772, occasioned by several problems posed to me by M. Monge, the capable professor of mathematics at the Engineering school in Mézières; I gave him my views then, at the same time sending them to M. Lagrange, and presented them to the Académie in February 1773. Since then the Marquis de Condorcet has published a beautiful memoir on this topic in the Académie's publications for 1771; but the route I am following is different from his." *Mémoires de mathématique et de physique* (1773), vol. 7, pp. 70–71 [*O.C.*, vol. 8, p. 103].
47. The dispute was in regard to a theorem that Laplace read before the Académie on 11 August 1784 and that he used in a 1785 publication (*Histoire et mémoires de l'Académie Royale des Sciences de Paris . . .* [Paris, 1782], pp. 13–196 [*O.C.*, vol. 10, pp. 341–419]) before Legendre could publish his ideas in 1787 (*Mémoires de mathématique et de physique . . . présentés par divers savans* [Paris, 1785], vol. 10, pp. 411–434). The priority was duly noted by Lalande in the *Journal des Savans* (October 1787): 667. See also Legendre, "Recherches sur la figure des planètes," in *Histoire et mémoires de l'Académie Royale des Sciences de Paris* (1784), note on p. 370. A similar observation was published in 1793 by Legendre for some

work presented before the Académie in 1790 that Laplace built on. See Andoyer, *L'œuvre scientifique de Laplace*, pp. 27–28.

48. Laplace chaired the review committee of Legendre's papers on probability theory submitted on 8 March 1780; on continuous fractions on 10 and 12 May 1780; and on spheroids on 22 January and 19 February 1783. These reports written in Laplace's hand are available in AAdS, dossiers 12 December 1781 and 15 March 1783.
49. Legendre is cited on pp. 96–97, but not in Laplace's subsequent memoir on the same subject that appeared in 1785.
50. Vulfran Warmé, *Eloge historique de M. Delambre* (Amiens, 1824); and Claude Louis Mathieu in *Biographie universelle*, vol. 10, p. 299.
51. See Laplace to Oriani, 5 March 1788, in Milan, Osservatorio di Brera, Corr 1788, 1788 11 05 PSB BO thanking him for his assistance and praising the work of Delambre. For Bouvard, see Jules Forni, *L'astronome Bouvard* (Chambéry, 1888).
52. In his autobiographical notes in the BI, MS 2042, fol. 410v, Delambre indicates that he volunteered to calculate new tables for Jupiter and Saturn on the day in 1787 that Laplace presented his findings to the Académie. The date he gives, which has been repeated by others, is mistaken. Laplace proposed his views on 10 May and 15 July 1786 as indicated in Gillispie, p. 291, item 52. In 1785 Delambre was already supplying data to Laplace on Herschel's planet. His first major publication was the *Tables de Jupiter et de Saturne* (Paris, 1789).
53. The earliest documented attempt at entry was in 1785. See Laplace to Delambre, 21 June 1785, in AAdS, dossier Laplace. Delambre was eventually elected on 15 February 1792.
54. Factual details are contained in Arthur F. O'd. Alexander, *The Planet Uranus: A History of Observation, Theory, and Discovery* (New York, 1965); more elaborate comments are in Eric Forbes, "The Pre-Discovery Observations of Uranus," in Garry Hunt, ed., *Uranus and the Outer Planets* (Cambridge, Eng., 1982), pp. 67–80; and Gillispie, p. 100.
55. Laplace provides his own version of the discovery of the orbit of Uranus, crediting Bochart de Saron with the key finding in his *Théorie du mouvement et de la figure elliptique des planetes* (Paris, 1784), pp. 28–32. The book was published with financial assistance from this same Bochart de Saron. See also *Journal des Savans* (June 1785): 345–349.
56. *Journal des Savans* (October 1787): 667–668; *Journal de Paris*, 5 January 1788, no. 5, and Montucla, *Histoire des mathématiques*, vol. 4, p. 121.
57. Montucla, *Histoire des mathématiques*, vol. 4, pp. 119–120.
58. "Théorie des attractions des sphéroïdes et de la figure des planètes," *Histoire et mémoires de l'Académie Royale des Sciences de Paris* (1782),

p. 115 [*O.C.*, vol. 10, p. 343], itself a continuation of his book completed in December 1783.

59. "Mémoire sur les inégalités séculaires des planètes et des satellites," *Histoire et mémoires de l'Académie Royale des Sciences* (1784), pp. 1–50 [*O.C.*, vol. 11, pp. 49–92]; Joseph Louis Lagrange, "Théorie des variations séculaires des élémens des planètes," *Nouveaux Mémoires de l'Académie Royale des Sciences et Belles-Lettres [Berlin]* (1782), p. 292 [*O.*, vol. 5, p. 344].
60. The best account of these thickly argued papers is in Gillispie, pp. 124–141; see also Morando, *Planetary Astronomy,* pp. 138–141.
61. "Mémoire sur les inégalités séculaires des planètes et des satellites," pp. 5 and 8 [*O.C.*, vol. 11, pp. 52 and 56]. That figure was raised to 919 years a few months later in the paper presented May 1786 in *Histoire et Mémoires de l'Académie Royale des Sciences* (1785), p. 35 [*O.C.*, vol. 11, p. 97].
62. *Histoire et mémoires de l'Académie Royale des Sciences de Paris* (1785), p. 33 [*O.C.*, vol. 11, p. 95]; *Histoire et mémoires de l'Académie Royale des Sciences de Paris* (1786), pp. 236 and 240 [*O.C.*, vol. 11, pp. 244 and 248].
63. *Histoire et mémoires de l'Académie Royale des Sciences de Paris* (1786), p. 241 [*O.C.*, vol. 11, pp. 248–249].

## 6. Exploring the Physical World

1. Laplace to Barnaba Oriani, 10 July 1790, in Milan, Osservatorio di Brera, Corrispondenza 1790, 1790 07 10 PSL BO.
2. Frederic Lawrence Holmes, *Lavoisier and the Chemistry of Life* (Madison, Wis., 1985).
3. Letter of 21 August 1783 in Lagrange, *O.*, vol. 14, p. 124.
4. Letter of 28 June 1783 in my possession.
5. Ibid.
6. Undated letter to Deluc (probably in June 1783) in my possession and partly reproduced in J. P. Muirhead, ed., *Correspondence of the Late James Watt on His Discovery of the Theory of the Composition of Water* (London, 1846), pp. 41–42. The exact words are "C'est mon premier essai en physique, et je vous avoue que j'en suis si peu content, que je suis tenté d'abandonner cette carrière et me restreindre uniquement à la géométrie."
7. See, for example, his 7 March 1782 letter to Lavoisier in Denis I. Duveen and Roger Hahn, "Deux lettres de Laplace à Lavoisier," *RHS* 11 (1958): 337–342 and in Lavoisier, *O., Correspondance,* vol. 3, pp. 712–713.
8. It was Vicq d'Azyr who was reported to have said, "Adam does not know how much effort Laplace and I have expended to forget what he taught us." Adolphe Jean Louis Marie Dufresne, *Notes sur la vie et les œuvres de Vicq d'Azyr (1748–1794)* (Bordeaux, 1906), pp. 13–14.

9. AAdS, dossier 1 February 1775.
10. AAdS, dossier séance 18 December 1776. The paper submitted through the intermediary of Lalande (see letter of Deluc to Lalande of 27 September 1776, now in AAdS, dossier Lalande) bore the title "Support de niveau et quelques autres instruments." It may have been part of the supplement to Deluc's *Recherches sur les modifications de l'atmosphère* referred to in the Fichier Charavay at the BN, MSS, vol. 106.
11. The entry in Gillispie, p. 289, item 22 can be supplemented by the announcements in the *Journal de Paris* for 12 April 1777 and the *Gazette de France* for 18 April 1777, p. 138, col. 2. There is a brief description of the talk in a letter of Pierre Prévost to Georges Louis Le Sage dated 13 April 1777, quoted in part in "Pierre Prévost: Notice relative à ses recherches sur la chaleur rayonnante," *Mémoires de la société physique et d'histoire naturelle de Genève*, supp., no. 2 (1890): 9. Henry Guerlac proposed a slightly different rendering of this paper and its origins in his comments in Antoine Laurent Lavoisier, *Memoir on Heat* (New York, 1982), p. xv.
12. AAdS, dossier 28 June 1777.
13. AAdS, dossier 26 June 1779. It was published in Lavoisier, *O.*, vol. 4, p. 327.
14. AAdS, dossier 11 August 1780.
15. Mathurin Jacques Brisson, *Dictionnaire raisonné de physique* (Paris, 1781), 3 vols. The favorable report, written by Laplace and co-signed by Bézout on 23 August 1780, is in BN, MS, n.a.f. 5152, pp. 208–209. Laplace also reported four years later on a supplement to this dictionary entitled "Observations sur les nouvelles découvertes aérostatiques, et sur la probabilité de pouvoir diriger les ballons." See AAdS, Procès-verbaux, 4 December 1784.
16. Laplace accompanied Franklin, Macquer, La Rochefoucauld, and Montigny. See Antoine Laurent Lavoisier, *Mémoires de chimie* (Paris, 1805), vol. 1, p. 348. Guerlac corrects the date given in this publication to 1777, when Lavoisier tested some of Joseph Priestley's findings. See Henry Guerlac, "Chemistry as a Branch of Physics: Laplace's Collaboration with Lavoisier," *HSPS* 7 (1976): 197, note 13.
17. Guerlac, "Chemistry," p. 196.
18. This issue is discussed in detail in ibid., pp. 205–216. Guerlac cites a number of references to their unpublished experiments.
19. AN, $O^1$679, no. 457.
20. Private residences are recorded in the annual *Almanach Royal* and the *Connoissance des Tems*.
21. Five hundred livres came from the coffers of the Académie, and two stipends of six hundred livres each were obtained when he left his teaching post at the Ecole Militaire.
22. Jean-Pierre Poirier, *Antoine Laurent de Lavoisier, 1743–1794* (Paris, 1993).

I prefer to refer to the English translation, *Lavoisier: Chemist, Biologist, Economist* (Philadelphia, 1996), which is a corrected and amended version of the original text.

23. On 21 August 1783, Laplace wrote to Lagrange, "As a considerably wealthy man, he spares nothing to provide his experiments with the precision indispensable for this delicate research program." Lagrange, *O.*, vol. 14, p. 124.
24. T. H. Lodwig and William A. Smeaton have pointed out that the accuracy of the experimental setup they carried out was quickly criticized across the Channel. See "The Ice Calorimeter of Lavoisier and Laplace and Some of Its Critics," *Annals of Science* 31 (1974): 1–18.
25. Lavoisier was in the habit of holding a salon on Saturdays after the meetings of the Académie. But Laplace might have also expected to be a special guest at the Lavoisiers' dinner table on Monday nights, which indeed he was. See Poirier, *Antoine Laurent de Lavoisier*, p. 96 and Birembaut, "L'Académie Royale des Sciences en 1780 vue par l'astronome suédois Lexell (1740–1784)," *RHS* 10 (1957): 161.
26. Archives du Calvados, F 6357, dated 24 June 1777. The first documents of this affair refer to a decree of 27 January 1775, when several doctors and apothecaries were asked to make analyses of the suspected cider. See Archives de Seine-Inférieure, C 945, dossier 6.
27. Details are in notarial documents dated 27 October 1775 and 18 March 1777 in Bancroft, box 18, folders 18 and 19.
28. These edicts are dated 22 July, 30 August and 18 September 1777. See the Bibliothèque Municipale de Dieppe, Archives Municipales, FF 11.
29. Lavoisier wrote a "Rapport concernant les cidres de Normandie" with several other academicians in *Histoire et mémoires de l'Académie Royale des Sciences de Paris* (1786), pp. 479–506 and Lavoisier, *O.*, vol. 3, pp. 536–561. See also Antoine François Hardy, *Expériences sur les cidres* (Rouen, 1785), and Antoine François Fourcroy, Antoine Laurent de Lavoisier, and Michel Augustin Thouret, "Rapport sur la falsification des cidres," Société Royale de Médecine, *Histoire et mémoires de l'Académie Royale des Sciences de Paris* (1786), pp. 159–166, and Lavoisier, *O.*, vol. 3, pp. 529–535. Thouret was also a contemporary scientist who studied natural philosophy at the University of Caen with Laplace.
30. Loan of three thousand livres by Lavoisier to Pierre Laplace dated 25 March 1777, registered before the Parisian notary Maître Dominique Destrein (Etude 99). The loan was paid back after Pierre Laplace's death by his son on 25 January 1789.
31. Laplace to Lavoisier, 14 March 1782, in AAdS, dossier Lavoisier, Chabrol donation, carton no. 1, 177.

32. In Laplace's marriage contract of 1788, there is reference to his investment of 12,600 livres in the powder administration. See Poirier, *Antoine Laurent de Lavoisier,* pp. 89–94.
33. Roger Hahn, "The Laplacean View of Calculation," in Tøre Frängsmyr et al., eds., *The Quantifying Spirit in the Eighteenth Century* (Berkeley, 1990), pp. 363–380.
34. The term is used in his exchanges with Turgot to denigrate popular demonstrators. See *Correspondance inédite de Condorcet et de Turgot 1770–1779,* ed. Charles Henry (Paris, 1882), p. 215.
35. Roger Hahn, "Lavoisier et ses collaborateurs: Une équipe au travail," in *Il y a 200 ans Lavoisier* (Paris, 1995), pp. 55–63.
36. Laplace to Lavoisier, undated but written "ce dimanche soir," in AAdS, dossier Lavoisier, Chabrol donation, carton no. 1, p. 176.
37. Laplace to Lavoisier, 7 March 1782, published by Denis I. Duveen and Roger Hahn, "Deux lettres de Laplace à Lavoisier," *RHS* 11 (1958): 337–342 and in Lavoisier, *O., Correspondance,* vol. 3, pp. 712–714. The results of their experiments on thermal expansion carried out in the last half of 1781 were not reported until after Lavoisier's death by his widow, who discussed its details with Laplace. See Lavoisier, *Mémoires de chimie,* vol. 1, pp. 295–311.
38. Laplace to Lavoisier, 7 March 1782, in Denis I. Duveen and Roger Hahn, "Deux lettres de Laplace à Lavoisier," *RHS* 11 (1958): 338–339, reproduced in Lavoisier *O., Correspondance,* vol. 3, p. 713.
39. Laplace, *O.C.,* vol. 10, pp. 203–205.
40. Laplace to Lavoisier, 14 March 1782. In this letter, Laplace lamented Lavoisier's absence at the meeting of the Académie where Volta displayed his electrometer, and praised the Italian scientist, suggesting they repeat their work in his presence using the new instrument. AAdS, dossier Lavoisier, Chabrol donation, carton no. 1, 177. A more detailed account is in Guerlac, "Chemistry," pp. 234–240; and Marco Beretta, "From Nollet to Volta: Lavoisier and Electricity," *RHS* 54 (2001): 41–47.
41. Details of the experiments are given in a letter of Volta to Van Marum dated 28 November 1782, published in Volta, *Opere,* vol. 3, p. 303, and discussed by Johannes Bosscha in *La correspondance de A. Volta et M. van Marum* (Leiden, 1905), pp. 5–7.
42. Giuliano Pancaldi, *Volta: Science and Culture in the Age of Enlightenment* (Princeton, 2003), pp. 238–243.
43. *Recherches sur les modifications de l'atmosphère* (Geneva, 1772).
44. Volta, *Epistolario,* vol. 2, pp. 162–166.
45. *Recherches sur les modifications de l'atmosphère* (Geneva, 1772), vol. 2, p. 368.

46. Conversation in May 1782, recorded on p. 213 of the copy book now in Yale University's Sterling Library, Deluc papers, MS 179, ser. 2, box 26. Deluc kept Le Sage abreast of many of his encounters in Paris.
47. Deluc abandoned the project of addressing these thoughts to Laplace several years later, when Laplace demurred at accepting this "honor." By this time, Deluc had begun to contest both the originality and the validity of the new chemistry as established by the synthesis of water—which Laplace agreed was a key feature of Lavoisier's chemical revolution. The substance of these letters was published by Deluc as *Idées sur la météorologie* (London, 1786–1787), 2 vols. See especially vol. 1, p. 5 for an explanation of their publication.
48. Deluc papers, MS 179, ser. 2, box 26, p. 272. Deluc is overjoyed: "Nothing is better for us than to see Laplace drawn into a branch of physics where one cannot see clearly without developing a notion about elastic fluids."
49. The issue had already been broached by Deluc in his *Recherches sur les modifications de l'atmosphère,* vol. 1, pp. 165–167 and vol. 2, p. 368. It appears on pp. 5–6 of the "Mémoire sur la chaleur lû à l'Académie Royale des Sciences, le 28 juin 1783 par Mrs Lavoisier et de La Place de la même Académie."
50. Deluc papers, MS 179, ser. 2, box 26, pp. 213 and 245.
51. *Théorie du mouvement et de la figure elliptique des planetes,* p. xii.
52. Lissa Roberts, "A Word and the World: The Significance of Naming the Calorimeter," *Isis* 82 (1991): 199–222 offers a sophisticated discussion of the way the newly invented "machine" served theoretical ends in the chemical revolution.
53. *Memoir on Heat,* pp. 6 and 8.
54. See Guerlac, "Chemistry"; Holmes, *Lavoisier and the Chemistry of Life;* as well as Robert Fox, *The Caloric Theory of Gases from Lavoisier to Regnault* (Oxford, 1971); Fabio Sebastiani, "Laplace e le teorie sulla natura del calore," *Giornale di Fisica* 22 (1981): 279–294; S. Tugnoli Pattaro, "Le ricerche calorimetriche di Lavoisier e Laplace," *Giornale di Fisica* 29 (1988): 333–361. Lavoisier generously credited Laplace with the concept of the calorimeter in his *Traité elémentaire de chimie* (Paris, 1789), vol. 1, p. 285, but paid for its confection. See Maurice Daumas, *Lavoisier: Théoricien et expérimentateur* (Paris, 1955), pp. 141–142.
55. Carleton E. Perrin, "Lavoisier, Monge, and the Synthesis of Water: A Case of Pure Coincidence?" *British Journal for the History of Science* 6 (1973): 424–428.
56. Maurice Daumas and Denis I. Duveen, "Lavoisier's Relatively Unknown Large-Scale Decomposition and Synthesis of Water, February 27 and 28, 1785," *Chymia* 5 (1959): 113–129.

57. Lavoisier, *O.*, vol. 2, p. 342.
58. Laplace lamented this fact in his letter to Deluc on 5 November 1783 (in my possession).
59. The importance of the salary, along with a generous assessment of both Bézout's work in algebra and Laplace's friendship with him, is mentioned in Laplace to Lagrange, 11 February 1784, Lagrange, *O.*, vol. 14, p. 130.
60. Laplace to Lavoisier, October 1783. In this letter composed one month after Bézout's passing, all Laplace's machinations are revealed. See Denis I. Duveen and Roger Hahn, "Laplace's Succession to Bézout's Post of Examinateur des Elèves de l'Artillerie," *Isis* 48 (1957): 416, and Lavoisier, *O., Correspondance*, vol. 3, pp. 750–752.
61. See Duveen and Hahn, "Laplace's Succession," p. 423.
62. See Laplace to the Minister of the Navy, the Marquis de Castries, in ibid., p. 425; and his reply on 13 November 1783 in Vincennes, Archives de la Marine, dossier Laplace, no. 2.
63. These amounts are listed in Laplace's 1788 marriage contract.
64. Report of Laplace to the Minister of War, 12 September 1785, in Vincennes, Archives de la Guerre, $X^D$ 249, vol. 1, pp. 22–25.
65. Arthur Chuquet, *La jeunesse de Napoléon: Brienne* (Paris, 1897–1899), vol. 1, p. 225.
66. *Essai d'une théorie sur la structure des crystaux* (Paris, 1784). The glowing report written by Laplace is in AAdS, *Procès-verbaux,* 26 November 1783. I discussed this briefly in "The Laplacean View of Calculation," in Frängsmyr et al., *Quantifying Spirit,* pp. 371–372.

## 7. Revolutionary Tumult

1. Laplace was on the review committees that commented on a disagreement about military tactics between Fourcroy de Ramecourt and Montalembert.
2. Bernard Bru, "Estimations Laplaciennes. Un exemple: La recherche de la population d'un grand empire, 1785–1812," in Jacques Mairesse, ed., *Estimation et sondages* (Paris, 1988), pp. 7–46; Jean-Claude Perrot and Stuart J. Woolf, *State and Statistics in France, 1789–1815* (Chur, 1984); and Eric Brian, *La mesure de l'état: Administrateurs et géomètres au XVIIIè siècle* (Paris, 1994). In 1790, Laplace also reported on various proposals by Louis François de Beaufleury, Emmanuel Etienne Duvillard de Durand, and Joachim Lafarge.
3. See Louis S. Greenbaum, "Jean-Sylvain Bailly, the Baron de Breteuil, and the Four New Hospitals of Paris," *Clio Medica* 8 (1973): 261–284; "Tempest in the Academy: Jean-Baptiste Le Roy, the Paris Academy of Sciences, and the Project of a New Hôtel-Dieu," *AIHS* 24 (1974): 122–140; and

"'Measures of Civilization': The Hospital Thought of Jacques Tenon on the Eve of the French Revolution," *Bulletin of the History of Medicine* 49 (1975): 43–56. The Bordeaux projects were authored by Guy-Louis Combes.
4. See the multiple receipts from Ysabeau for payment of rental that are in Bancroft, box 10, folder 21. My guess is that he lived there from August 1784 until mid-1788.
5. Laplace sat on the review committee that considered the removal of slaughterhouses from the center of the capital. See *Histoire et mémoires de l'Académie Royale des Sciences* (Paris, 1787), pp. 19–43.
6. A veiled reference to help from Madame Lavoisier in December 1783 is the only evidence that could be construed as relating to an arranged marriage. See Lavoisier, *O., Correspondance,* vol. 3, p. 758.
7. She died a widow at age ninety-four on 20 July 1862. A necrology appeared in the *Journal Général de l'Instruction Publique* 31 (23 July 1862): 547 and in the *Revue de l'Instruction Publique de Littérature et des Sciences en France et dans les Pays Etrangers* 22 (24 July 1862): 270.
8. Jean Baptiste Courty died in 1824 at age ninety. See Jean-Tiburce de Mesmay, *Dictionnaire historique, biographique et généalogique des anciennes familles de Franche-Comté* (Paris, 1958), vol. 14, p. 294, mimeographed copy consulted at the Archives Départementales du Doubs, and (Paris, 1987), vol. 1, p. 786; and Jean-François Solnon, *215 bourgeois gentilshommes au XVIIIe siècle: Les secrétaires du roi à Besançon* (Paris, 1980) [*Cahiers d'études comtoises,* no. 28], pp. 221–222. According to François Lassus, in his *Métallurgistes Franc-Comtois du XVIIe et XVIIIe siècles: Les Rochet* (Besançon, 1980), vol. 1, p. 387, the Courty dynasty was one of the few that progressed from miner to clerk to iron master and then to entrepreneur.
9. The Mollerat family, though principally living in Poissons (near Joinville) in the Champagne region, also built its fortune through managing ironworks. The extended family had ties with Besançon through its office holders in Moutiers-sur-Saulx. See Archives Départementales de la Haute-Marne, 22 J 8, C 91 and C 306. They were also well-off and recently ennobled.
10. Guy Richard, "Les forges champenoises à la fin du XVIIIe siècle," in *Actes du 95e Congrès des Sociétés Savantes: Reims, 1970* (Paris, 1974), vol. 2, pp. 115–118.
11. The loan by Louis Mollerat was made on 10 January 1788 and registered by the Parisian notary Maître Guillaume jeune (AN, Minutier des Notaires, Etude 78 [Guillaume]).
12. Contract dated 14 March 1788 at AN, Minutier des Notaires, Etude 78 (Guillaume), no. 933.

13. Power of attorney dated 12 February 1788 in Bancroft, box 25, folder 20.
14. Since the figures for Laplace included travel reimbursements for his examinations for the artillery corps, one must reduce the annual salary to 8,400 livres. His assets were estimated at 16,600 livres in savings and 4,000 livres for personal effects. Calculation for the dowry is complicated because the family promised funds from an expected inheritance in return for her renunciation of rights to the estate. Her personal effects were evaluated at 4,000 livres.
15. A copy made from the Paris "Registre des Actes de Mariage" indicates that banns were posted on 24 February 1788, and the marriage was witnessed by Berthollet and a lawyer named Bouton de Souville for the groom; and for the bride, her father, maternal uncle Mollerat de Riaucourt, and Colin, also a lawyer. See AN, BB30 1044, dossier Charles Emile Pierre Joseph Laplace, Canal de Loing.
16. In a letter to Delambre dated 25 May 1788, Lalande reports that Laplace is about to leave the following Tuesday for his homeland. (London, Wellcome Historical Medical Library, Lalande folder).
17. The Bancroft Library, box 30, has a "Journal pour l'administration des biens, fonds et rentes de Normandie dont est propriétaire Monsieur le Marquis de Laplace, Pair de France, commencé en 1788 par M. Lieutaud."
18. Contract signed between Pierre Simon and Marie Anne dated 13 July 1788 and signed at Beaumont, in Bancroft, box 10, folder 12.
19. Pierre Laplace died on 29 June 1788 and was interred at Beaumont the next day without any of his family in attendance. Archives Départementales du Calvados, 4E 124.
20. Among Laplace's papers kept at the Bancroft Library (box 3, folder 30, and box 10, folder 21), there are records of extensive repairs made to the apartment once the couple moved in. His living standards changed considerably after becoming head of a family. Contract between Robert Arthur and René Grenard, owners of the building, and Laplace dated 1 January 1789, in Bancroft, box 10, folder 21. The Brechainvilles occupied the third floor, and the Laplaces rented the fourth floor.
21. 5 April 1789. See Léonce de Brotonne, *Tableau historique des Pairs de France, 1789–1814–1848* (Paris, 1889), p. 34.
22. The birth date is given as 15 April 1792. Her godparents were René Grenard, merchant and co-owner of the building where the Laplaces lived, and Marie Suzanne Arthur, probably the wife of the other co-owner. Copy of Register of the Parish, Bancroft, box 19, folder 10.
23. Receipt signed by Jean Baptiste Courty on 7 June 1790 in Bancroft, box 10, folder 12. The son Charles Emile was already referred to simply as Emile.
24. Already in a letter to Lacroix dated 7 March 1789, Laplace envisaged leav-

ing the capital for the quiet of the countryside. See René Taton, "Laplace et Silvestre-François Lacroix," *RHS* 6 (1953): 351.
25. Receipts signed by the paper manufacturers and decorators Robert Arthur and René Grenard on 22 March 1793 in Bancroft, box 10, folder 21.
26. Receipt from Etienne Guérin de Vaux dated 27 April 1793 and certificate dated 12 November 1792 issued by the Directory of the Melun district in Bancroft, box 10, folder 21. Another record indicates the rental contract for 550 livres for fifteen months was written on 27 September 1792 and registered with the notary Maître Duverger. See Archives Départementales de Seine-et-Marne, 101C 149, p. 44 verso.
27. Receipt dated 14 brumaire an 3 [4 November 1794] from Courty in Bancroft, box 10, folder 12. Courty acquired several properties in the region starting in early 1792. These transactions are recorded with various notaries in Versailles, Langres, and Paris and their sale can be traced back through the Melun notary Maître Antoine Derosière.
28. There were 112 candidates examined in August 1789; 87 in January and February 1792; and a smaller number in May and June 1793. See Vincennes, Archives de la Guerre, $X^D 249$; and the letter of Laplace of 14 April 1793 to the minister of war at Rouen, Bibliothèque Municipale, Duputel 530.
29. Letter from Jancé of the naval ministry dated 28 ventôse an 2 [18 March 1794] and 30 thermidor an 2 [18 July 1794] in Bancroft, box 1, folders 15–16; and the draft of a letter to Tierzot in November 1794 indicating that he was still expecting to examine naval engineering students, in Bancroft, box 8, folder 11.
30. Receipts dated 16 October 1789 and 27 June 1791 in Bancroft, box 2, folder 28.
31. Affidavits of 21 January and 24 February 1793 in Bancroft, box 1, folder 23; and 26 December 1793 in Bancroft, box 1, folder 23.
32. Many of these are in Bancroft, box 10, folder 21.
33. See documents of 22 frimaire and 6 fructidor an 2 [12 December 1793 and 23 August 1794] in Bancroft, box 2, folder 13, and box 8, folder 10; and 16 prairial an 2 [4 June 1794] in Archives Départementales de Seine-et-Marne, L 706b.
34. See Dorinda Outram, "The Ordeal of Vocation: The Paris Academy of Sciences and the Terror, 1793–1795," *History of Science* 21 (1983): 251–273; and Roger Hahn, "The Triumph of Scientific Activity: From Louis XVI to Napoléon," *Proceedings of the Annual Meeting of the Western Society for French History* 16 (1989): 204–211.
35. Simon Pierre Mérard de Saint-Just, *Eloge Historique de Jean Sylvain Bailly* (London, 1794), pp. 139–140; and Gabriel Leroy, *Histoire de Melun*

(Melun, 1887), p. 438. In a later work by Leroy, *Le Vieux Melun* (Melun, 1904), pp. 354–362, he indicates that Bailly had rented a different house from Guérin de Vaux starting 1 July 1793. Bailly did not arrive until the first week of September that year.

36. Gabriel Leroy, "Une aventure de M. de La Place," in *Almanach historique, topographique et statistique du Département de Seine-et-Marne et du Diocèse de Meaux, année 1867*, pp. 143–149. The register noting the arrest indicates he was released by 4 A.M. the next day. See Archives de la ville de Melun, ser. 4-I-1, bundle 2.
37. For the text, see *CIPCN, 3*, 239. The decree dated 3 nivôse an 2 [24 December 1793] was introduced by Prieur de la Côte d'Or who, according to Delambre, held a grudge against Lavoisier and the circle of his friends who shared his political views. See Delambre, *Grandeur et figure de la Terre*, ed. Guillaume Bigourdan (Paris, 1912), pp. 212–214. Georges Bouchard, in his *Prieur de la Côte-d'Or* (Paris, 1946), pp. 296–298, offers another explanation for this culpable act: it may have stemmed from the rebuff Prieur felt when his nomenclature for the new units was rejected.
38. Lacroix replaced him on 1 October 1793. See BI, MS 2398. See also Roger Hahn, "Le rôle de Laplace à l'Ecole Polytechnique," in Bruno Belhoste et al., eds., *La formation polytechnicienne, 1794–1994* (Paris, 1994), pp. 50–51.
39. See item 29 of the sale catalogue by J. A. Stargardt in Marburg for 5 June 1962.
40. Boston Public Library, Department of Rare Books and Manuscripts, MS 2837.
41. Milan, Osservatorio di Brera, Corrispondenza 1790, 1790 07 10 PSL BO.
42. See the draft of November 1793 letter to the tax collector in Melun, citoyen Liger, in Bancroft, box 8, folder 5. His request to reduce the tax from 3,000 to 1,000 livres was granted; see Archives Départementales de Seine-et-Marne, L 49, fol. 61 verso and 571.
43. Hahn, chapters 6–7.
44. The story of the metric system has been told many times over; see, for example, Guillaume Bigourdan, *Le système métrique des poids et mesures: Son établissement et sa propagation graduelle* (Paris, 1901); Adrien Favre, *Les origines du système métrique* (Paris, 1931); Ruth Inez Champagne, "The Role of Five Eighteenth-Century French Mathematicians in the Development of the Metric System," Ph.D. diss., Columbia University, 1979; John L. Heilbron, "The Measure of Enlightenment," in Heilbron, *Weighing Imponderables and Other Quantitative Science around 1800*, (Berkeley, 1993), pp. 243–277; and Ken Alder, *The Measure of All Things* (New York, 2002).
45. Gillispie, pp. 151–153 offers a more elaborate hypothesis for Laplace's

preference for the geodetic approach. I follow the explanation given by Delambre in *Grandeur et figure de la Terre,* pp. 202–203.

46. Arthur Birembaut, "Les deux déterminations de l'unité de masse du système métrique," *RHS* 12 (1959): 25–54.
47. Along with five other academicians, Laplace was ousted from the committee in late December 1793 until his reinstatement in May 1795 along with Borda, Coulomb, Brisson, and Delambre. The other purged member of the original committee, Lavoisier, had in the meantime been executed.
48. Laplace, *Discours prononcé [au Conseil des Anciens] par le citoyen Laplace au nom de l'Institut National des Sciences et des Arts* (1er jour complémentaire an 4) [17 September 1796].
49. Session 9 of his Ecole Normale lectures and ten pages of chapter 12 of book 1 of the first edition of the *Exposition* (pp. 124–133) are devoted solely to this topic. For Laplace's lecture, see *Journal de l'Ecole Polytechnique,* vol. 2, cahiers 7 et 8 (June 1812): 126–139.
50. Laplace follows this radical proposal in the first edition of the *Exposition* (1796). See the comments by Heilbron in his "Measure of Enlightenment," pp. 211–215; and Paul Smith, "La division décimale du jour: L'heure qu'il n'est pas," in Bernard Garnier et al., eds., *Genèse et diffusion du système métrique,* (Caen, 1990), pp. 123–134.
51. *Annales de Chimie* 16 (March 1793): 250–255, published separately as *Recueil de pièces relatives à l'uniformité des poids et mesures* (Paris, 1793), pp. 26–31.
52. *Annales de Chimie* 16 (March 1793): 267–282.
53. Charles Ballot, "Procès-verbaux du Bureau de Consultation des Arts et Métiers," *Bulletin d'Histoire Economique de la Révolution* (1913): 15–160. In a letter dated 27 pluviôse an 2 [15 February 1794] to the Bureau, he specifically referred to his assiduity in carrying out the tasks assigned him. See Paris, Conservatoire des Arts et Métiers, Archives du Musée National des Techniques, ser. 10–49, no. 1. See also Dominique de Place, "Le Bureau de Consultation pour les Arts, Paris, 1791–1796," *History and Technology* 5 (1988): 139–178.
54. In the 1780s, Laplace had a few squabbles with Brissot de Warville over Marat's scientific merit. After the Académie approved of the creation of a corporation of scientific instrument-makers in 1787, Laplace insisted that they demonstrate a mathematical aptitude. See Hahn, pp. 71 and 154, and Gillispie, pp. 154–155.
55. Nicole Dhombres and Jean Dhombres, *Naissance d'un nouveau pouvoir: Sciences et savants en France, 1793–1824* (Paris, 1989), chapter 1.
56. Ibid., especially pp. 69–91; and Bronislaw Baczko, *Comment sortir de la Terreur: Thermidor et la Révolution* (Paris, 1989).

57. Robert R. Palmer, *The Improvement of Humanity: Education and the French Revolution* (Princeton, 1985); and Dominique Julia, *Les trois couleurs du tableau noir: La Révolution* (Paris, 1981).
58. A detailed analysis of these debates is in the notes by James Guillaume in *CIPCN*. See also Bronislaw Baczko, ed., *Une éducation pour la démocratie* (Paris, 1982).
59. Paul Marie Dupuy, *Le centenaire de l'Ecole Normale, 1795–1895* (Paris, 1895).
60. *CIPCN*, vol. 5, p. 185.
61. Letter of Tierzot, president of the district of Melun, dated 22 brumaire an 2 [12 November 1794] in Bancroft, box 8, folder 13.
62. Draft reply in Bancroft, box 8, folder 11.
63. Nominally, he was there to replace the medical doctor Hallé. See *CIPCN*, vol. 5, p. 364. This left Laplace less than a month to prepare his lectures. Laplace's salary for his lectures was 4,700 livres according to records in AN, $F^{17}$ 1562.
64. Dhombres and Dhombres, *Naissance d'un nouveau pouvoir*, pp. 69–71.
65. However mythical was this phrase supposedly uttered by Fouquier-Tinville, the scientists shared the belief that they needed to prove their usefulness to the new regime. See James Guillaume, "Un mot légendaire: 'La République n'a pas besoin de savants,'" *Etudes Révolutionnaires* 1 (Paris, 1908): 136–155; and Janis Langins, *La République avait besoin de savants* (Paris, 1987).
66. Even though some 1,200 students had been selected, the lecture hall held only six to seven hundred seats. See Dupuy, *Le centenaire de l'Ecole Normale*, p. 94.
67. The opening event was given a prominent place in the official newspaper, the *Moniteur Universel* of 9 pluviôse an 3 [28 January 1795], p. 530. It was also featured in the *Décade Philosophique*, 4 (20 pluviôse an 3 [9 February 1795]), pp. 276–277; and *L'Abréviateur Universel* of 1 ventôse an 3 [19 February 1795], p. 603. Details are in Dupuy, *Le centenaire de l'Ecole Normale*, pp. 156–158.
68. *Séances des Ecoles Normales recueillies par des sténographes, revues par les professeurs* (Paris, 1795), 10 vols. Student questions and responses by the lecturers were also recorded and published as *Débats* (Paris, 1795), 3 vols.
69. *Séances des Ecoles Normales*, vol. 1, pp. 16–17, reproduced in Dupuy, *Le centenaire de l'Ecole Normale*, p. 106; and Jean Dhombres, ed., *L'Ecole Normale de l'an III: Leçons de mathématique* (Paris, 1992), p. 45.
70. Laplace advised his students to "favor general methods in your teaching, dedicate yourselves to presenting them in the simplest manner and you will see that they are almost always the easiest." *Séances des Ecoles*

*Normales,* vol. 4, pp. 49–50 and *Leçons de mathématiques,* p. 90. This passage was added in the printed version of the seventh lesson.

71. See the important articles by Jean Dhombres, "L'enseignement des mathématiques par la 'méthode révolutionnaire': Les leçons de Laplace à l'Ecole Normale de l'an III," *RHS* 33 (1980): 315–348; and "Enseignement moderne ou enseignement révolutionnaire des sciences?" *Histoire de l'Education* 42 (May 1989): 55–78 for more details.
72. The first six of the ten lectures are devoted to numbers and algebra, which are given a noticeable primacy over geometry. Geometry was covered by Monge in another set of lectures. In the second edition of these lectures, printed in the *Journal de l'Ecole Polytechnique,* cahiers 7 and 8 (June 1812), pp. 24 and 47, Laplace added a note referring readers to the brilliant new solutions by Gauss. For the fundamental theorem of algebra, see Hellmuth Kneser, "Laplace, Gauss und der Fundamentalsatz der Algebra," *Deutsche Mathematik* 4 (1939): 318–322; and annex 6 by Jean Dhombres, in *Leçons de mathématiques,* pp. 477–484.
73. The quotation is from Jean Baptiste Biot, *Essai sur l'histoire générale des sciences pendant la Révolution française* (Paris, 1803), p. 68. See also what Lacroix inserted in Jean-Baptiste Joseph Delambre, *Rapport historique sur les progrès des sciences mathématiques depuis 1789, et sur leur état actuel* (Paris, 1810), pp. 35 and 65. Sylvestre François Lacroix, in his *Essais sur l'enseignement en général, et sur celui des mathématiques en particulier* (Paris, 1805), pp. 35–36, suggests that these lectures gave a "prodigious impulse" to the teaching of mathematics by reviving a method of presentation set by Clairaut and adopted at the écoles centrales (pp. 281–282 and 146).
74. *Séances des Ecoles Normales,* vol. 6, p. 32 and *Leçons de mathématiques,* p. 125.
75. The German translator Johann Karl Friedrich Hauff was in touch with Laplace in 1799. See Weimar, Nationale Forschungs-und Gedenkstatten der Klassischen Deutschen Literatur, Goethe Autographen, no. 1046. Mary Somerville also obtained permission from Laplace to paraphrase his work for her *Mechanism of the Heavens* (London, 1831). See Elizabeth C. Patterson, *Mary Somerville, 1780–1872* (Oxford, 1979), p. 49; Patterson, *Mary Somerville and the Cultivation of Science* (The Hague, 1983), and Oxford, Bodleian Library, Dep c. 371, no. 21.
76. A textbook of lessons offered at the Athénée and the Ecole Polytechnique written by Jean Henri Hassenfratz—*Cours de physique céleste; ou, Leçons sur l'exposition du système du monde* (Paris, 1803)—is explicitly based on Laplace's work.
77. *Exposition,* vol. 2, p. 7.

78. Ibid., pp. 310 and 27.
79. This historical chapter was later published separately as the *Précis de l'histoire de l'astronomie* (Paris, 1821), and has stimulated some revealing comments by Joseph Agassi. See Agassi, *Toward an Historiography of Science* (The Hague, 1963).
80. Laplace's exact words are "Avec la défiance que doit inspirer tout ce qui n'est point un résultat de l'observation ou du calcul." (*Exposition,* vol. 2, p. 303).
81. Remigius Stölzle, "Ist die Bezeichnung Kant-Laplacesche Hypothese berechtigt?" *Philosophisches Jahrbuch* 20 (1907): 324–327.
82. For the most recent literature on the legacy he left, see Ronald L. Numbers, *Creation by Natural Law: Laplace's Nebular Hypothesis in American Thought* (Seattle, 1977); Stanley L. Jaki, *Planets and Planetarians: A History of Theories of the Origin of Planetary Systems* (New York, 1977); Jacques Merleau-Ponty, *La science de l'univers à l'age du positivisme* (Paris, 1983); and Stephen G. Brush, *Nebulous Earth: The Origin of the Solar System and the Core of the Earth from Laplace to Jeffreys* (New York, 1996).
83. An excellent introduction to these ideas is Jean Seidengart, "Genèse et structure de la cosmologie Kantienne précritique," in the French translation of Kant's *Allgemeines Naturgeschichte* (Paris, 1984), pp. 9–59.
84. See *Séances des Ecoles Normales,* vol. 6, pp. 52–57 or *Leçons de mathématiques,* pp. 132–134.
85. Ibid.
86. Jean Seidengart proposes an important other distinction between French and Anglo-Saxon treatments of cosmology related to theology in his "Genèse et structure de la cosmologie Kantienne précritique," pp. 21–24. See also the excellent section on the relation between stability and cosmology in Merleau-Ponty, *La science de l'univers à l'age du positivisme,* pp. 39–57.
87. *Mémoires de mathématique* (Paris, 1773), vol. 7, p. 503. [O.C., vol. 8, p. 279].
88. Daniel Bernoulli, "Recherches physiques et astronomiques sur le problème proposé pour la seconde fois par l'Académie Royale des Sciences 'Quelle est la cause physique de l'inclinaison des plans des orbites des planètes' . . ." in *Recueil des pièces qui ont remporté des prix de l'Académie Royale des Sciences* (Paris, 1735), vol. 3, Prix de 1734, p. 99.
89. Bailly, *Histoire de l'astronomie moderne,* vol. 2, pp. 719–721 and vol. 3, pp. 219–224.
90. Buffon, *Les époques de la nature,* ed. Jacques Roger (Paris, 1962), pp. 25–39; and Jean Seidengart, "Le traitement du problème cosmologique dans l'œuvre de Buffon," *Buffon* 88 (1992): 309–325.

91. Dionis du Séjour, *Traité analytique des mouvements apparents des corps célestes* (Paris, 1786–1789), vol. 2, p. 605. In his *Essai sur les comètes* (Paris, 1775), pp. 195–196, Dionis had already objected to Buffon's theory, but with a gracious and polite tone. Laplace agreed, first in *Exposition*, vol. 2, pp. 298–301.
92. Dionis du Séjour, *Traité analytique*, vol. 2, p. xl.
93. Jaki, *Planets and Planetarians*, pp. 122–134, gives a spirited rendition of the successive changes Laplace brought to his theory.
94. Bailly, *Histoire de l'astronomie moderne*, vol. 3, pp. 216–218.
95. Dionis du Séjour, *Traité analytique*, vol. 2, p. vi.
96. *Exposition*, vol. 2, p. 305.
97. Laplace did not calculate the gravitational effects on light under these theoretical conditions in the *Exposition* because he wrote it for laymen. A calculation was published, however, in German in *Allgemeine Geographische Ephemeriden* 4 (July 1799): 1–6. I suspect it was sent for Laplace by the German translator of the *Mécanique,* Johann Karl Burckhardt, who was in contact with the editor of this monthly journal, Baron Franz von Zach.
98. Simon Schaffer, "John Michell and Black Holes," *Journal for the History of Astronomy* 10 (1979): 42–43.
99. "Astronomical Observations Relating to the Construction of the Heavens . . . ," *Philosophical Transactions* 101 (1811): 269–336; in J. L. E. Dreyer, ed., *The Scientific Papers of Sir William Herschel* (London, 1912), vol. 2, pp. 459–497.
100. *Moniteur Universel,* 7 July 1812, pp. 739–740.
101. The enthusiasm was not shared by all members of the Bureau; some realized that there were many new objections to Laplace's theory, including the existence of newly discovered asteroids. See Olbers to Bessel, 10 July 1812, in *Briefwechsel zwischen W. Olbers und F. Bessel* (Leipzig, 1852), vol. 1, p. 337; on 9 July Laplace had forwarded Olbers a copy of the *Moniteur* article. See Bremen, Staatsbibliothek, MS 6, La (Laplace) 1. On the same day, he sent one to Oriani; see Milan, Osservatorio di Brera, Corrispondenza 1812, 07 09 PSL BO.
102. In Olbers to Bessel, 10 July 1812, Olbers refers sneeringly to Laplace's "love of his own hypothesis." Yet in the letter of thanks he sent Laplace after his return to Bremen, he is quite obsequious, referring to him as "the greatest mathematician of Europe." See the draft of a letter in Bremen, Staatsbibliothek, MS 5, Ol (Olbers) 438.
103. *Moniteur Universel,* 9 pluviôse an 3 [28 January 1795], p. 530; *Séances des Ecoles Normales,* vol. 1, pp. 30–31; and *Leçons de mathématiques,* pp. 51–52. This remark is discussed critically in a letter by Gregorio Fontana to Oriani from Pavia on 23 April 1795. See Milan, Osservatorio di Brera, Corrispondenza 1795.

## 8. The Politics of Science

1. A notable exception was Laplace's behavior at the time of elections of new members. Lalande was especially upset that he did not back Charles le Géomètre, who was a disciple of Bossut and Monge. See Lalande to Deluc, 12 May 1785, letter in my possession.
2. Emilie Trembley, "Un savant genevois: Jean Trembley-Colladon 1749–1811. Son jugement sur le monde scientifique de Paris en 1786," *Bulletin de la Société d'Histoire et d'Archéologie de Genève* 9 (1948): 110.
3. Laplace was notably attached to Paul Charpit de Villecourt, a precocious mathematician who died young, introduced to him by Arbogast. See copy of letter to his mother on 30 December 1784 in Florence, Biblioteca Medicea-Laurenziana, Ashburnham-Libri, add. 1838, vol. 3, fol. 60v. He carried on a long and fruitful mentoring relationship with the Genevan Jean Frédéric Théodore Maurice starting in 1796; and befriended the young Charles Chisson in 1804, who died young as well. For Chisson, see *Inventaire des autographes et des documents historiques composant la collection de M. Benjamin Fillon* (Paris, 1877), vol. 2, p. 79.
4. Hahn, pp. 167–173.
5. See Dorinda Outram, "The Ordeal of Vocation: The Paris Academy of Sciences and the Terror, 1793–1795," *History of Science* 21 (1983): 251–273; and Nicole Dhombres and Jean Dhombres, *Naissance d'un nouveau pouvoir: Sciences et savants en France, 1793–1824* (Paris, 1989), pp. 11–91. For Bochart, see Félix Louis Christophe Ventre de la Touloubre Montjoie, *Eloge historique de Jean-Baptiste-Gaspard Bochart de Saron* (Paris, [1799]), pp. 131–143.
6. In apologizing to his Genevan colleague Le Sage on 17 germinal an 5 [6 April 1797], Laplace said: "The *Exposition* is the fruit of my country's retreat during the lamentable period of revolutionary governance that moved so many true friends of France and humanity to tears. The impressions of these sad events which flashed by so rapidly while I was busy on it necessarily affected my work.' See Geneva, Bibliothèque Publique et Universitaire, MS D.O. Autographes.
7. See Nicole Dhombres and Jean Dhombres, "Laplace: Mathématicien, astronome, physicien, . . . et ministre français," *Crux Mathematicorum* 5 (1979): 32–40, 69–75, 92–102.
8. Letters of 17 and 22 December 1794 in Joseph Lakanal, *Exposé sommaire des travaux de Joseph Lakanal* (Paris, 1838), pp. 207–209. In the first letter, Laplace says he made a special trip to Paris to confer with Lakanal.
9. *CIPCN*, vol. 6, p. 62. More details can be found in the same volume and in Bruno Morando, "La création du Bureau des Longitudes en 1795," *Conférences de la Société Philomatique de Paris* 3 (1993): 23–44.

10. Morando, "La création du Bureau des Longitudes."
11. Paris, Bibliothèque de l'Observatoire, MS 1022; and several copybooks at Archives du Bureau des Longitudes. The meetings were held until 1804 in the Louvre, where the former Académie des Sciences had met until 1793. See Guillaume Bigourdan, "Le Bureau des Longitudes: Son histoire et ses travaux," *Annuaire du Bureau des Longitudes* (1928): A8–A9.
12. In addition to Alexis Bouvard, who became his major assistant in the calculation of tables, Laplace supported Burckhardt, Arago, Biot, and Damoiseau for various appointments. He also attempted to have the young Augustin Cauchy named librarian, but failed because of an irregularity in the election procedures. See Bigourdan, "Le Bureau des Longitudes" (1928), pp. A30–A31.
13. Laplace promoted the preparation of various tables for the position of the sun (Delambre), the moon (Bürg, Burckhardt, and Damoiseau), Jupiter and Saturn (Bouvard), satellites of Jupiter (Delambre), and atmospheric refractions (Delambre). All received the approval and some were funded by the Bureau des Longitudes. See Guillaume Bigourdan, "Le Bureau des Longitudes," *Annuaire du Bureau des Longitudes* (1932): A2–A11.
14. BI, MS 2042, fol. 411–412. Much of this section is crossed out because Delambre did not wish Laplace to read it while he was still alive.
15. See also the rather critical private report that Olbers sent to Bessel after attending a meeting of the Bureau: *Briefwechsel zwischen W. Olbers und F. Bessel* (Leipzig, 1852), vol. 1, p. 337.
16. The decree creating the Institut was one of the last acts of the National Convention on 26 October 1795. For details see Hahn, chapter 10.
17. Roger Hahn, "From the Académie des Sciences to the Institut National," in *Annalen der Internationalen Gesellschaft für Dialektische Philosophie: Societas Hegeliana* 6 (1989): 202–219. The inaugural session took place on 4 April 1796.
18. *Discours prononcé [au Conseil des Anciens] par le citoyen Laplace au nom de l'Institut National des Sciences et des Arts (premier jour complémentaire an 4)* [17 September 1796], pp. 4–5. The text of the speech was reported in *Moniteur Universel,* cinquième jour complémentaire de l'an 4 [22 September 1796], p. 1460, and the reply by Pastoret printed the next day.
19. See Laplace to Delambre, 24 and 29 January 1798, in Columbia University, Rare Book Library, David Eugene Smith Historical Collection and in *RHS* 14 (1961): 287–290.
20. Maurice P. Crosland, "The Congress on Definitive Metric Standards, 1798–1799," *Isis* 60 (1969): 226–231.
21. Claude Nicolas Amanton, *Recherches biographiques sur le Professeur d'Artillerie, Lombard* (Dijon, 1802), pp. 28–29. See also Jean Louis Lombard, *Ta-*

bles du tir des canons et des obusiers (n.p., 1787) and *Traité du mouvement des projectiles appliqué au tir des bouches à feu* (Dijon, 1797).
22. Joachim Fischer, *Napoleon und die Naturwissenschaften* (Stuttgart, 1988), p. 30.
23. *Correspondance de Napoléon Premier* (Paris, 1858), vol. 1, pp. 392–393, 5 prairial an 4 [24 May 1796].
24. *Moniteur Universel*, 22 messidor an 4 [10 July 1796], p. 1165.
25. Paul V. Aubry, *Monge: Le savant ami de Napoléon Bonaparte 1746–1818* (Paris, 1954), chapter 10. Monge was away from Paris for seventeen months on this mission.
26. Charles C. Gillispie, "Historical Introduction," in his *Monuments of Egypt: The Napoleonic Edition* (Princeton, 1987), pp. 1–29; Henry Laurens et al., *L'expédition d'Egypte, 1798–1801* (Paris, 1989); and Yves Laissus, *L'Egypte, une aventure savante, 1798–1801* (Paris, 1998).
27. At a reception in December 1797 attended by Lagrange and Laplace, Bonaparte spoke with the scientists about a minor treatise on geometry by Mascheroni he had brought back from his Italian campaign. Laplace quipped: "We expected everything from you, but not a mathematics lesson!" *Moniteur Universel*, 25 frimaire an 6 [15 December 1797], pp. 341–342; and in *Narrateur Universel*, 23 frimaire an 60 [13 December 1797], cited in François Alphonse Aulard, ed., *Paris pendant la réaction thermidorienne et sous le Directoire* (Paris, 1902), vol. 4, pp. 490–491.
28. A tell-tale example is that Cabanis asked Laplace on 5 February 1798 to intervene on his behalf to allow Cabanis to present his *Du degré de certitude de la médecine* to the general. See BN, MS, Fichier Charavay, vol. 33, fol. 11, no. 8. Bonaparte also invited him to dinner in a casual manner, as seen in his letter of 27 vendémiaire an 8 [19 October 1799] in *Correspondance de Napoléon Premier* (Paris, 1860), vol. 6, p. 1.
29. For a list of candidates, see Institut de France, *PV, Institut*, vol. 1, p. 296.
30. Georges Lacour-Gayet, *Bonaparte: Membre de l'Institut* (Paris, 1921); and Fischer, *Napoleon*, pp. 55–57.
31. Registres des Séances du Bureau des Longitudes, 19 nivôse an 6 [8 January 1798] and 19 floréal an 6 [8 May 1798]. See also Laplace to Delambre, 14 floréal an 6 [3 May 1798], in which Laplace relates his meeting with Bonaparte, in AAdS, dossier Laplace.
32. Guillaume Bigourdan, "Le Bureau des Longitudes," *Annuaire du Bureau des Longitudes* (1931): A91–A92.
33. AN, AFiv1. The order is signed by the Consuls Ducos and Siéyès, but not by Bonaparte. On that very day, Laplace attended a meeting of the Institut with Bonaparte where they presented their report, read by Lacroix, on a mathematical memoir by Biot. *PV, Institut*, vol. 2, pp. 30–32. Biot discusses

this meeting in "Une anecdote relative à Laplace," in his *Mélanges scientifiques et littéraires* (Paris, 1858) vol. 1, pp. 1–9.
34. Jean Savant, *Les ministres de Napoléon* (Paris, 1959).
35. Roger Hahn, "L'autobiographie de Lacepède retrouvée," *Dix-Huitième Siècle* 7 (1975): 72. Another author claims that Volney was offered the post as well. See Antoine Guillois, *Le Salon de Madame Helvétius, Cabanis et les idéologues,* 2d ed. (Paris, 1894), pp. 136–137. Both were members of the Institut.
36. He was appointed examiner of students at Polytechnique on 25 May 1795; a member of the "jury" for Paris schools in December 1795; and to the same post for the school in Fontainebleau in April 1796. See Ambroise Fourcy, *Histoire de l'Ecole Polytechnique* (Paris, 1828), p. 106; letter from the Department of Seine, 23 frimaire an 4 [14 December 1795] (Bancroft, box 1, folder 5); and from the Department of Seine-et-Marne, 17 April 1796 (Bancroft, box 1, folder 8).
37. Paul Bouteiller, *Histoire du ministère de l'intérieur de 1790 à nos jours* (Paris, 1993).
38. [Gaspard Gourgaud], *Mémoires pour servir à l'histoire de France sous Napoléon, écrites à Sainte-Hélène* (Paris, 1823), vol. 1, pp. 111–112. See also the different assessments by Fischer, *Napoleon,* pp. 109–110 and Dhombres and Dhombres, *Naissance,* pp. 749–751.
39. *Moniteur Universel,* 6 frimaire an 8 [27 November 1799], no. 66, p. 258. There is a dramatic story about the pension, probably romanticized, reported by Arago in his biography of Bailly. See Dominique François Arago, *O., Notices biographiques* (Paris, 1854), vol. 2, pp. 425–426. See also report of 2 frimaire an 8 [23 November 1799] in AN, AFiv1, 3.
40. *PV, Institut,* vol. 2, p. 53.
41. 12 frimaire an 8 [3 December 1799] in AN, AFiv925, dossier 1.
42. Letter of 15 frimaire an 8 [6 December 1799] from Delambre in AN, AA 64(267). A reply indicating the assignment of funds from the Ministry to compensate Borda's family and subsidize the publications of his *Tables trigonométriques décimales* was read to the Bureau des Longitudes on 24 frimaire an 8 [15 December 1799] by Delambre and published the following year.
43. Letter of 6 frimaire an 8 [27 November 1799] in Leiden, Universitetsbibliotheek, MS BPL 755. Laplace also lobbied for the law of 19 frimaire that adopted the prototypes for the meter and kilogram and placed the metric system in the Constitution.
44. *Recueil des lettres circulaires, instructions, arrêtés et discours publics, émanés des citoyens Quinette, Laplace, Lucien Bonaparte et Chaptal, ministres de l'intérieur* (Paris, 1799–1802), vol. 3, p. 118. There are several other references in AN, AFiv1, 4 for 4 frimaire an 8 [25 November 1799].

45. AN, AFiv925, dossier 1, 12 frimaire an 8 [3 December 1799].
46. Fourcy, *Histoire de l'Ecole Polytechnique,* pp. 192–193; and Roger Hahn, "Le rôle de Laplace à l'Ecole Polytechnique," in Bruno Belhoste et al., eds., *La formation polytechnicienne, 1794–1994* (Paris, 1994), p. 54.
47. Letter of 27 brumaire an 8 [18 November 1799] in BI, MS 2398.
48. AN, AFiv3, 12, no. 11.
49. Letter to Gueret, 27 brumaire an 8 [18 November 1799], in Washington, Smithsonian Institution, Bern Dibner library, MS 823A. A handful of these Departmental *Statistique* appeared in print in 1802.
50. See Laplace to Dieudonné, 30 frimaire an 10 [21 December 1801], in Philadelphia, Historical Society of Pennsylvania, Dreer Collection.
51. Laplace to Van Swinden, 6 frimaire an 8 [27 November 1799], in Leiden, Universitetsbibliotheek, MS BPL 755.
52. Laplace to Delambre, 25 ventôse an 5 [15 March 1797], in Berlin, Staatsbibliothek, Slg. Darmstaedter J 1796.
53. Laplace to Dieudonné, 28 germinal an 9 [18 April 1801], BN, MS, n.a.fr. 1305, fols. 99r–100v.
54. AN, AFiv4, no. 15, 3 nivôse an 8 [24 December 1799]. The colleagues named to the Senate included Berthollet, Bougainville, Cabanis, Cousin, Darcet, Daubenton, Destutt de Tracy, Garat, Lacepède, Lagrange, Monge, and Volney. See Jean Thiry, *Le Sénat de Napoléon (1800–1814),* 2d ed. (Paris, 1949), chapter 4.
55. AN, AFiv 2, 4è jour complémentaire an 11 [21 September 1803].
56. He rented an apartment from a M. Astruc, according to the account books kept by Madame Laplace, now in Bancroft, box 6.
57. The official change of domicile is dated 9 frimaire an 9 [30 November 1800] in Bancroft, box 2, folder 14.
58. In 1808 Laplace received rents of around 11,000 francs from a property in Westphalia. See AN, AFiv6 and 32.
59. According to Paul Marmottan, *Lettres de Madame Laplace à Elisa Napoléon* (Paris, 1897), pp. 8–9, this corresponds today to no. 6 rue de Tournon.
60. Bancroft, box 6.
61. Arago recounts an anecdote about the key to the sugar box kept by the master of the house, which indicates it was considered a precious commodity. Arago, "Histoire de ma jeunesse," *O., Notices biographiques* (Paris, 1854), vol. 1, p. 17.
62. Bancroft, box 6.
63. Many of the contemporary memoirs mention the Laplaces' participation in this social world. For example, see Johann Friedrich Reichardt, *Un hiver à Paris sous le Consulat, 1802–1803* (Paris, 1896), pp. 29–30; [Antoine Thibaudeau], *Mémoires sur le Consulat* (Paris, 1827), pp. 17–21; Louis Antoine Bourrienne, *Mémoires de M. Bourrienne* (Paris, 1829), pp. 94–95;

Louis Marie Bajot, *Lettres rétrospectives sur la marine* (Paris, [1852]), 2d letter; [Marie Julie Cavaignac], *Les mémoires d'une inconnue* (Paris, 1894), p. 55; Alexandrine Sophie Bawr, *Mes souvenirs* (Paris, 1853), pp. 226–228.

64. There are several notes in box 14 of the Laplace papers referring to dinner parties with members of the Bonaparte family.
65. In 1800, Bonaparte asked Laplace how to handle Guyton de Morveau's publications on the disinfection of squalid airs in prisons, hospitals, and vessels, along with Guyton de Morveau's request to be appointed Senator. See Laplace's draft reply in Bancroft, box 14, folder 9.
66. See letters to Oriani on 12 fructidor an 8 and 20 brumaire an 9 [30 August and 11 November 1800] in Milan, Osservatorio di Brera, Corrispondenza 1800, 08 30 PSL BO and 11 11 PSL BO.
67. *PV, Institut*, vol. 2, pp. 7–8, 103, 175, and 189.
68. Gavin R. De Beer, "The Relations between Fellows of the Royal Society and French Men of Science when France and Britain Were at War," *Notes and Records of the Royal Society* 9 (1952): 268–269.
69. *Monatliche Correspondenz zur Beförderung der Erd- und Himmels-Kunde* 8 (November 1803): 448.
70. Ibid., 7 (March 1803): 182.
71. Laplace to Charles Blagden, 2 April 1806, in London, Royal Society, BLA L19.
72. *Monatliche Correspondenz zur Beförderung der Erd- und Himmels-Kunde* 6 (1802): 272–278.
73. The original draft is in Bancroft, box 10, folder 10. A slightly different version was quoted in Arthur Jules Morin, "Notice historique sur le système métrique, sur ses développements, et sur sa propagation," *Annales du Conservatoire des Arts et Métiers* 9 (1870): 573–640.
74. *Moniteur,* 4 July 1837.
75. Jean Baptiste Biot, "Note relative à l'habitation de M. Laplace à Arcueil," in *Mélanges Scientifiques et Littéraires* (Paris, 1858) vol. 1, pp. 9–10. The house was sold by Jean François Reubell according to Raymond Guyot, *Documents biographiques sur J.-F. Reubell, membre du Directoire Exécutif* (Paris, 1911), no. 489. The cost was 60,000 francs, as indicated in the contract dated 22 October 1806, notarized by Maître Charles Lebrun, in the Minutier des Archives Notariales de Paris, Etude 92.
76. Bancroft, box 14, folder 28.
77. Some intimate letters of Amédée to Emile written in 1809 and 1810, the last one shortly before Amédée committed suicide, are in Bancroft, box 26, folder 15.
78. *Théorie du mouvement et de la figure elliptique des planètes* (Paris, 1784), p. xii. His exact words were "Knowledge [of the laws of affinity] is the

principal concern of chemistry, and it will only be when they are sufficiently observed to be treated mathematically that this science will reach the degree of perfection astronomy has achieved through the discovery of universal gravitation."

79. *Séances des écoles normales* (Paris, 1800), vol. 9, p. 45.
80. Michelle Sadoun-Goupil, *Le chimiste Claude-Louis Berthollet, 1748–1822: Sa vie—son œuvre* (Paris, 1977), vol. 2, chapter 6; and Maurice Crosland, *The Society of Arcueil* (London, 1967), p. 237.
81. *Histoire et Mémoires de l'Académie Royale des Sciences* (Paris, 1782), pp. 534–535.
82. For details consult Sadoun-Goupil, *Le Chimiste Claude-Louis Berthollet*, pp. 62–80 and Crosland, *Society of Arcueil*, pp. 248, 254, and 130–131.
83. *Mémoires de physique et de chimie de la Société d'Arcueil* (Paris, 1807), vol. 1, pp. iii–iv.

## 9. Celestial Mechanics

1. Laplace to Deluc, 18 March 1785, in Geneva, Bibliothèque Publique et Universitaire, MS D.O.
2. Louis Pierre Eugène Amélie Sédillot, *Les professeurs de mathématiques et de physique générale au Collège de France* (Rome, 1869), pp. 186–187.
3. The text is in the AAdS, dossier 28 February 1788.
4. Laplace to Oriani, 2 July 1789, in Milan, Osservatorio di Brera, Corrispondenza 1789, 1789 07 02 PSL BO. The exact term was "un ouvrage de longue haleine."
5. The term "mécanique céleste" actually shows up in a paper on tides that was read before the Académie in December 1790 but appeared only in 1797, by which time Laplace had the opportunity to alter the words of the original paper he had given orally.
6. Draft of letter from Le Sage to Laplace, 31 March 1797, in Geneva, Bibliothèque Publique et Universitaire, MS Suppl. 518, fols. 54–55.
7. Laplace to Le Sage, 17 germinal an 5 [6 April 1797] in Geneva, Bibliothèque Publique et Universitaire, MS D.O., Autographes.
8. Bancroft, box 18, folder 9.
9. Laplace to Oriani, 15 vendémiaire an 7 [6 October 1798] in Milan, Osservatorio di Brera, Corrispondenza 1798, 10 06 PSF BO.
10. Maurice to Pierre Prévost, 31 March and 30 April 1797, in Geneva, Bibliothèque Publique et Universitaire, MS Suppl. 1051, fols. 10r and 12r.
11. Laplace to Biot, 30 germinal an 7 [19 April 1799], in Sotheby's sale catalogue, 17/18 November 1988, no. 250.
12. *Analyse du "Traité de Mécanique Céleste de P. S. Laplace"* (Paris, an 9

[1801]), in 93 pages. Biot had also published parts of this work previously in the *Magasin Encyclopédique* 3 (1799): 433–459; 6 (1799): 497–520; and 1 (1800): 7–38; as well as in the new edition of *Séances des écoles normales* (Paris, 1800–1801), vol. 7, pp. 5–141.
13. Laplace to Oriani, 10 floréal an 10 [30 April 1802], in Milan, Osservatorio di Brera, Corrispondenza 1802, 04 30 PSL BO.
14. Laplace to Oriani, 8 August 1806, in Milan, Osservatorio di Brera, Corrispondenza 1806, 08 08 PSL BO.
15. *Mechanik des Himmels* (Berlin, 1800–1802), 2 vols.
16. *A Treatise upon Analytical Mechanics: Being the First Book of the* Mécanique Céleste, trans. John Toplis (Nottingham, 1814); *A Treatise of Celestial Mechanics,* trans. Henry H. Harte (Dublin, 1822–1827), 2 vols.; *Mécanique Céleste,* trans. Nathaniel Bowditch (Boston, 1829–1839), 4 vols. Another minor English commentator in 1821 was Captain Henry Kater, who wrote *A Treatise on Mechanics* (London, 1830) with Dionysius Lardner. In *Elementary Illustrations of the Celestial Mechanics of Laplace* (London, 1821), Thomas Young tried to provide "the connecting link between the geometrical and algebraic modes of representation" (p. iii), but ended by writing a work largely independent of the *Mécanique céleste.*
17. Review by John Playfair in *Edinburgh Review* 11 (1808): 249–283. The issue of final causes is taken up on pp. 278–279.
18. Ivor Grattan-Guinness, "Before Bowditch: Henry Harte's Translation of Books 1 and 2 of Laplace's *Mécanique Céleste,*" *NTM* 24, no. 2 (1987): 53–55.
19. Bancroft, box 7, folder 31. The beginnings of this projected work appear in draft form in box 9, folder 19.
20. Ibid.
21. *Mécanique,* vol. 1, pp. 84–85.
22. Bancroft, box 7, folder 31.
23. *Mécanique,* vol. 1, pp. 121–122.
24. "Extrait des registres de l'Académie des Sciences, du 28 mars 1787," in Jacques Antoine Cousin, *Introduction à l'étude de l'astronomie physique* (Paris, 1787), p. 319. Laplace echoes this idea in *Mécanique,* vol. 3, p. x. For background, consult John Bennett Shank, "Before Voltaire: Newtonianism and the Origins of the Enlightenment in France, 1687–1734," Ph.D. diss., Stanford University, 2000.
25. Bancroft, box 7, folder 35. The list covers the years 1726–1772.
26. The son of Nathaniel Bowditch took Laplace to task for his unacknowledged borrowing from contemporary mathematicians, but tried to excuse the habit by calling it an "error of judgment" on Laplace's part, not an "intentional or unfair appropriation of the labor of others." See Henry I.

Bowditch, "Memoir of the Translator by His Son," in volume 4 of the translation, pp. 65–67.
27. *Mémoires de mathématiques* (Paris, 1773), vol. 7, p. 163 [O.C., vol. 8, p. 201].
28. In 1773, Laplace wrote that this was "an original work that shines throughout with the brilliance of ingenuity, and that one can regard as containing the germ of all that has been done since on the mechanics of solid bodies." Ibid.
29. Curtis A. Wilson, "The Precession of the Equinoxes from Newton to d'Alembert and Euler," in René Taton and Curtis A. Wilson, eds., *Planetary Astronomy from the Renaissance to the Rise of Astrophysics. Part B: The Eighteenth and Nineteenth Centuries* (Cambridge, Eng., 1995), pp. 50–54.
30. *Mécanique*, vol. 1, pp. 61–64 and 317–320. These ideas were first introduced in a paper read to the Académie four days after the Bastille was captured, on 18 July 1789, and published at another inopportune time in 1793. They were reiterated succinctly to the Institut National in 1796 and published there and again in 1798 in the *Journal de l'Ecole Polytechnique*, cahier 5 (1798), pp. 155–159 [O.C., vol. 14, pp. 3–7], where they were better appreciated.
31. *Tables de Jupiter et de Saturne* (Paris, 1789).
32. Hahn, pp. 163–164.
33. Laplace to Deluc, 1 February 1793, in London, Royal Astronomical Society, W.1/13 L.32, 1–2.
34. Gillispie, pp. 194–195.
35. Laplace to Deluc, 1 February 1793.
36. Laplace to Delambre, 25 ventôse an 5 [15 March 1797], in Berlin, Staatsbibliothek, Sgl. Darmstaedter J 1796.
37. *Mécanique*, vol. 4, pp. 231–293.
38. Ramond published a paper summarizing these features a few years later as *Mémoires sur la formule barométrique de la Mécanique Céleste* (Clermont-Ferrand, 1811). It is curious that Laplace initially seems to have ignored Christian Kramp's *Analyse des réfractions astronomiques et terrestres* (Strasbourg, 1798), which was a work known at the Institut National.
39. Laplace to Deluc, 1 February 1793, in London, Royal Astronomical Society, W. 1/13 L.32, 1–2.
40. *Mécanique*, vol. 4, pp. 1–192.
41. Morando, *Planetary Astronomy*, p. 142, points out that in the 1860s, some of Laplace's ideas were resuscitated with some success.
42. The original drafts and several revisions are in Bancroft, box 7, folders 22 and 24.
43. MS 182, published by Caussin in "Le livre de la grande table Hakémite,"

*Notices et Extraits des Manuscrits de la Bibliothèque Nationale et Autres Bibliothèques* 7 (1804): 16–240. For comments about Laplace, see p. 230, n. 1.
44. PV, *Institut* (6 prairial an 8) [26 May 1800], vol. 2, p. 172.
45. See Laplace to Delambre, 25 ventôse an 5 [15 March 1797] in Berlin, Staatsbibliothek, Sgl. Darmstaedter J 1796; and Laplace to Bouvard, 24 vendémiaire [15 October 1797] and 15 fructidor [1 September 1798], in Washington, Smithsonian Institution, Dibner Library, MS 823A and Leipzig, Handschriftenabteilung, Universitätsbibliothek.
46. Giovanni Antonio Amedeo Plana, *Théorie du mouvement de la lune* (Turin, 1832) and Damoiseau, *Tables de la lune* (Paris, 1828). See also Guido Tagliaferri and Pasquale Tucci, "Carlini and Plana on the Theory of the Moon and Their Dispute with Laplace," *Annals of Science* 56 (1999): 221–269.
47. Pontécoulant, *Théorie analytique du système du monde* (Paris, 1829–1843), 4 vols.
48. *Mécanique Céleste,* trans. Nathaniel Bowditch, vol. 3, p. 149.
49. *Mécanique,* vol. 4, pp. 204–207.
50. Ibid., vol. 3, p. x.
51. Clifford J. Cunningham, *The First Asteroid: Ceres, 1801–2001* (Surfside, Fla., 2001).
52. Von Zach to Laplace, 24 June 1801, in Paris, Bibliothèque de l'Observatoire.
53. Laplace to von Zach, 14 fructidor [1 September 1801] sent to Bürg in Vienna, in Österreichische Nationalbibliothek, Autogr. 4/54–2.
54. Eugene Frankel, "Career-Making in Post-Revolutionary France: The Case of Jean-Baptiste Biot," *British Journal for the History of Science* 11 (1978): 42–45.
55. Jean-Baptiste Biot, *Traité elémentaire d'astronomie physique* (Paris, 1810–1811), 3 vols.
56. Jean-Baptiste Biot, "Une anecdote relative à Laplace," in *Mélanges Scientifiques et Littéraires* (Paris, 1858), vol. 1, pp. 1–10. See also the touching dedication in his *Traité* cited in ibid.
57. Dominique François Arago, "Poisson," in *O. Notices Biographiques,* vol. 2 (Paris, 1854), pp. 596–598; and Bancroft, box 15, folder 17. In letters to Madame Laplace in August and September 1799, Billy refers to himself as an "apprentice mathematician," and an "unworthy student" of Laplace. In one dated 24 ventôse an 10 [15 March 1802], he refers to Poisson as "a fortunate and surprising young man who owes so much to the friendship of Mr. Laplace."
58. David H. Arnold, "The *Mécanique Physique* of Siméon Denis Poisson: The Evolution and Isolation in France of His Approach to Physical Theory (1800–1840)," *AHES* 28 (1983): 243–287.

59. Roger Hahn, "Le rôle de Laplace à l'Ecole Polytechnique," in Bruno Belhoste et al., eds., *La formation polytechnicienne, 1794–1994* (Paris, 1994), pp. 45–57.
60. "Mémoire sur la détermination d'un plan . . . ," in cahier 5 (1798), pp. 155–159; "Sur la mécanique," in cahier 6 (1799), pp. 343–344; and "Leçons de mathématiques . . . ," in cahier 7 (1800), pp. 1–172.
61. Michel Métivier et al., eds., *Siméon-Denis Poisson et la Science de son Temps* (Paris, 1981), pp. 213, 233–234; and Ivor Grattan-Guinness, *Convolutions in French Mathematics, 1800–1840* (Basel, 1990), vol. 1, pp. 371–374.
62. Jean Henri Hassenfratz, *Cours de physique de céleste; ou, Leçons sur l'exposition du systême du monde, données à l'Ecole Polytechnique en l'an dix* (Paris, an 11 [1803]).
63. Bruno Belhoste, *La formation d'une technocratie: L'Ecole Polytechnique et ses élèves de la Révolution au Second Empire* (Paris, 2003).
64. Université Paris, Faculté des Sciences, *Thèse de mécanique soutenue le 9 mars 1811 . . . suivie du programme de la thèse d'Astronomie . . . par Pierre-Marie Bourdon* (Paris, 1811); and *Thèses de mécanique et d'astronomie pour le doctorat . . . par Lefébure de Fourcy,* (Paris, 1811).
65. Université Paris, Faculté des Sciences, *Thèse d'astronomie sur quelques points des théories de la lune et des planètes . . . par Alfred Gautier* (Paris, 1817), published at the same time as *Essai historique sur le problème des trois corps.* The quotation is from pp. 215–216.
66. *DSB*, vol. 15, pp. 225–226. He wrote a "Coup d'œil philosophique sur la mécanique céleste" that was neither published nor refuted. See *PV, Institut* (9 September 1811), vol. 4, p. 522.
67. See *Address of M. Hoene Wronski to the British Board of Longitude upon the Actual State of the Mathematics, Their Reform, and upon the New Celestial Mechanics* (London, 1820); *Trois lettres à Sir Humphry Davy sur l'imposture publique des savants . . .* (London, 1822); and *Accomplissement de la réforme de la Mécanique céleste, contenant son universelle loi fondamentale pour l'établissement a priori de la rationalité de l'univers . . .* (Metz, 1851).
68. Laplace to Gauss, 20 November 1812, in Karin Reich, *Im Umfeld der 'Theoria Motus': Gauss' Briefwechsel mit Perthes, Laplace, Delambre und Legendre* (Göttingen, 2001), pp. 99–100, in which both the 1809 and 1812 papers of Ivory are mentioned, but somewhat dismissed because of an error detected by Laplace.
69. Laplace to Davy, 5 January 1824, in Washington, Smithsonian Institution, Bern Dibner Library, MS 823A.
70. Grattan-Guinness, *Convolutions,* vol. 2, pp. 1190–1195.
71. Albert Maquet, *L'astronome royal de Turin: Giovanni Plana, 1781–1864* (Brussels, 1965), pp. 37–42, 82–105, and 222–231; and Tagliaferri and Tucci, "Carlini and Plana."

72. In *Allgemeine Geographische Ephemeriden* 3 (1799): 50, Zach refers to "Laplace, der Newton unser Zeit," a term that he uses frequently.
73. Legendre to Plana, 28 July 1826, copy in Milan, Osservatorio di Brera, Corrispondenza.
74. Grattan-Guinness, *Convolutions,* vol. 2, pp. 1154–1157.
75. Nicole Capitaine, ed., *Deux siècles d'évolution du système du monde: Hommage à Laplace* (Paris, 1997).
76. Robert Fox, "The Rise and Fall of Laplacian Physics," *HSPS* 4 (1975): 89–136.
77. *Essai de statique chimique* (Paris, an 11 [1803]), vol. 1, pp. 1–2.
78. In the Anglo-Saxon world, these are known as Boyle's laws. See ibid., pp. 245–247, note 5 and pp. 522–523, note 18. Michelle Sadoun-Goupil, *Le Chimiste Claude-Louis Berthollet, 1748–1822: Sa vie—Son œuvre* (Paris, 1977), vol. 2, chapter 6, has more details about the content of Berthollet's *Essai.*
79. René Just Haüy, *Traité élémentaire de physique* (Paris, an 12 [1803]), p. xxvii.
80. This feature is popularly known as the eclipse of the "Ecole de Monge" by the "Ecole de Laplace," terms coined by Théodore Olivier in his *Mémoire de géométrie descriptive: Théorique et appliquée* (Paris, 1851). See the important comments by Charles C. Gillispie in his "Un enseignement hégémonique: Les mathématiques," in *La formation polytechnicienne, 1794–1994,* pp. 35–40; and Bruno Belhoste, "Ecole de Monge, école de Laplace: Le débat autour de l'Ecole Polytechnique," in François Azouvi, ed., *L'Institution de la raison* (Paris, 1992), pp. 101–112.
81. No relation to the academician Mathurin Jacques Brisson, the author of older textbooks and dictionaries on physics and natural history.
82. Ernst Gottfried Fischer, *Physique mécanique,* trans. Jean Baptiste Biot (Paris, 1806), vol. 1, p. iii.
83. This belief has been championed by the contemporary school of French historians of science, who published "La mathématisation, 1780–1830" in *RHS* 42 (1989): 1–172; and John L. Heilbron, *Weighing Imponderables and Other Quantitative Science around 1800* (Berkeley, 1993).
84. "Sur la théorie du son," *Journal de Physique, de Chimie et d'Histoire Naturelle* 55 (fructidor an 10 [August–September 1802]), pp. 173–182. For Fox's views, see Gillispie, pp. 199–202.
85. "Mémoire sur l'intégral complet et fini de l'équation de la propagation du son," and "Sur l'intégration générale complète des équations sur les lois de la propagation du son, l'air étant considéré avec ses trois dimensions," in *PV, Institut,* vol. 1, pp. 524 and 546; and vol. 2, pp. 373 and 392. Though these papers were presented in 1799 and 1800, one of them did not appear in print until 1806. Biot's work was completed in 1801.

86. Thomas S. Kuhn, "The Caloric Theory of Adiabatic Compression," *Isis* 49 (1958): 132–140; Bernard S. Finn, "Laplace and the Speed of Sound," *Isis* 55 (1964): 7–19; and Grattan-Guinness, *Convolutions,* vol. 1, pp. 452–460.
87. Mathurin Jacques Brisson, *Dictionnaire raisonné de physique* (Paris, 1800), vol. 1, p. 172.
88. According to Gillispie, pp. 108 and 290, the work became available only in 1793. It was probably timed to buttress the measurements needed to establish the unit measure of weight based on a unit of length, not for any fundamental theoretical purposes. Its preparation around 1803 for Madame Lavoisier's publication of her late husband's *Mémoires de physique* (see vol. 1, pp. 246–280), coincides with Laplace's revived concern with heat theory. See Lavoisier, *O.,* vol. 2, pp. 739–759, and for the apparatus, vol. 6, pp. 711–712. Biot took a special interest in this work, reconstructing the results from participants' recollections, and presenting it in his 1816 *Traité de physique expérimentale et mathématique,* vol. 1, pp. 146–158, as well as in a table of calculations.
89. It was in fact added to volume 4 as the first supplement to book 10, the year following its initial appearance. Biot's review is in *Moniteur Universel,* 24 May 1806, pp. 709–719.
90. Lucan, *De Bello Civili* (Cambridge, Eng., 1992), book 2, lines 744–745. The original text differs slightly from the one used by Biot, but has the same meaning.
91. Fischer, *Physique Mécanique,* vol. 1, p. 149.
92. Laplace to an unknown scientist, 20 January 1806, in BN, Fichier Charavay, vol. 106, no. 86.
93. "Théorie mathématique de l'action capillaire," *Journal de l'Ecole Polytechnique,* cahier 16 (May 1813): 1–40. Petit begins by assuming short-range forces, but quickly turns to mathematical representations rather than to discussions of the physical models entailed by the theory.
94. Here I follow Jean Dhombres, "La théorie de la capillarité selon Laplace: Mathématisation superficielle ou étendue," *RHS* 42 (1989): 43–77. See also John L. Heilbron, "Laplace's School," *Historical Studies in the Physical and Biological Sciences* 24, supp. (1993): part 1, pp. 139–184.
95. Thomas Young, "An Essay on the Cohesion of Fluids," *Philosophical Transactions of the Royal Society* (1805): 65–87, read 20 December 1804. In his "Additional" remarks printed as an appendix to his *A Course of Lectures on Natural Philosophy and the Mechanical Arts* (London, 1807), vol. 2, p. 670, Young contends that "the first mathematicians on the continent have exerted great ingenuity in involving the plainest truth of mechanics in the intricacies of algebraic formulas, and in some instances have even lost sight of the real state of an investigation, by attending only to the symbols, which they have employed for expressing its steps."

96. "Review of Laplace's Memoir 'Sur la loi de la réfraction extraordinaire dans les cristaux diaphanes,'" *Quarterly Review* 2 (November 1809): 337–348, quotation from p. 338. The anonymous author mentioned was one Thomas Knight, who had written *An Examination of M. La Place's Theory of Capillarity* (London, 1809). I suspect the true author was Thomas Young himself.
97. "Review of Laplace's Memoir," p. 343.
98. Eugene Frankel, "Corpuscular Optics and the Wave Theory of Light: The Science and Politics of a Revolution in Physics," *Social Studies of Science* 6 (1976): 141–184; Geoffrey N. Cantor, *Optics after Newton: Theories of Light in Britain and Ireland, 1704–1840* (Manchester, 1983); André Chappert, *Etienne Louis Malus (1775–1812) et la théorie corpusculaire de la lumière* (Paris, 1977); and Jed Z. Buchwald, *The Rise of the Wave Theory of Light: Optical Theory and Experiment in the Early Nineteenth Century* (Chicago, 1989).
99. Eugene Frankel, "The Search for a Corpuscular Theory of Double Refraction: Malus, Laplace, and the Price Competition of 1808," *Centaurus* 18 (1974): 223–245.
100. Etienne Malus, "Théorie de la double réfraction," *Mémoires Présentés à l'Institut des Sciences, Lettres et Arts par Divers Savants* 2 (January 1811): 505.
101. *Journal de Physique, de Chimie, d'Histoire Naturelle et des Arts* 68 (January 1809): 107–111; and *Mémoires de physique et de chimie de la Société d'Arcueil* (Paris, 1809), vol. 2, pp. 111–142. Laplace was also setting forth his priority claim, much to the dismay of Malus, who had discussed some of the concepts behind this discovery with Laplace before publishing them himself. See Dominique François Arago, "Malus," *O., Notices Biographiques* (Paris, 1855), vol. 3, pp. 153–154.
102. Eugene Frankel, "Corpuscular Optics and the Wave Theory of Light: The Science and Politics of a Revolution in Physics," *Social Studies of Science* 6 (1976): 147–152.
103. Ibid., pp. 156–157.
104. Augustin Jean Fresnel, *O. C.* (Paris 1868), vol. 1, p. 261; *PV, Institut* (15 March 1819), vol. 6, p. 427.
105. Fresnel, *O. C.*, vol. 1, pp. lxxxvi–lxxxvii.
106. See the comments by Grattan-Guinness in *Convolutions*, vol. 1, pp. 867–868.
107. Ibid., pp. 594–595.
108. See the comments of Ivor Grattan-Guinness to the critical edition of Fourier's "Théorie de la propagation de la chaleur dans les solides" (1807) in *Joseph Fourier, 1768–1830: A Survey of His Life and Work* (Cambridge, Mass., 1972), pp. 440–452 and 476n.

109. John Herivel, *Joseph Fourier face aux objections contre sa théorie de la chaleur* (Paris, 1980); and Jean Dhombres and Jean-Bernard Robert, *Joseph Fourier, 1768–1830: Créateur de la Physique-Mathématique* (Paris, 1998).
110. Prévost, a disciple of Le Sage and Deluc, visited Laplace in 1808 and corresponded with him regarding the research leading to his *Du calorique rayonnant* (Geneva, 1809). See Burghard Weiss, *Zwischen Physikotheologie und Positivismus: Pierre Prévost (1751–1839) und die korpuskularkinetische Physik der Genfer Schule* (Frankfurt am Main, 1988).
111. Robert Fox, *The Caloric Theory of Gases from Lavoisier to Regnault* (Oxford, 1971), pp. 165–177.

## 10. Probability and Determinism

1. For a modern refutation, see Richard Green, *The Thwarting of Laplace's Demon Arguments against the Mechanistic Worldview* (New York, 1995).
2. *Essai philosophique* (Paris, 1814), p. 1. Laplace repeats this terminology in the body of the work, and maintained it in successive editions, even after the Restoration in 1815. See the critical edition by Bernard Bru (Paris, 1986), pp. 79 and 31, from which I will mostly be quoting [*O.C.*, vol. 7, pp. xlviii and vi].
3. "A NAPOLEON-LE-GRAND. Sire, la bienveillance avec laquelle Votre Majesté, a daigné accueillir l'hommage de mon Traité de Mécanique Céleste, m'a inspiré le désir de Lui dédier cet Ouvrage sur le Calcul des Probabilités. Ce calcul délicat s'étend aux questions les plus importantes de la vie, qui ne sont en effet, pour la plupart, que des problèmes de probabilité. Il doit, sous ce rapport, intéresser Votre Majesté dont le génie sait si bien apprécier et si dignement encourager tout ce qui peut contribuer au progrès des lumières, et de la prospérité publique." *Théorie analytique des probabilités* (Paris, 1812), pp. i–ii.
4. Letter from Vitebsk, dated 1 August 1812. *Correspondance de Napoléon Premier* (Paris, 1868), vol. 24, p. 131.
5. Keith A. Baker, *Condorcet: From Natural Philosophy to Social Mathematics* (Chicago, 1975); Jean-Claude Perrot and Stuart J. Woolf, *State and Statistics in France, 1789–1815* (Chur, 1984); Eric Brian, *La mesure de l'état: Administrateurs et géomètres au XVIIIè siècle* (Paris, 1994).
6. *Séances des Ecoles Normales Recueillies par des Sténographes, et Revues par les Professeurs* 6 (Paris, 1795): 32–73, same pagination (Paris, 1800) and (1810); *Journal de l'Ecole Polytechnique* cahiers 7 and 8 (1812): 140–172.
7. *Discours prononcé au deux conseils par le citoyen Laplace au nom de l'Institut National des Sciences et des Arts* (Paris, an 4 [1796]).
8. François Picavet, *Les idéologues* (Paris, 1891); Sergio Moravia, *Il tramonto dell'illuminismo: Filosofia e politica nella società francese (1770–1810)* (Bari,

1968); Martin S. Staum, *Cabanis: Enlightenment and Medical Philosophy in the French Revolution* (Princeton, 1980); François Azouvi, ed., *L'institution de la raison: La révolution culturelle des idéologues* (Paris, 1992).

9. Martin S. Staum, *Minerva's Message: Stabilizing the French Revolution* (Montreal, 1996).
10. Xavier Bichat, *Recherches physiologiques sur la vie et la mort* (Paris, an 8 [1800]), pp. 93–98.
11. *O.C. de Cabanis* (Paris, 1823–1825), vol. 3, p. 9, n. 1.
12. Cabanis to Laplace, 17 pluviôse an 6 [5 February 1798], BN, Fichier Charavay, vol. 33, fol. 11, no. 18. Cabanis asked that the two other copies be presented by Laplace to Lagrange and General Bonaparte. Receipt of the book was recorded by the Institut on 21 pluviôse [9 February] as is indicated in *PV, Institut,* vol. 1, p. 343.
13. A much fuller historical account of this issue is given by Bernard Baertschi, *Les rapports de l'âme et du corps* (Paris, 1992), particularly on pp. 235–256. Cabanis's essential contribution to the discourse is given on pp. 253–256.
14. Roger Hahn, "Laplace and the Vanishing Role of God in the Physical Universe," in *The Analytic Spirit,* ed. Harry Woolf (Ithaca, 1981), pp. 85–95.
15. *The Scientific Papers of Sir William Herschel,* ed. J. L. E. Dreyer (London, 1912), vol. 1, p. lxii.
16. Edmond Pictet, "Journal d'un genevois [Marc-Auguste Pictet] à Paris sous le Consulat," *Mémoires Publiées par la Société d'Histoire et d'Archéologie de Genève* 5, 2d ser. (1893): 101–102; Antoine Augustin Cournot, *Souvenirs (1760–1860)* (Paris, 1913), p. 86; and *L'Intermédiaire des Chercheurs et Curieux* 29 (1894): columns 87–88.
17. Pierre Sylvain M[aréchal], *Dictionnaire des athées anciens et modernes,* ed. Joseph Jérôme Lalande (Paris, an 8 [1800]), pp. 231–232.
18. Paris, Bibliothèque Victor Cousin, MS 99, unpaginated.
19. François Victor Alexandre Alphonse Aulard, "Napoléon et l'athée Lalande," *Etudes et leçons sur la Révolution française* (Paris, 1908), ser. 4, pp. 303–316.
20. *Essai philosophique,* pp. 82 and 16 of the first edition (pp. 165 and 128 of the Bru edition; and *O.C.,* vol. 7, pp. cxix and lxxxvii).
21. Roger Hahn, "Laplace's Religious Views," *AIHS* 8 (1955): 38–40. The edited text is in an appendix, pp. 215–232.
22. An exception is a single reference to a work published anonymously, but written by Abraham Louis Bréguet, entitled *Essai sur la force animale et sur le principe du mouvement volontaire* (Paris, [1811]). But since the reference is squeezed into Laplace's text as a footnote, it is entirely possible it was added after the main body of the manuscript was composed.
23. François Magendie, "Quelques idées générales sur les phénomènes par-

ticuliers aux corps vivans," *Bulletin des Sciences Médicales* 4 (1809): 145–170.

24. For a general discussion, see Paul Hazard, *La crise de la conscience Européenne* (Paris, 1935) and Alan Charles Kors, *Atheism in France, 1650–1729* (Princeton, 1990). The passage Laplace quotes from Henry Dodwell, *Dissertationes 1ª in Irenaeum* (Oxford, 1689), paragraphs 38–39, is at times miscopied. Bertram Eugene Schwarzbach points out in "Les lectures d'un curé de campagne," *La Lettre Clandestine* 9 (2000): p. 269, that Dodwell was often read in the mid-1750s.

25. Frank E. Manuel, *The Eighteenth Century Confronts the Gods* (Cambridge, Mass., 1959).

26. Even though there is no internal evidence that Laplace was familiar with Hume's writings, the appearance of a new French edition of Hume's works in 1788 makes it likely that he was aware of the Scottish philosopher's views.

27. Here and in his analysis of the functioning of will, Laplace is addressing issues broached directly by Cabanis. See Elizabeth A. Williams, *The Physical and the Moral: Anthropology, Physiology, and Philosophical Medicine in France, 1750–1850* (Cambridge, Eng., 1994).

28. AAdS, dossier Laplace, "De l'idée du pouvoir." A similar position is taken in the segment entitled "De la cause."

29. *Essai philosophique* (1814), pp. 77–83.

30. In the body of the first edition of the *Essai philosophique,* pp. 16–17, Laplace takes Pascal to task for his use of miracles in support of religion [pp. 126–127 of the Bru edition; *O.C.,* vol. 7, p. lxxxvi]. In 1833, Augustin Louis Cauchy turned Laplace's argument on its head in his "Sept leçons de physique générale faites à Turin," in Cauchy, *O.C.,* 2d ser., vol. 15, p. 419.

31. Laplace borrowed this book from the Imperial Library on 16 October 1812. See BN, MS, n.a.fr. 854, fols. 301–302. See also Richard Nash, *John Craige's Mathematical Principles of Christian Theology* (Carbondale, Ill., 1991); and Stephen M. Stigler, "John Craig and the Probability of History: From the Death of Christ to the Birth of Laplace," *Journal of the American Statistical Association* 81 (1986): 881–887.

32. *PV, Institut* (16 messidor an 8) [5 July 1800], vol. 2, p. 186.

33. Dugald Stewart, *Elémens de la philosophie de l'esprit humain* (Geneva, 1808), 2 vols. See the letter of thanks from Laplace dated 19 April 1808 in Geneva, Bibliothèque Publique et Universitaire, MS supp. 1050, fols. 223–224.

34. There is no evidence to link Laplace with either the Genevans Jean Trembley or Daniel de la Roche, but they were also involved in writings that would have carried some weight with him. See Isaac Todhunter, *A His-*

*tory of the Mathematical Theory of Probability from the Time of Pascal to that of Laplace* (Cambridge, Eng., 1865), pp. 411–431 and p. 459 for details. Another Swiss writer who treated similar issues and who cited de la Roche was Charles Victor de Bonstetten, *Recherches sur la nature et les lois de l'imagination* (Geneva, 1807), 2 vols. He has a chapter entitled "De la volonté" that arrives at conclusions quite different from those of Laplace.

35. Gauss, *Werke* (Göttingen, 1929), vol. 12, pp. 241–252; Martin Brendel, "Über die astronomischen Arbeiten von Gauss," in Gauss, *Werke* (Göttingen, 1927), vol. 11, part 2, essay 3, esp. pp. 150–187; Clifford J. Cunningham, *The First Asteroid: Ceres, 1801–2001* (Surfside, Fla., 2001); and Karin Reich, *Im Umfeld der "Theoria Motus": Gauss' Briefwechsel mit Perthes, Laplace, Delambre und Legendre* (Göttingen, 2001), p. 15.

36. Comparing probability theory to number theory, he said they were "un des meilleurs exercices de l'esprit par la finesse et la justesse qu'elles demandent dans les raisonnements." Reich, *Im Umfeld der "Theoria Motus,"* p. 89.

37. "Mémoire sur divers points d'analyse," *Journal de l'Ecole Polytechnique* 8 (1809): 229–265 [*O.C.*, vol. 14, pp. 178–214].

38. See Gillispie, pp. 216–223, and *Mémoires de la Classe des Sciences Mathématiques et Physiques de l'Institut de France* 10 (1809): 353–415 [*O.C.*, vol. 12, pp. 301–345]; the issue is discussed at greater length in the *Moniteur Universel,* 10 January 1814, pp. 37–38.

39. As explained earlier, Laplace was encouraged to make this move because of Herschel's observations of the evolution of nebulae.

40. *Mémoires de la Classe des Sciences Mathématiques et Physiques de l'Institut de France* 10 (1809): 353–415 [*O.C.*, vol. 12, pp. 301–345]; Paris, Bureau des Longitudes, *Procès-verbaux des séances,* 20 and 27 September 1809.

41. Letter of 7 July 1810 from Laplace to Gauss, in which he refers to him as a "rare génie" and tells him of his plans to assemble and rearrange his earlier papers on probability into a general treatise. See Reich, *Im Umfeld der "Theoria Motus,"* p. 93.

42. Stephen M. Stigler, *The History of Statistics: The Measurement of Uncertainty before 1900* (Cambridge, Mass., 1986), pp. 140–148.

43. *Mémoires de l'Académie* 10 (1809): 559–565 [*O.C.*, vol. 12, pp. 349–353]. In it he mentions Gauss, but without a precise citation.

44. Stigler, *History of Statistics,* pp. 143–148.

45. See letters to Friedrich Schubert on 5 November 1811 in *Nauchnoe Nasledstvo* 1 (1948): 801–802; to Jöns Svanberg on 16 November 1811 in Stockholm, Kungliga Svenska Vetenskapsakademiens Bibliotek; and Barnaba Oriani on 9 July 1812 in Guido Tagliaferri and Pasquale Tucci, "P. S. de Laplace e il grado meridiano d'Italia," *Giornale di Fisica* 34 (1993): 275–276.

46. "Notice sur les probabilités," *Annuaire Publié par le Bureau des Longitudes* (1811): 98–125; "Mémoire sur les intégrales définies, et leur application aux probabilités et spécialement à la recherche du milieu qu'il faut choisir entre les résultats des observations," *Mémoires de l'Académie* 11 (1810): 279–347 [*O.C.*, vol. 12, pp. 357–412]; and "Du milieu qu'il faut choisir entre les résultats d'un grand nombre d'observations," *Connaissance des Temps* (1813): 213–223 [*O.C.*, vol. 13, p. 78 and vol. 12, pp. 401–412].
47. *Mémoires de l'Académie* 11 (1810): 283–284, [*O.C.*, vol. 12, pp. 360–361].
48. Ibid. pp. 286–287 [*O.C.*, vol. 12, p. 364].
49. Ibid., p. 284 [*O.C.*, vol. 12, p. 360].
50. Sylvestre François Lacroix, *Traité élémentaire du calcul des probabilités* (Paris, 1816); and Siméon Denis Poisson, *Recherches sur la probabilité des jugements en matière criminelle et en matière civile, précédées des règles générales du calcul des probabilités* (Paris, 1837).
51. See Laplace to Oriani, 30 November 1813, in which Laplace refers to Cauchy as "a young mathematician full of merit." In Tagliaferri and Tucci, "P. S. de Laplace e il grado meridiano d'Italia," p. 297.
52. The original version he prepared for students at the Ecole Normale is even less methodical.
53. In the lectures at the Ecole Normale, he often refers to them as theorems.
54. Condorcet, "Sur la probabilité des faits extraordinaires," in Condorcet, *Arithmétique politique: Textes rares ou inédits (1767–1789)* (Paris, 1994), pp. 431–436.
55. *Essai philosophique* (1814), pp. 11–12 [pp. 42–43 of the Bru edition; *O.C.*, vol. 7, p. xv].
56. Fourth edition of the *Essai Philosophique* (1819), p. 56 [p. 66 of the Bru edition; *O.C.*, vol. 7, p. xxxv].
57. Gillispie, "Probability and Politics: Laplace, Condorcet, and Turgot," *Proceedings of the American Philosophical Society* 116 (1972): 1–20, and Gillispie, pp. 93–95.
58. Madeleine Mazars-Chadeau, "Le savant, le social et le politique: A propos de 'l'Essai philosophique des probabilités' de Laplace," *L'Année Sociologique* (1986): 75–92, especially pp. 80–81.
59. Bernard Bru, "De la Michodière à Moheau: L'évalution de la population par les naissances," in Moheau, *Recherches et considérations sur la population de la France (1778)*, ed. Eric Vilquin (Paris, 1994), pp. 493–516; and "Estimations Laplaciennes. Un exemple: La recherche de la population d'un grand Empire, 1785–1812," in Jacques Mairesse, ed., *Estimation et sondages* (Paris, 1988), pp. 7–46.
60. Before the Revolution, Laplace had been placed on committees of the Académie to review several works presented to them dealing with finances. See Brian, *La mesure de l'état*, pp. 419 and 429–430.

61. Jean Charles Borda, "Mémoire sur les élections au scrutin," *Mémoires de l'Académie royale des Sciences* (1781): 657–665; and Roger Hahn, "L'Académie Royale des Sciences et la réforme de ses status en 1789," *RHS* 18 (1965): 15–28.
62. Pierre-Claude-François Daunou, *Mémoire sur les élections au scrutin* (Paris, an 11 [1803]), p. 50.
63. *Essai philosophique* (1816), pp. 115–118 [pp. 104–107 of the Bru edition; *O.C.*, vol. 7, pp. lxxiii–lxxv]; "Mémoire sur le flux et le reflux de la mer," *Mémoires de l'Académie Royale des Sciences de l'Institut de France* 3 (1818–1820): 1–90 [*O.C.*, vol. 12, pp. 473–546].
64. Details may be found in Gillispie, pp. 252–256.
65. *Essai philosophique* (1814), pp. 95–97 [pp. 79–80 of the Bru edition; *O.C.*, vol. 7, pp. xlviii–xlix].
66. *Essai philosophique* (1819), p. 80 [p. 80 of the Bru edition; *O.C.*, vol. 7, pp. xlvii–xlix].
67. *Essai philosophique* (1816), p. 124 [pp. 117–118 of the Bru edition; *O.C.*, vol. 7, p. lxxviii]. The last sentence was added only in the fourth edition (1819), p. 134.
68. *Essai philosophique* (1814), pp. 109–110 [pp. 115–116 of the Bru edition; *O.C.*, vol. 7, pp. lxxvi–lxxvii]. In later editions this passage is included as an example of the application of probability theory to the natural sciences.
69. *PV, Institut* (24 vendémiaire an 12) [17 October 1803], vol. 3, pp. 11–17 for a report on the annual galvanism prize instituted by Napoleon in 1802. Laplace was periodically named to sit on the committee assessing the submissions. For a more detailed account, see Pierre Sue, *Histoire du galvanisme* (Paris, 1802–1805), 4 vols.
70. According to Bru in his edition, p. 215, n. 137, the term *"physiologie intellectuelle"* was used in the third edition, but I was unable to locate the exact reference. For the renewal of the study of experimental physiology, see John E. Lesch, *Science and Medicine in France: The Emergence of Experimental Physiology, 1790–1855* (Cambridge, Mass., 1984), pp. 117–122; and Niklaus Egli, *Der "Prix Montyon de Physiologie expérimentale" im 19. Jahrhundert* (Zürich, 1970).
71. *Essai philosophique* (1814), pp. 77–83 [pp. 157–189 in the Bru edition; *O.C.*, vol. 7, pp. cxii–cxxxviii]. Laplace greatly enlarged this section from its original version in the first edition.
72. Laplace quotes from a long passage in Pascal's *Pensées* that emphasizes the need first to believe and then to destroy false ideas learned from a young age: see *Essai philosophique* (1819), pp. 235–237 [pp. 185–187 in the Bru edition; *O.C.*, vol. 7, pp. cxxxiv–cxxxvi].

73. *Essai philosophique* (1819), p. 242 [p. 189 in the Bru edition; *O.C.*, vol. 7, p. cxxxviii].
74. *Riflessioni critiche sopra il Saggio filosofico intorno alle probabilità del Sig. Conte Laplace* (Modena, 1821). Ruffini died a year later, before Laplace had responded to it. The work had been translated into French, but, to my knowledge, never published. See Patrizia Accordi, "Le 'Riflessioni Critiche . . .' di Paolo Ruffini (1821)," *Atti e Memorie dell'Accademia Nazionale di Scienze Lettere e Arti, Modena,* ser. 7, vol. 9 (1991–1992): 31–51.
75. *Della immaterialità dell'anima* (Modena, 1806).
76. Francesco Barbieri and Franca Cattelani Degani, *Catalogo della corrispondenza di Paolo Ruffini* (Pisa, 1997); and Francesco Barbieri and Carla Fiori, "L'ultima lettera di Ruffini a Cauchy," *Nuncius* 4 (1989): 161–163.
77. See the critical edition by Bernard Bru (Paris, 1984); and Thierry Martin, *Probabilités et critique philosophique selon Cournot* (Paris, 1996), p. 38.
78. For an assessment of the influence of Laplace's work, see the postface by Bernard Bru in his critical edition of the *Essai philosophique,* pp. 245–296. Bru also points to other features of the book not treated here.
79. *Essai philosophique,* pp. 95–96 [pp. 206–207 of the Bru edition; *O.C.*, vol. 7, p. cliii].

## 11. The Waning Years

1. Laplace mentioned his weakened eyesight to John Herschel in a letter dated 4 November 1824 in Royal Society of London, HS 11.103; and his other problem [*une incommodité*] on 24 January 1826 in a letter at the Smithsonian Institution, Bern Dibner Library, MS 823A.
2. Alexis Eymery, ed., *Dictionnaire des girouettes* (Paris, 1815), p. 274; see also *Dictionnaire des protées modernes* (Paris, 1815), pp. 159–160.
3. Jean Antoine Claude Chaptal, *Mes souvenirs sur Napoléon* (Paris, 1893), p. 342; see also Jean Jacques Cambacérès, *Lettres inédites à Napoléon, 1802–1814* (Paris, 1973), vol. 2, p. 1058, which indicates that the emperor was aware of Laplace's plight, but feigned surprise.
4. Madame Laplace to son Emile, 29 March 1814, in Bancroft, box 17, folder 11.
5. Letters of Madame Laplace to Emile, 2 April 1814, in Bancroft, box 17, folder 7; Laplace to his secretary Groslier, 2 April 1814, in Paris, Observatoire MS 1001.
6. Berthollet discussed his absence in a letter to Charles Blagden, dated 25 March 1815, in Royal Society of London, BLA. B. 141. See also Michelle

Sadoun-Goupil, *Le chimiste Claude-Louis Berthollet, 1748–1822: Sa vie— son œuvre* (Paris, 1977), pp. 85–87.
7. Laplace to Emile, 12 April 1814, in Bancroft, box 12, folder 2.
8. Louis Philippe d'Orléans to Emile, 30 November 1815, consulted at Château de Mailloc, papers of Colbert-Laplace family.
9. Letters of 12 January 1824 and 6 April 1825 in Czech Central State Archives, Acta Clementina 7, part 125, carton 3.
10. The administrative note appointing Laplace president is in Vincennes, Archives de la Guerre, $X^D 452$. The minutes of the meeting of the commission are in the Palaiseau archives of the Ecole Polytechnique, vol. 3, p. 1.
11. *Cours d'analyse de l'Ecole Royale Polytechnique* (Paris, 1821), p. i.
12. Bruno Belhoste, *Augustin-Louis Cauchy: A Biography* (New York, 1991), pp. 80–85.
13. *Quelques réflexions sur l'Ecole Polytechnique* (Paris, 1816).
14. Laplace to Countess Rumford, 26 September 1816, in AAdS, Lavoisier, Chabrol donation, no. 1, p. 185.
15. Josef W. Konvitz, *Cartography in France, 1660–1848* (Chicago, 1987); Henri Marie Auguste Berthaut, *La carte de France, 1750–1898: Etude historique* (Paris, 1898–1899), 2 vols.; Patrice Bret, "Le dépôt général de la guerre et la formation scientifique des ingénieurs-géographes militaires en France (1780–1830)," *Annals of Science* 48 (1991): 113–157.
16. Correspondence between Gauss and Laplace, in Karin Reich, *Im Umfeld der 'Theoria Motus': Gauss' Briefwechsel mit Perthes, Laplace, Delambre und Legendre* (Göttingen, 2001), pp. 102–104.
17. Laplace to Blagden, 2 April 1806, in London, Royal Society, BLA19; and Laplace to John Herschel, 18 April 1823 and 4 November 1824, in London, Royal Astronomical Society, HS 11.102 and 11.103.
18. Alfred Fierro, *La Société de Géographie, 1821–1946* (Geneva, 1983).
19. See the list of published speeches in Gillispie, pp. 303–306.
20. Letter of 8 October 1826 in BN, MS, n.a.fr. 26525, fols. 241–242.
21. In a letter to Charles Blagden dated 14 April 1816, Laplace pointedly remarked that he did not seek the post, but was flattered by his election. New Haven, Yale University, Beinecke Library, Osborn files, Laplace.
22. Archives de l'Académie Française, Collection Moulin, dossier Laplace; and Emile Gassier, *Les cinq cents immortels* (Paris, 1906), pp. 147–150. Vaublanc does not mention Laplace specifically in his *Souvenirs* (Paris, 1838), vol. 1, pp. 47–52.
23. "Carnet d'invitations," Bancroft, box 23, folder 52.
24. See André Morellet, *Mémoires inédits de l'abbé Morellet*, 2d ed. (Paris, 1882), vol. 2, chapters 29 and 31, and the correspondence that follows; as

well as Charles Nisard, *Mémoires et correspondances historiques et littéraires inédits* (Paris, 1858), chapter 7.
25. "Discours de M. le Marquis de La Place, chancelier de l'Académie Française, 25 août 1821," in *Recueil des discours, rapports et pièces diverses lus dans les séances publiques et particulières de l'Académie Française, 1820–1829* (Paris, 1843), pp. 551–556; and *Les prix de vertu fondés par M. de Montyon* (Paris, 1864), vol. 1, pp. 14–18.
26. Archives de l'Académie Française, 2B7, fols. 47–51; and Abel François Villemain, *La tribune moderne* (Paris, 1858), vol. 1, pp. 398–399.
27. Gassier, *Cinq cents immortels*, p. 149.
28. *L'Etoile*, 17 January 1827, p. 3, and repeated on 23 January.
29. Gillispie, p. 243. This section was written by Robert Fox.
30. Amable Guillaume Prosper Brugière, Baron de Barante, *La vie politique de M. Royer-Collard: Ses discours et ses écrits* (Paris, 1861), vol. 2, p. 344.
31. Louis Guimbaud, *Un grand bourgeois au XVIIIe siècle: Auget de Montyon (1733–1820)* (Paris, 1909); Eric Brian, "Le prix Montyon de statistique à l'Académie des Sciences pendant la Restauration," *Revue de Synthèse*, 4th ser., no. 2 (April–June 1991): 208–236; Niklaus Egli, *Der "Prix Montyon de Physiologie Expérimentale" im 19. Jahrhundert* (Zürich, 1970), pp. 7–18.
32. In Montyon's testament written in 1819, Laplace is asked to assist the executor to establish several prizes and is compensated with a gift of a diamond intended for Angélique de Portes. See articles 10–15 of the testament in René André Polydore Alissan de Chazet, *Des mœurs, des lois et des abus* (Paris, 1829), pp. lxxviii–lxxx.
33. "Nécrologie," *Revue Encyclopédique* 33 (1827): 880.
34. See, for example, *Recommendations in Favour of Charles Babbage, Esq., Candidate for the Professorship of Mathematics in the University of Edinburgh* (1819); and Biot to Babbage, 9 August 1819, in London, British Library, Add. MS 37182, fol. 137.
35. Mary Somerville, *Personal Recollections: From Early Life to Old Age* (London, 1874), pp. 108–109; and Elizabeth Chambers Patterson, "A Scotswoman Abroad: Mary Somerville's 1817 Visit to France," *Light of Nature* (Dordrecht, 1985), pp. 360–361.
36. Laplace to Deluc, 7 November 1789 and 5 July 1790, quoted in Chapter 7.
37. Letters of Laplace to Gauss in 1804, in Reich, *Im Umfeld der 'Theoria Motus,'* pp. 84–86.
38. *Memoirs of the Life of Sir Humphry Davy*, ed. John Davy (London, 1839), pp. 167–168.
39. Johann Friedrich Benzenberg, *Briefe geschrieben auf einer Reise nach Paris im Jahn 1804* (Dortmund, 1805), vol. 1, p. 192. Benzenberg's exact words

were that Laplace's manner gave a sense "of feeling of self-assuredness [sustained] by those who acknowledged his intellectual superiority" ("das Gefühl der Sicherheit und der von anderen anerkannten Ueberlegenheit des Geistes").

40. Von Zach to Oriani, Lyon, 3 August 1814, in Milan, Osservatorio di Brera, Corrispondenza. The exact words are "son orgueil, sa morgue et son insolence."
41. Johann Friedrich Reichardt, *Un hiver à Paris sous le Consulat, 1802–1803*, trans. Arthur Laquiante (Paris, 1896), p. 29.
42. Robert Christison, *The Life of Sir Robert Christison* (Edinburgh, 1885–1886), vol. 1, pp. 241–242.
43. Bancroft, box 2, folder 11.
44. Benzenberg, *Briefe geschrieben auf einer Reise nach Paris im Jahr 1804*, vol. 1, pp. 240–241. A fuller account of their scientific relationship is in Domenico Bertoloni Meli, "St. Peter and the Rotation of the Earth: The Problem of Fall around 1800," in *The Investigation of Difficult Things*, ed. P. M. Harman and Alan E. Shapiro (Cambridge, Eng., 1992), pp. 421–447, especially the last four pages. See also Giambattista Guglielmini, *Carteggio: De Diurno Terrae Motu*, ed. Maria Teresa Borgato and Allesandra Fiocca (Florence, 1994).
45. *Memoirs of the Life of Sir Humphry Davy*, p. 168.
46. In his "Eloge historique de M. le Marquis de Laplace," p. lxxxiii, Fourier comments on his "constance imperturbable des vues."
47. In her rue du Bac bedroom, Madame Laplace had two engravings under glass of the Virgin Mary; in the living room at Arcueil, there were eight religious engravings under glass handsomely framed. See the 72-page inventory after death registered in Paris, Minutier des Notaires, Etude 97 (Agasse), 12 March 1827.
48. This correspondence is in Bancroft, box 19.
49. The fourth supplement to the *Théorie analytique* was submitted to the Académie by both Laplace and his son, Emile. See *PV, Institut* (7 February 1825), vol. 8, p. 182. After his father's death, Emile also sent a manuscript supplement to volume 5 of the *Mécanique*. See *PV, Institut* (23 July 1827), vol. 8, p. 571.
50. Anne Marie Godlewska, *Geography Unbound: French Geographic Science from Cassini to Humboldt* (Chicago, 1999), pp. 76–85, offers a detailed explanation of the nature of their differences. Delambre shared the same views about Cassini IV.
51. Letters from Madame Laplace to Emile, 21 and 26 April 1823, in Bancroft, box 17, folder 3.
52. Laplace to John Herschel, 18 April 1823, in London, Royal Society, HS.

11102. See also Roger Hahn, "L'image de Newton reconstruite par Laplace," in Ulla Kölving and Irène Passeron, eds., *Sciences, musiques, lumières* (Ferney-Voltaire, 2002), pp. 463–466.
53. Thomas Chalmers, *Des preuves et de l'autorité de la Révélation Chrétienne,* trans. Jacques Louis Samuel Vincent (Paris, 1819), a translation of *The Evidence and Authority of the Christian Revelation* (Edinburgh, 1816). According to Antoine Augustin Cournot, in *Matérialisme, vitalisme, rationalisme* (Paris, 1875), p. 368, n. 1, Laplace was quite familiar with Louis de Potter's arguments, but not the *Histoire philosophique, politique et critique du Christianisme,* which was published after Laplace's death (this contradicts Cournot, who thought Laplace had read the *Histoire philosophique*).
54. William Hanna, *Memoirs of the Life and Writings of Thomas Chalmers* (Edinburgh, 1852), vol. 4, pp. 48–49.
55. See Geneva, Archives Maurice. The family let me copy some documents, which are cited here. Others have been transferred to Archives d'Etat de Genève, Archives Privées 12.
56. This opinion is confirmed by an independent remark of Adam Sedgwick in his *A Discourse on the Studies of the University of Cambridge,* 5th ed. (London, 1850), appendix pp. 129–130; Sedgwick had also met Laplace during his last illness.
57. "Relation de Monsieur le Baron Maurice sur le Marquis de Laplace," Geneva, Bibliothèque Publique et Universitaire, Archives Tronchin 349, no. 18. Emphasis in the original.
58. Full text in the appendix.
59. This account of Newton's self-effacing evaluation was reported by Jean-Baptiste Biot in his biographical account of Newton published in the 1822 edition of *Biographie universelle ancienne et moderne,* vol. 31, p. 192, col. 1, a work most likely known to Fourier.
60. "Relation de Monsieur le Baron Maurice."
61. Archives Maurice, private archives. Text in the appendix.
62. *La Quotidienne* (7 March 1827), no. 66, p. 2, col. 2.

## Conclusion

1. Joseph Fourier, "Eloge historique de M. le Marquis de Laplace, prononcé dans la séance publique de l'Académie Royale des Sciences, le 15 juin 1829," *Mémoires de l'Académie Royale des Sciences de l'Institut de France* 10 (1827): xcviii.
2. In his "Histoire de ma jeunesse," published posthumously, Arago took umbrage at Laplace's opposition to candidates Arago favored in various elections. He also took occasion to deride Laplace for his niggardliness

with regard to the supply of sugar in his home. See Arago, O., *Notices biographiques* (Paris, 1854), vol. 1, p. 17. By contrast, in a public talk to support the first publication of Laplace's *Œuvres,* Arago was full of praise for his senior contemporary. See Arago, O., *Notices biographiques* (Paris, 1855), vol. 3, pp. 456–545.
3. "Une anecdote relative à Laplace," in *Mélanges Scientifiques et Littéraires* (Paris, 1858), vol. 1, pp. 1–10.
4. Laplace's last will and testament leaves everything to his widow. See Minutier Central des Archives Notariales, Etude 98, no. 764 (8 March 1827).
5. Jean Dhombres and Roger Hahn, "La première rencontre entre Laplace et d'Alembert en 1769," *Revue de Métaphysique et de Philosophie* (in press).
6. *DSB,* vol. 3, p. 375.

## Four Nonscientific Manuscripts by Laplace

1. Henry Dodwell (1641–1711), erudite English philologist and theologian who held the mistaken belief that the Gospels had been written during Trajan's reign (98–117). The long passage quoted below comes from *Dissertationes in Irenæum* (Oxford, 1689), pp. 66–69. Laplace miscopied the text in numerous ways, but without altering the sense of Dodwell's message. He also punctuated the text differently from the original version from which he was quoting. Most of these insignificant deviations from the printed version will not be noted here.
2. Roman Emperors Trajan (98–117) and Hadrian (117–138).
3. The word "Scriptorum" in the Dodwell text was omitted by Laplace.
4. Saint Irenus (130–208), theologian whose *Adversus hæreses* is the first systematic exposition of Catholic belief.
5. A phrase in the Dodwell text is omitted without indication by Laplace. It reads "Sacrorum Canon, nec receptus aliquis in Ecclesia Catholica Librorum."
6. The word "Ecclesiasticos" in the Dodwell text was omitted by Laplace.
7. Clement of Rome (end of first century); Saint Barnabas (first century); Hermas (second century); Ignatius (second century), Bishop of Antioch; and Polycarp (69–115), bishop of Smyrna, all commonly known as the Apostolic Fathers.
8. The word "Scriptis" in the Dodwell text was omitted by Laplace.
9. Both the Epistle of Jude and the three Epistles of John are now believed, from internal evidence, to have been written later than the other books of the New Testament.
10. The word "Libros" in the Dodwell text was omitted by Laplace.

11. The Greek word "ἐναντιοφανόη" in the Dodwell text was replaced by Laplace by "apparentos contradictiones."
12. The Greek word "Αυτοπτόων" in the Dodwell text was replaced by Laplace by "testi [eam] e visû."
13. The Greek word "Αυτόπταζ" in the Dodwell text was replaced by Laplace by "testes è visû" twice in this sentence.
14. Matthew 1:18–25; Luke 1:26–35, 2:1–7.
15. Matthew 2:1–12, 2:13–15, 2:16–18.
16. Luke 2:22.
17. Matthew 1:1–17; Luke 3:23–38.
18. Luke 3:23.
19. John 13; Matthew 26:26–29; Mark 14:22–25; Luke 22:19–20.
20. John 13:1–5.
21. This paragraph is an amended version of the previous one.
22. Matthew 27:38–40.
23. Luke 23:39–42.
24. Matthew 28:1–10.
25. Luke 24:1–6 and 13–15.
26. Mark 14:19 and Luke 24:51.
27. Matthew 4:1–10.
28. *Annals* [of Imperial Rome], trans. John Jackson (Cambridge, Eng., 1937), vol. 4, book 15, chapter 44, p. 283. He reports that Christ founded the Christian sect and was executed by Pontius Pilate.
29. Matthew 27:57–60; Mark 15:43–46; Luke 23:50–53; John 19:38–42.
30. Matthew 27:62–64.
31. John 19:34.
32. Marguerite Perrier, Pascal's niece, afflicted with an incurable fistula in her left eye, was reported to have been cured instantly in 1656 by kissing a sacred relic. See *Abrégé de l'Histoire de Port-Royal* (Paris, 1908), part 1, pp. 77–87.
33. A skeptical satirist of mid-second-century Rome, Lucian discusses how Peregrinus fooled the Christians.
34. Justin, author of the influential *Apologies for the Christian Religion,* died as a martyr in the second century A.D. Simon the Magician, a founder of gnosticism in the first century A.D., was a Jewish sectarian considered a prime heretic by Christians.
35. Saint Clement of Alexandria (150–213); Saint Cyril (315–386); Tertullian (late second century); Eusebius (265–339), Church historian and apologist; Theodoret (390–458), bishop of Cyrrhus and historian of the Greek Church, and Saint Augustine (354–430), author of *City of God.*
36. Edward Gibbon (1737–1796) lists in *The Decline and Fall of the Roman Em-*

*pire* (London, 1776–1788), chapter 15, five reasons for the success of Christianity: the intolerant zeal of Christians; the doctrine of a future life; the miraculous powers ascribed to the primitive church; the pure and austere morals of the Christians; and the union and discipline of the Christian republic.

37. St. Augustine, *The City of God,* trans. Marcus Dods (Edinburgh, 1881), vol. 1, book 11, chapters 26–27; and Bossuet, *Discours sur l'histoire universelle* (Paris, 1681), part 2, chapter 19, "Jésus Christ et sa doctrine."
38. The French word "intelligence" has been translated as "intellectual faculties."
39. I have translated "soutien" as basic properties.
40. Jacques Bénigne Bossuet (1627–1704), renowned theologian and historian; and Antoine Arnauld (1612–1694), leader of Jansenist movement within the Church. Among other issues, Arnauld cast doubts on Descartes's argument that animals have no souls.
41. Leibniz discusses his theory of preestablished harmony most fully in *Essais de théodicée* (Amsterdam, 1710), which went through eight French editions in the eighteenth century. He also mentions it in the *Système nouveau de la nature* and in the *Monadologie.*
42. The original document had this footnote here: "Someone at the First Class of the Institute read a memoir in which he claimed that willpower is a force to be added to mechanical forces. The commissioners charged with examining the memoir asked its author to lift by means of a pulley a weight heavier than his body, and completely free. When subjected to this test, he recognized, after fruitless efforts, the falsity of his assertion." The someone was Abraham Louis Bréguet reading his *Essai sur la force animale, et sur le principe du mouvement volontaire* (Paris, 1811).
43. *Theogony; and Works and Days,* trans. M. L. West (Oxford, 1988).
44. For Leibniz's theory of preestablished harmony, see his *Essais de théodicée* and *Système nouveau de la nature* and the *Monadologie.*
45. See Bréguet, *Essai sur la force animale.* In the original document there appeared another footnote here: "At the First Class of the Institute of France, a memoir was read to establish the fact that the will adds a force to mechanical forces. The commissioners charged with examining the memoir asked its author to lift by means of a pulley a weight heavier than his body and completely free. When subjected to this test, he recognized, after fruitless efforts, the falsity of his assertion."
46. Examples are to be found in the following series of well-known tracts: David Hume, *Natural History of Religion* (1757), with French translation in 1759; Charles de Brosses, *Du culte des dieux fétiches* (1760); Paul Henri Thiry d'Holbach, *La contagion sacrée; ou, Histoire de la superstition* (1768);

Jacques-Henri Meister, *De l'origine des principes religieux* (1768); Charles-François Dupuis, *Origine de tous les cultes; ou, Religion universelle* (1794) and *Abrégé de l'origine de tous les cultes* (1798); and Antoine Destutt de Tracy, *Analyse raisonnée de l'origine de tous les cultes* (1799).

47. *Theogony; and Works and Days.*

# Index

Académie Française, Paris, 49, 113, 195–196, 200
Académie Royale de Marine, Brest, 24, 29
Académie Royale des Sciences, Paris, 14, 37, 45, 56, 60, 62, 68, 76, 83–84, 98–99, 105, 107–108, 120–121, 124, 148, 157, 183, 190, 197–198, 201, 213; elections, 38–43, 45, 76, 127, 194, 201; review commissions, 54, 65–66, 68–69, 75–76, 84–85, 96, 98, 141, 146, 167, 197; weights and measures, 99, 104, 106–107, 112, 121, 125–126, 129, 134–135, 148, 150
Académie Royale des Sciences et Belles-Lettres (Berlin), 36, 39, 42–43, 51, 55, 74, 147
Accademia delle Scienze (Turin), 36, 40, 45, 147
Adam, Jean, 18–25, 27–30, 48, 50, 84
Aepinus, Franz Ulrich Theodosius, 96
Agnesi, Maria Gaetana, 38
Akademiia Nauk (St. Petersburg), 36, 39, 42, 147
Alembert, Jean Le Rond d', 21–22, 26–35, 38–39, 41–42, 44–45, 48–51, 53–56, 58, 60–62, 64, 66, 68, 71–74, 95, 97, 120, 141, 144, 147, 152, 155, 170, 181, 195, 206, 208
Ampère, André-Marie, 166
Andoyer, Henri, vii
André, Yves Marie, 23, 54, 56
Andrieux, François Guillaume Jean Stanislas, 193, 196
Antelmy, Pierre Thomas, 38, 42, 69
Arago, Dominique François Jean, 84, 134, 137, 165, 167, 205
Arbuthnot, John, 182
Argand, Jean Robert, 111

Arrests, 103–104, 106, 121
Arsenal laboratory, Paris, 88, 94, 136–137
Asteroids, 153–154, 177–178
Aubry de la Boucharderie, Claude Charles, 103
Augez de Villers, Clément, 67, 213
Azaïs, Pierre-Hyacinthe, 199

Babbage, Charles, 198
Baciocchi, Elisa, *née* Bonaparte, 133
Bailly, Jean-Sylvain, 66, 69, 98, 103, 113, 116–117, 121, 129, 195, 201
Baker, Keith Michael, 62
Banks, Joseph, 133–134
Batteux, Charles, 53
Baudin, Thomas Nicolas, 134
Bayes, Thomas, 59
Beaumont-en-Auge, Normandy, 3–9, 15, 18, 87, 100–101; Belle-Croix Inn, 5–6, 8, 17; Collège de, 9–14, 16, 19, 36, 42, 111; Le Mérisier, 6, 101
Benedictine Order. *See* Congregation of St. Maur (Benedictine Order)
Benzenberg, Johann Friedrich, 199–200
Bernoulli, Daniel, 59–60, 72, 92, 116–117
Bernoulli, Jacob, 57
Berthelot, Claude François, 35, 38
Berthier, Louis-Alexandre, 128
Berthollet, Amédée Barthelmy, 136–137
Berthollet, Claude-Louis, 83–84, 99–101, 110, 119, 126–127, 135–137, 159–160, 191–192, 199, 201
Berthollet, Marie Marguerite, 136
Bertier, Joseph Etienne, 54
Bézout, Etienne, 22, 34, 36–37, 39, 66, 95–97, 111
Bichat, Marie François Xavier, 171, 174

303

Bicquilley, Pierre Marie, 103
Billy, A. L., 155
Biot, Françoise Gabrielle, *née* Brisson, 160
Biot, Jean-Baptiste, 84, 111, 134, 136, 142–143, 154–155, 160–163, 165–167, 198, 205, 207
Blagden, Charles, 104, 117–118, 137, 151, 199
Blanchard, Charles Antoine, 13, 16, 19
Bochart de Saron, Jean Baptiste Gaspard, 66–67, 69, 77, 95, 103, 121, 201
Bonaparte, Lucien, 128, 131
Bonaparte, Napoleon, 90, 96, 125–128, 131, 133–137, 169, 172–173, 184, 191–193, 195, 199, 205, 207
Bonnet, Charles, 187
Borda, Jean Charles, 41, 106, 183, 201
Bošković, Ruđer Josip, 42, 58, 60, 67, 77–78, 92, 152
Bossut, Charles, 24–25, 34, 36, 70, 84, 95–96, 111, 146, 154, 201
Bougainville, Louis-Antoine de, 195
Bouisset, Jean, 18–19
Boullier, David Renaud, 54, 56
Bourdon, Louis Pierre Marie, 155
Bourdon, Marc Antoine, 128
Bouvard, Alexis, 76, 143, 148, 150–152, 183, 204, 234
Bowditch, Nathaniel, 143–144, 152
Bradley, James, 76
Brahe, Tycho, 118
Brest, Britanny, 134, 149, 183
Bretocq family, 5, 8
Brisson, Mathurin Jacques, 65, 85, 129
Buffon, Georges-Louis Leclerc, Comte de, 26, 28, 61–62, 113, 116, 195
Burckhardt, Johann Karl, 142–143, 148, 150
Bureau de Consultation des Arts et Métiers (Paris), 107, 121
Bureau des Longitudes (Paris), 109, 118, 121–124, 127, 129–130, 133, 148, 178, 183, 190, 207
Bürg, Johann Tobias, 76, 134, 148
Burnet, Gilbert, 202

Cabanis, Pierre Jean Georges, 171–172, 174, 186
Caen, Normandy, 16–18, 23, 30, 53; Eglise du Saint-Sépulcre, 19, 22; Jesuits, 20, 22–23. *See also* University of Caen

Cagnoli, Antonio, 38
Calorimetry, 86, 94, 206
Camus, Charles Etienne Louis, 13, 36, 111
Capillarity, 93, 162–164, 207
Caritat de Condorcet, Jacques Marie de, 8
Caritat de Condorcet, Marie Jean Antoine. *See* Condorcet, Marie Jean Antoine Nicolas de Caritat, Marquis de
Carlini, Francesco, 152, 157–158
Carnot, Lazare Nicolas Marguerite, 14, 107
Carrey, Robert, 4–5
Cassini, Jacques, 117
Cassini, Jean Dominique (Cassini IV), 69, 103, 121, 201
Cassini de Thury, César François, 69
Cauchy, Augustin Louis, 111, 137, 158, 179, 188, 193, 198
Cauchy, Louis François, 137, 179, 193
Causality, 28, 49, 53–59, 63, 71–72, 80, 91, 112, 115–117, 142, 152, 165–166, 168, 171–172, 174, 176, 179, 181, 184–186, 208–209, 220, 222, 225, 227–232,
Caussin de Perceval, Jean Baptiste Jacques Antoine, 152
Cavallo, Tiberius, 90
Chalmers, Thomas, 202
Chamber of Peers (French), 182, 192, 194, 196, 200, 206–207
Chaptal, Jean-Antoine Claude, 135, 137, 191
Charles, Jacques, 42, 76
Charles, Jacques Alexandre César, 129
Charles X, 196
Chemical revolution, 81–82, 85, 88–90, 94
Chemistry, 56, 66, 92, 136, 160, 162–163
Clairaut, Alexis-Claude, 41
Clarke, Samuel, 55
Colbert, Jean Baptiste, 206
Colbert-Laplace, Jean Baptiste Charles Auguste, 3
Collège des Quatre-Nations (Collège Mazarin), Paris, 14–15, 39, 99
Collège Royal (Collège de France, Paris), 14, 22, 35, 38, 42–43, 72, 141, 152, 154
Collin d'Harleville, Jean François, 124
Comets, 28–29, 60, 68, 77–78, 116–117, 145, 152–153, 177–178
Comte, Auguste (Isidore-Auguste-Marie-François-Xavier), 176, 208
Concordat of 1801, 173

Condillac, Etienne Bonnot de, 64, 171, 221
Condorcet, Marie Jean Antoine Nicolas de Caritat, Marquis de, 14–15, 34, 39, 41–42, 44–45, 47–48, 50–53, 58, 60–64, 66, 73–74, 88–89, 98, 103, 107, 112–114, 119, 121, 169–171, 176, 181–183, 195, 201, 206
Congregation of St. Maur (Benedictine Order), 9–14, 16, 18, 30, 111
Cordier, Louis François, 4
Coulomb, Charles Augustin de, 96
Cournot, Antoine Augustin, 188
Courty de Romange, Jean-Baptiste Joseph, 100–102, 121
Courty de Romange, Marie Anne Charlotte. See Laplace, Marie Anne Charlotte, née Courty de Romange
Cousin, Jacques Antoine Joseph, 34–35, 38, 41, 44, 48, 67, 72, 76, 141, 146, 201
Craig, John, 173, 176
Crawford, Adair, 91
Cuvier, Jean-Léopold-Nicolas-Frédéric (Baron Georges), 113, 234

D'Alembert. See Alembert, Jean Le Rond d'
Damoiseau, Marie Charles Théodore, 152, 157, 198
Daru, Pierre Antoine Noël Bruno, 196
Daubenton, Louis Jean-Marie, 96, 110
Daunou, Pierre-Claude-François, 183
Davy, Humphry, 156, 198–200
Delambre, Jean-Baptiste Joseph, 14, 25, 76–77, 81, 104, 106, 111, 120–121, 123, 125, 135, 142, 148–151, 173, 197
Dellebarre, Louis François, 84
Deluc, Jean André, 82–84, 90–92, 104, 117, 141, 149–150, 200
Demollien, 132
De Morgan, Augustus, 61, 72
Descartes, René, 29, 49, 54–56, 70, 111, 155, 187, 224–225, 229
Descotils, Hippolyte Victor, 137
Desmarest, Nicolas, 42
Destutt de Tracy, Antoine Louis Claude, 171–172
Determinism, 1, 34, 52, 58, 70, 78, 80, 93, 158, 168, 171, 173–175, 179, 188, 204
Dicquemare, Jacques François, 113
Dieppe, Picardy, 86–87
Dietrich, Philippe Frédéric, 103

Dieudonné, Christophe, 129–130
Dionis du Séjour, Achille Pierre, 66–69, 77–78, 116–117
Dodwell, Henry, 174, 215, 217, 221
Dolomieu, Dieudonné Sylvain Guy Tancrède Gratet de (known as Déodat), 133–134
Dortous de Mairan, Jean Jacques. See Mairan, Jean Jacques Dortous de
Dulong, Pierre-Louis, 137, 198
Du Marsais, César Chesneau, 27
Dumouchel, Jean Baptiste, 129
Duprat, J. B. M., 142–143
Dupuis, Charles François, 174
Duvaucel, Charles, 42

Ecole Centrale des Travaux Publiques (Paris), 109
Ecole Normale de l'an III (Paris), 108–112, 115, 119, 122, 125, 136, 155, 170–171, 173, 179–180, 183, 207
Ecole Polytechnique (Paris), 103–104, 109, 112, 127–128, 130, 133, 136–137, 142–143, 154–155, 157, 160, 166, 177, 188, 192–194
Ecole Royale Militaire (Paris), 34–39, 41–43, 53, 60, 62, 65–66, 69, 85, 98, 109, 111, 126, 195
Ecoles centrales, 30, 128, 155, 159
Electricity, 89–91, 96, 159, 161–163, 186, 198, 226, 230
Encontre, Daniel, 176
Encyclopédie, 21, 26–27, 35, 45, 49, 120
Euler, Leonhard, 39–40, 42, 44, 49, 70, 72, 84, 95, 97, 111, 141, 144, 146–147, 152, 208

Ferme Générale, 66, 82, 86
Fischer, Ernst Gottfried, 160, 163
Flamsteed, John, 76
Follebarbe family, 4
Fontaine des Bertin, Alexis, 39
Fontana, Gregorio, 125
Fontenelle, Bernard Le Bovier de, 25, 29, 113, 195
Fontette, François Orceau de, 19
Forbin, Gaspard François Anne de, 54
Fortin, Nicolas, 85
Fouché, Joseph, 128
Fouchy, Jean Paul Grandjean de, 45
Fourcroy, Antoine François, 108, 119, 121

Fourier, Jean-Baptiste-Joseph, 32–33, 84, 166–167, 197–198, 203–204, 206
Fox, Robert, 161
Foy, Maximilien Sébastien Frédéric, 103
Francœur, Louis Benjamin, 198
François de Neufchâteau, Nicolas Louis, 196
Free will, 51, 188
French Revolution: Committee of Public Instruction, 109, 122; Committee of Public Safety, 109; Directorate, 127, 131; legislatures, 106, 109–110, 122, 125; Senate, 129, 131, 134, 136, 179, 182, 191–192, 195, 207
Fresnel, Augustin Jean, 84, 165–167
Fyot, François Marie, 42

Gabriel, Jacques Ange, 35
Gadbled, Christophe, 18–19, 22–29, 38, 42, 48, 69
Galilei, Galileo, 114
Garat, Dominique Joseph, 171
Gaudin, Martin-Michel-Charles, 128
Gauss, Carl Friedrich, 44, 111, 153–154, 156–157, 176–179, 198–199, 207
Gautier, Alfred, 155, 158
Gay-Lussac, Joseph Louis, 137, 150
Geneva, Switzerland, 82, 84, 90–91, 141, 155, 167, 176, 186–187, 192, 202
Gibbon, Edward, 174, 222
Gillispie, Charles Coulston, vii, 149, 177
Girault de Kéroudou, Mathurin Georges, 42
God, 25–26, 28, 51–56, 67, 119, 172–173, 188, 201–203, 213, 221–222
Gosset des Aunes, 65
Grandi, Guido, 173
Grattan-Guinness, Ivor, 166
Gribeauval, Jean Baptiste Vaquette de, 95–96
Griois, Charles Pierre Lubin, 103
Guerlac, Henry, 85
Guernier, Marin Simon, 6
Guyton de Morveau, Louis Bernard, 108

Halévi, Ran, vii
Halley, Edmund, 76, 153
Halley, Pierre, 5, 7–8
Harte, Henry Hickman, 143
Hassenfratz, Jean Henri, 155
Haüy, René Just, 15, 83, 96, 106, 110, 159–160, 163

Heat theory, 1, 82, 85–86, 92–94, 117, 150, 161–162, 166–167, 176, 206–207
Helvétius, Claude Adrien, 18
Henry, Maurice, 134
Héricy, Jacques Armand d', 18, 29
Herschel, Friedrich Wilhelm (Sir William), 77–78, 81, 104, 117–118, 151,172
Herschel, John Frederick William, 198
Herschel's planet (Uranus), 77–78, 81, 151
Hobbes, Thomas, 63
Hoëné-Wroński, Jósef Maria, 156, 199
Holbach, Paul Henri Thiry d', 18, 43, 66, 120
Humboldt, Alexander von, 137, 150, 198, 202
Hume, David, 174
Huygens, Christiaan, 165

Idéologues, 171–174
Imperial Academy of St. Petersburg. *See* Akademiia Nauk (St. Petersburg)
Imponderable fluids, 43, 82–83, 89, 161, 206, 226, 230
Institut d'Egypte (Cairo), 127
Institut National des Sciences et des Arts (France), 76, 109, 124–125, 127–128, 130–131, 134, 136–137, 152, 156, 170–172, 176–178, 183, 195, 199, 207
Ivory, James, 156–157, 198, 207

Jars, Antoine Gabriel, 213
Jeaurat, Edme Sébastien, 38, 69
Jesuit Order. *See* Society of Jesus
Jupiter/Saturn problem, 72, 76, 78–79, 81, 142–143, 148–149, 157, 183, 234
Jussieu, Antoine Laurent, 81, 201

Kant, Immanuel, 114–116, 156
Kepler, Johannes, 145
Keranflech, Charles Hercule de, 54, 56
Kirwan, Richard, 95
Koyré, Alexandre, vii

Labey, Jean Baptiste, 130
Lacepède, Bernard-Germain-Etienne de la Ville-sur-Illon de, 128
La Chapelle, Jean Baptiste de, 13, 16, 23, 25, 42
Lacretelle, Jean Charles Dominique, 196

Lacroix, Sylvestre François, 111, 121, 130, 161, 179
Lagrange, Joseph Louis, 39–42, 44–45, 49, 60, 67, 71–75, 78–79, 81–82, 86, 94–95, 110–111, 122–124, 130, 141, 144, 146, 152–156, 160, 181, 201, 206
Lair, Pierre Noël Aimé, 17
Lakanal, Joseph, 122, 125
Lalande, Joseph-Jérôme Lefrançois, 34, 38, 42, 68–69, 76–78, 113, 120, 122–123, 153, 172–173, 178
La Londe, François Richard de, 21
Lamanon, Jean Honoré Robert Paul de, 85
Lambert, Johann Heinrich, 60
Lamennais, Hugues Félicité Robert de, 193
La Michodière, Jean Baptiste François de, 98
Lamotte, Marie Madeleine Isabel, 5–6
Landriani, Marsilio, 91
Lanney, Marie Magdeleine de, 5, 7
Laplace, Charles Emile (called Emile), 11, 101–102, 133, 135–137, 155, 179, 191–192, 200–201, 204, 206, 233–234
Laplace, François de, 4
Laplace, Jacques Pierre, 6
Laplace, Julie Marguerite, 6
Laplace, Louis de, 7–11, 14, 30
Laplace, Marie de, 4
Laplace, Marie Anne de, 6–7, 10, 101
Laplace, Marie Anne Charlotte, *née* Courty de Romange, 11, 100–103, 131, 133, 136, 151, 191, 198, 200, 203–204, 206, 233–234
Laplace, Olivier de, 4, 6–7, 11, 87
Laplace, Pierre de, 3–9, 11, 14–15, 30, 37, 43, 86–88, 100–101, 133; cider affair, 86–88, 101
Laplace, Sophie Suzanne. *See* Portes, Sophie Suzanne, *née* Laplace
Laplace, Thomas François de, 4
La Rochefoucauld d'Enville, Louis Alexandre, 103
Lassone, Joseph Marie François de, 95
Lavoisier, Antoine-Laurent, 14–15, 43, 66–67, 86–90, 92–96, 99, 101, 103, 106–107, 120–121, 136–137, 162, 167, 200–201, 206
Lavoisier, Marie Anne, *née* Paulze, 82, 86, 193, 213
Law of gravitation, 53, 55–56, 70–72, 79–80, 83, 91–93, 113, 140, 142, 144–145, 148–149, 157–158, 162, 206, 224, 228
Le Canu, Pierre, 18–19, 24, 29–30, 32–33, 35, 38, 48
Le Carpentier family, 4
La Chevalier, Marie, 5
Le Clerc, Pierre, 113
Le Coq, Olivier, 5
Lefébure de Fourcy, Louis Etienne, 155
Le Gaigneur, René Jacques, 23–24
Legendre, Adrien Marie, 15, 39, 42, 44, 48, 72, 75, 96, 141, 146, 154–155, 157, 178–179, 233
Le Guay, Thomas François, 22
Leibniz, Gottfried Wilhelm, 54–56, 58, 64, 119, 173, 225–226, 229–230
Lelièvre, Pierre, 30
Le Mée (Île de France), 102–103, 131, 136, 155
Lemercier, Louis Nicolas, 196
Le Monnier, Pierre, 22, 34, 38
Leperchey, 7
Leroy, Jean Baptiste, 84
Le Sage, Georges Louis, 71, 90–92, 141–142
Le Turc, Bonaventure, 65
Lévêque lectures, 25
Lexell, Anders Johan, 67, 77, 153
Lhuilier, Simon Antoine Jean, 176
Light theory, 93, 149, 159, 164–167
Locke, John, 25, 64
Lombard, Jean Louis, 126
Louis XVI, 67, 99
Louis XVIII, 190–191
Louize, François Jacques, 6
Louvel, Jacques, 19, 30
Lucan, Marcus Annaeus, 163

Mabon, Jacques, 5
Mabon, Jean, 87
Maclaurin, Colin, 146
MacParlan, Milesius, 20
Magendie, François, 174–175, 186–187, 198, 204
Mairan, Jean Jacques Dortous de, 117, 195
Malebranche, Nicolas, 54–56
Mallet, Edme, 27
Malus, Etienne Louis, 137, 165
Marat, Jean-Paul, 199
Marguerie, Jean Jacques de, 24, 29, 42, 48

Marie, Joseph François, 42
Marie-Antoinette de Lorraine, 100
Mariotte, Edme, 159
Marmont, Auguste Frédéric Louis Viesse de, duc de Raguse, 103
Martinne, Henri, 6
Marum, Martin Van, 91
Maskelyne, Nevil, 76, 151
Materialism, 67, 213–214
Matter, 34, 51–52, 55, 70, 93, 142, 145, 161–162, 222–223, 226, 230–231
Mauduit, Antoine René, 42
Maupertuis, Pierre-Louis Moreau de, 56, 58, 195, 208
Maurice, Jean Frédéric Théodore, 141–142, 155, 186, 199, 202–204, 206
Mayer, Johann Tobias, 76
Méchain, Pierre-François-André, 76–77, 106, 148
Melun (Île de France), 102–103, 108, 110, 121, 131
Meridian, 106, 121, 134, 146, 148, 193–194
Messier, Charles Joseph, 76, 118, 153
Metric system, 107, 112, 123, 125, 129, 134–135, 149, 207
Metternich, Clemens Wenzel Lothar, 192
Meusnier de la Place, Jean Baptiste Marie Charles, 95
Michaud, Joseph-François, 196
Michell, John, 117–118
Milan (city), 104–105, 142, 153, 192
Milan Observatory (Brera), 133, 199
Ministry of the interior, 127–131, 134, 182, 193, 195, 207
Mogensen, Nels Wayne, 5
Moivre, Abraham de, 57, 59, 181
Mollerat, Marie Hélène Angélique, 100–101
Mollerat de Brechainville, François Nicolas, 101
Monge, Gaspard, 42, 48, 72, 75–76, 94–96, 108, 110, 119, 121, 123, 126–127, 136, 160, 193, 201
Montesson, Charlotte Jeanne Béraud de la Haie, 100
Montmort, Pierre Rémond de, 57, 182
Montucla, Jean-Etienne, 113
Montyon, Antoine-Jean-Baptiste-Robert Auget de, 196–197
Moysant, François, 18–19, 30

Napoleon I. *See* Bonaparte, Napoleon
Nebular hypothesis, 29, 114–116, 118, 151, 177, 208
Newton, Isaac, 1, 22, 29, 33, 44, 48–51, 53–56, 70, 72, 77–79, 83, 92, 111, 113–114, 119, 140, 144–147, 154, 157–159, 161, 167–168, 184, 198, 201, 204, 206; religious views, 201–202
Noël, François Joseph Michel, 128–129
Nollet, Jean Antoine, 22, 25

Oriani, Barnaba, 76–77, 81, 104–105, 125–126, 133, 142–143, 153, 157, 188, 199
Orléans, Louis Philippe, duc d', 16
Orléans, Louis Philippe Joseph, duc d', 67, 98, 100
Orléans family, 9–10, 100, 192
Ørsted, Hans Christian, 198

Paris de Meyzieu, Jean Baptiste, 35
Paris Observatory, 103, 121, 123, 127
Parseval, Marc Antoine des Chênes, 161
Pascal, Blaise, 54–55, 176, 187, 221
Patronage, 2, 32, 50, 58, 62, 67, 73, 121, 124, 135, 137, 142–143, 147, 205, 208
Personal finances, 10, 35, 43, 85–86, 96, 98–102, 124, 131,
Personal traits, 7, 10–12, 17–18, 29–31, 33–34, 36–38, 41–43, 45, 47, 50, 63, 65–68, 73–75, 99, 101, 104–105, 108, 120–121, 133, 151, 157, 173, 176, 184, 190–191, 197, 199–200, 202–204
Petit, Alexis Thérèse, 163
Peyronnet, Pierre Denis, 196–197
Piazzi, Giuseppe, 153–154, 177
Piel, Marie, 5
Pingré, Alexandre-Guy, 66, 69, 77, 117
Pius VII, 173
Plana, Giovanni, 152, 157–158, 188, 208
Pluche, Antoine, 113
Poincaré, Henri, 152, 158, 208
Poinsot, Louis, 157–158
Poisson, Siméon Denis, 84, 143, 154–155, 157, 163, 167, 203, 207, 234,
Pontécoulant, Gustave de, 152
Portal, Antoine, 201

Portes, Adolphe François René de, 191, 233
Portes, Angélique Joséphine Charlotte, 198, 200, 206
Portes, Sophie Suzanne, *née* Laplace, 11, 101, 191, 206
Portraits (Laplace), 46, *132, 138*
Potter, Louis Joseph Antoine, 202
Prades, Jean-Martin, 26
Prévost, Pierre, 84, 91, 167, 176, 186
Priestley, Joseph, 91, 118
Principle of inertia, 33–34, 50
Prony, Gaspard Clair François Marie Riche de, 107, 123, 154
Psychology, 82, 174–175, 186–188, 227
Public service, 102–103, 105, 107, 119, 125, 192, 194
Puiseux, Léon François, 19
Puissant, Louis, 194

Quesnot, François, 31
Quinette, Nicolas Marie, 128

Racine, Jean, 205
Ramond de Carbonnières, Louis François Elisabeth, 150, 183
Raynouard, François-Juste-Marie, 196
Reale Accademia delle Scienze. *See* Accademia delle Scienze (Turin)
Réaumur, René-Antoine Ferchault de, 85
Régnault de Saint Jean d'Angély, Michel Louis Etienne, 195
Reinhard, Charles Frédéric, 128
Religious conflict, 25–27, 50, 173
Religious criticism: on miracles, 173, 176, 180, 202, 219, 221, 223; on the New Testament, 174, 215–224; on Resurrection, 176, 180, 217–220; on transubstantiation, 176, 223–224
Residences, 17, 35–36, 85–86, 99, 101–103, 110, 131, 136
Restoration politics, 190–196
Revolutionary calendar, 99, 106–107
Robespierre, Maximilien Marie Isidore de, 108, 110, 122
Rochon, Alexis-Marie, 42
Royal Engineers (Génie Militaire), 14, 36–37, 96

Royal Society of London, 36, 59, 91, 104, 117–118, 133, 137, 151, 156, 199–200
Royer-Collard, Pierre Paul, 197
Ruffini, Paolo, 188, 208

Sabatier, Raphaël Bienvenu, 100
Sage, Balthazar Georges, 201
Saint Germain, Claude Louis, Comte de, 43
Saury, Jean, 42, 54
Saussure, Horace Bénédict de, 91, 199
Saxe-Gotha, Charlotte, Duchesse de, 142
Say, Jean-Baptiste, 194
Scanégatti thermometer, 84
Siegfried, André, 30
Sochon, Antoine, 5
Sochon, Jean Jacques, 5
Sochon, Marie Anne, 4–5, 7, 10
Société d'Arcueil, 135–137, 139, 151, 154–155, 160, 164–166, 179, 183, 191, 193, 198–200, 204–205, 234
Société de Géographie (Paris), 194
Société Maternelle (Paris), 129
Société Royale de Médecine (Paris), 87
Society of Jesus, 12, 14, 20–23, 26, 42, 58, 60, 67, 162
Sommerville, Mary, 198
Speed of sound, 161–162
Stability of solar system, 78–81, 115–116, 140, 158, 168, 206
Stewart, Dugald, 176
Stigler, Stephen Mack, 178
Swinden, Jan Hendrik Van, 91, 126, 129, 152

Talleyrand-Périgord, Charles Maurice de, 106, 125, 191
Taton, René, vii
Taylor, Brook, 181
Tenon, Jacques René, 98
Thénard, Louis-Jacques, 137
Thomassin, 66
Thouin, André, 110
Tinseau d'Amondans, Charles Marie, 42
Tisserand, François Félix, 158
Toplis, John, 143
Trémery, Jean Louis, 163
Turgot, Anne Robert Jacques, 62, 66

University of Caen, 13–15, 18–19, 25, 28–30, 54, 87, 111, 172, 174; Collège des Arts, 18–19, 22; Collège du Bois, 16–20, 24; Collège du Mont, 18–20, 22–23, 25, 29; religious controversies, 19, 24–29
University of Paris, 14, 22, 129; Faculty of Sciences, 155

Vandermonde, Alexandre Théodore, 41
Varignon, Pierre, 20
Vaublanc, Vincent Marie Viénot, 195
Vaucanson, Jacques de, 65
Vicq d'Azyr, Félix, 14, 24, 87, 100, 121, 195, 201

Villers, Charles François Domique de, 115
Volta, Alessandro, 83, 90–91, 186, 198
Voltaire, François Marie Arouet de, 113

Watt, James, 91
Wollaston, William Hyde, 165

Young, Thomas, 163–167
Yunus, Ibn, 152

Zach, Franz Xaver von, 134, 153–154, 157, 199